Gamification for Human Factors Integration:

Social, Education, and Psychological Issues

Jonathan Bishop
Centre for Research into Online Communities and E-Learning Systems, European Parliament, Belgium

A volume in the Advances in Human and Social Aspects of Technology (AHSAT) Book Series

Managing Director: Lindsay Johnston
Production Manager: Jennifer Yoder
Development Editor: Christine Smith
Assistant Acquisitions Editor: Kayla Wolfe
Typesetter: Henry Ulrich
Cover Design: Jason Mull

Published in the United States of America by
Information Science Reference (an imprint of IGI Global)
701 E. Chocolate Avenue
Hershey PA 17033
Tel: 717-533-8845
Fax: 717-533-8661
E-mail: cust@igi-global.com
Web site: http://www.igi-global.com

Library of Congress Cataloging-in-Publication Data

Gamification for human factors integration : social, education, and psychological issues / Jonathan Bishop, editor.
pages cm
Includes bibliographical references and index.
ISBN 978-1-4666-5071-8 (hardcover) -- ISBN 978-1-4666-5072-5 (ebook) -- ISBN 978-1-4666-5073-2 (print & perpetual access) 1. Internet--Social aspects. 2. Internet--Psychology. 3. Interpersonal relations. I. Bishop, Jonathan, 1979-
HM741.G36 2014
302.23'1--dc23
2013037492

This book is published in the IGI Global book series Advances in Human and Social Aspects of Technology (AHSAT) (ISSN: 2328-1316; eISSN: 2328-1324)

British Cataloguing in Publication Data
A Cataloguing in Publication record for this book is available from the British Library.

For electronic access to this publication, please contact: eresources@igi-global.com.

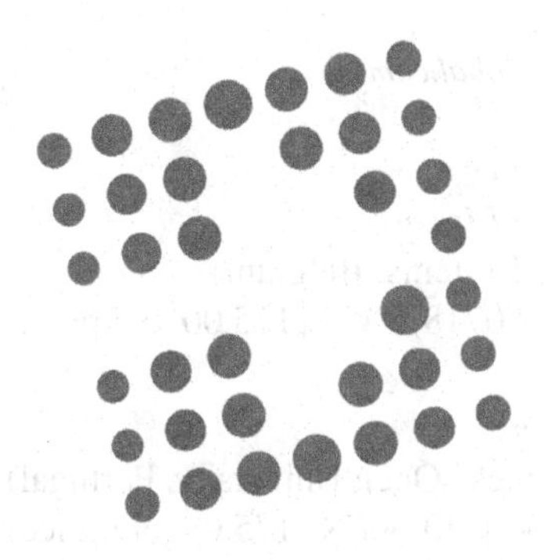

Advances in Human and Social Aspects of Technology (AHSAT) Book Series

ISSN: 2328-1316
EISSN: 2328-1324

Mission

In recent years, the societal impact of technology has been noted as we become increasingly more connected and are presented with more digital tools and devices. With the popularity of digital devices such as cell phones and tablets, it is crucial to consider the implications of our digital dependence and the presence of technology in our everyday lives.

The **Advances in Human and Social Aspects of Technology (AHSAT) Book Series** seeks to explore the ways in which society and human beings have been affected by technology and how the technological revolution has changed the way we conduct our lives as well as our behavior. The AHSAT book series aims to publish the most cutting-edge research on human behavior and interaction with technology and the ways in which the digital age is changing society.

Coverage

- Activism & ICTs
- Computer-Mediated Communication
- Cultural Influence of ICTs
- Cyber Behavior
- End-User Computing
- Gender & Technology
- Human-Computer Interaction
- Information Ethics
- Public Access to ICTs
- Technoself

IGI Global is currently accepting manuscripts for publication within this series. To submit a proposal for a volume in this series, please contact our Acquisition Editors at Acquisitions@igi-global.com or visit: http://www.igi-global.com/publish/.

The Advances in Human and Social Aspects of Technology (AHSAT) Book Series (ISSN 2328-1316) is published by IGI Global, 701 E. Chocolate Avenue, Hershey, PA 17033-1240, USA, www.igi-global.com. This series is composed of titles available for purchase individually; each title is edited to be contextually exclusive from any other title within the series. For pricing and ordering information please visit http://www.igi-global.com/book-series/advances-human-social-aspects-technology/37145. Postmaster: Send all address changes to above address.

Titles in this Series

For a list of additional titles in this series, please visit: www.igi-global.com

Gamification for Human Factors Integration Social, Education, and Psychological Issues
Jonathan Bishop (Centre for Research into Online Communities and E-Learning Systems, Belgium)
Information Science Reference • copyright 2014 • 378pp • H/C (ISBN: 9781466650718) • US $175.00 (our price)

Emerging Research and Trends in Interactivity and the Human-Computer Interface
Katherine Blashki (Noroff University College, Norway) and Pedro Isaias (Portuguese Open University, Portugal)
Information Science Reference • copyright 2014 • 580pp • H/C (ISBN: 9781466646230) • US $175.00 (our price)

Creating Personal, Social, and Urban Awareness through Pervasive Computing
Bin Guo (Northwestern Polytechnical University, China) Daniele Riboni (University of Milano, Italy) and Peizhao Hu (NICTA, Australia)
Information Science Reference • copyright 2014 • 440pp • H/C (ISBN: 9781466646957) • US $175.00 (our price)

Gender Divide and the Computer Game Industry
Julie Prescott (University of Bolton, UK) and Jan Bogg (The University of Liverpool, UK)
Information Science Reference • copyright 2014 • 334pp • H/C (ISBN: 9781466645349) • US $175.00 (our price)

User Behavior in Ubiquitous Online Environments
Jean-Eric Pelet (KMCMS, IDRAC International School of Management, University of Nantes, France) and Panagiota Papadopoulou (University of Athens, Greece)
Information Science Reference • copyright 2014 • 325pp • H/C (ISBN: 9781466645660) • US $175.00 (our price)

Uberveillance and the Social Implications of Microchip Implants Emerging Technologies
M.G. Michael (University of Wollongong, Australia) and Katina Michael (University of Wollongong, Australia)
Information Science Reference • copyright 2014 • 368pp • H/C (ISBN: 9781466645820) • US $175.00 (our price)

Innovative Methods and Technologies for Electronic Discourse Analysis
Hwee Ling Lim (The Petroleum Institute-Abu Dhabi, UAE) and Fay Sudweeks (Murdoch University, Australia)
Information Science Reference • copyright 2014 • 546pp • H/C (ISBN: 9781466644267) • US $175.00 (our price)

Advanced Research and Trends in New Technologies, Software, Human-Computer Interaction, and Communicability
Francisco Vicente Cipolla-Ficarra (ALAIPO – AINCI, Spain and Italy)
Information Science Reference • copyright 2014 • 361pp • H/C (ISBN: 9781466644908) • US $175.00 (our price)

www.igi-global.com

701 E. Chocolate Ave., Hershey, PA 17033
Order online at www.igi-global.com or call 717-533-8845 x100
To place a standing order for titles released in this series, contact: cust@igi-global.com
Mon-Fri 8:00 am - 5:00 pm (est) or fax 24 hours a day 717-533-8661

Editorial Advisory Board

Holly Tootell, *University of Wollongong, Australia*
Cristian Vaccari, *University of Bologna, Italy*
Andreas Veglis, *Aristotle University of Thessaloniki, Greece*
Nicole Velasco, *Lee University, USA*
Gyanendra Kumar Verma, *Indian Institute of Information Technology Allahabad, India*
Lucia Vesnic-Alujevic, *Ghent University, Belgium*
Shefali Virkar, *University of Oxford, UK*
Timothy D. Webster, *Genist Systems Inc., Canada*
Joseph Wilson, *University of Maiduguri, Nigeria*
Lincoln C. Wood, *Auckland University of Technology, New Zealand*
Prodromos Yannas, *Technological Institute (TEI) of Western Macedonia, Greece*

Table of Contents

Section 2
Health and Cognition

Section 3
Pedagogical Issues

Detailed Table of Contents

The growth in Internet use is not only placing pressure on service providers to maintain adequate bandwidth but also the people who run the Websites that operate through them. Called systems operators, or sysops, these people face a number of different obligations arising out of the use of their computer-mediated communication platforms. Most notable are contracts, which nearly all Websites have, and in the case of e-commerce sites in the European Union, there are contractual terms they must have. This chapter sets out to investigate how the role contract law can both help and hinder sysops and their users. Sysop powers are limited by sysop prerogative, which is everything they can do which has not been taken away by statute or given away by contract. The chapter finds that there are a number of special considerations for sysops in how they use contracts in order that they are not open to obligations through disabled or vulnerable users being abused by others.

The design of serious games does not always address players' empathy in relation to their cognitive capacity within a demanding game environment. Consequently players with inherent limitations, such as limited working memory, might feel emotionally drained when the level of empathy required by a game hinders their ability to cognitively attain the desired learning outcome. Because of the increasing attention being given to serious games that aim to develop players' empathy along with their cognitive competencies, such as Darfur is Dying (Ruiz et al., 2006), there is a need to investigate the empirical relationship between players' cognitive load and empathy development capacity during serious game play. Therefore this chapter examines cognitive load theory and empirical work on empathy development to propose a conceptual framework to inform the research and design of serious games that have empathy as part of the learning outcomes. Future research should focus on implementation and empirical validation of the proposed framework.

Much as been written about e-government within a growing stream of literature on ICT for development, generating countervailing perspectives where optimistic, technocratic approaches are countered by far more sceptical standpoints on technological innovation. This body of work is, however, not without its limitations: a large proportion is anecdotal in its style and overly deterministic in its logic, with far less being empirical, and there is a tendency for models offered up by scholarly research to neglect the actual attitudes, choices, and behaviour of the wide array of actors involved in the implementation and use of new technology in real organisations. Drawing on the theoretical perspectives of the Ecology of Games framework and the Design-Actuality Gap model, this chapter focuses on the conception and implementation of an electronic property tax collection system in Bangalore (India) between 1998 and 2008. The work contributes to not just an understanding of the role of ICTs in public administrative reform, but also towards an emerging body of research that is critical of managerial rationalism for an organization as a whole, and which is sensitive to an ecology of actors, choices, and motivations within the organisation.

As the responsibilities of modern business expand to multiple stakeholders, there is an increased need to understand how to manage conflicting normative expectations of different stakeholders. Corporate responsibilities to stakeholders are based on the need to minimize or correct harm from operations (respect negative injunctions) while contributing to the social welfare of communities the firm operates in (engage in positive duties). By comparing multiple decision scenarios in the traditional and online publishing industry, the chapter explores the tensions that arise between these imperatives. Based on these tensions, the chapter outlines a framework and a practical industry-independent heuristic decision making process, embracing normative expectations, the consequences to a company and to stakeholders, and potential mitigating actions. The proposed heuristic approach allows balancing the tensions among stakeholder expectations to ensure selection of the appropriate alternative. The discussion is finished by pointing out the usefulness and applicability of the proposed heuristics in other industries and settings of the contemporary business environment.

The quantitative-qualitative and subjectivity-objectivity debates plague research methods textbooks, divide academic departments, and confuse post-modernists as to their existence. Those from the objective-quantitative camps will usually demand methods assume parametric principles from the start, such as homogeneity and normal distribution. Many of the subjective-qualitative camps will insist on looking and the individual meanings behind what someone is saying through their narratives and other discourses. The objective-quantitative camps on the other hand think anything that does not involve systematic acquisition and analysis or data cannot be valid. This chapter presents an approach to derive a parametric user model for understanding users that makes use of the premises and ideals of both these camps.

Section 3
Pedagogical Issues

Play has been an informal approach to teach young ones the skills of survival for centuries. With advancements in computing technology, many researchers believe that computer games[1] can be used as a viable teaching and learning tool to enhance a student's learning. It is important that the educational content of these games is well designed with meaningful game-play based on pedagogically sound theories to ensure constructive learning. This chapter features theoretical aspects of game design from a pedagogical perspective. It serves as a useful guide for educational game designers to design better educational games for use in game-based learning. The chapter provides a brief overview of educational games and game-based learning before highlighting theories of learning that are relevant to educational games. Selected theories of learning are then integrated into conventional game design practices to produce a set of guidelines for educational games design.

In recent years, researchers and classroom teachers have started to explore purposefully designed computer/video games in supporting student learning. This interest in video and computer games has arisen in part, because preliminary research on educational video and computer games indicates that leveraging this technology has the potential to improve student motivation, interest, and engagement in learning through the use of a familiar medium (Gee, 2005; Mayo, 2009; Squire, 2005; Shaffer, 2006). While most of this early research has focused on the impact of games on academic and social outcomes, relatively few studies have been conducted exploring the influence of games on civic engagement (Lenhart et al, 2008). This chapter will specifically look at how *Quest Atlantis*, a game designed for learning, can potentially be utilized to facilitate the development of ecological stewardship among its players/students, thereby contributing to a more informed democratic citizenry.

Is it possible to enhance the learning of sociology students by staging simulated field studies in a MMOLE (massively multi-student online learning environment) modeled after successful massively multiplayer online games (MMOG) such as Eve and Lineage? Lacking such a test option, the authors adapted an existing MMOG—"The Sims Online"—and conducted student exercises in that virtual environment during two successive semesters. Guided by questions keyed to course objectives, the sociology students spent 10 hours observing online interactions in TSO and produced essays revealing different levels of analytical and interpretive ability. The students in an advanced course on deviance performed better than those in an introductory course, with the most detailed reports focusing on scamming, trashing, and tagging. Although there are no technical obstacles to the formation and deployment of a sociology MMOLE able to serve hundreds of thousands of students, such a venture would have to solve major financial and political problems.

Foreword

As Jonathan Bishop invited me to write this foreword, I had the privilege to dive into the chapters even before the book was published, and maybe even before the authors had the opportunity to read each others' contributions. Being in such a lucky position, I take the opportunity to lead you around in the wider scope of the book.

For about four decades we have welcomed new metaphors for interactive applications at least every three to four years. I think this book offers more than a new metaphoric framework. Its essence is that it is no longer reasonable to regard the virtual and the real as different sides of the same coin. What Jonathan's book shows is that the interactive "application" has taken a leading role in the question on how to address societal and existential issues, not long ago too large concerns to even imagine how media could mitigate between the strict intimate and what we used to call "the collective consciousness."

Let us go through the main topics as reflected in the titles of the chapters. First of all, it is interesting to see that the division in societal sectors like education, care, industry, and entertainment: they do not matter too much; it seems that our current societal needs have washed away our scruples to admit that in essence all these "sectors" cope with the same essential dilemmas and challenges.

1. Relentless optimism and at the same time realism that the human mind can transcend from everyday concerns and feel attached to the big social challenges of today like accepting that thinking about "disabilities" is essential for making progress with the "normal."
2. The awareness that before expecting our neighbours to "awake" from "not in my backyard," we need the harsh side of life, like Jonathan describes in his chapter on the poor boy who was willing to give his life in order to prevent his sister from being raped by their grandpa. Mass media has exploited our weak spot in order to elicit donations. We continue to understand that for conquering abstract concepts like relativity theory or quantum mechanics, a similar condition is needed as well.
3. Bringing "Cognitive Load" and "Empathy" in one sentence together is a big step forward as it confirms that we have been too obsessed in upgrading everyday life phenomena to the level of "cognitive," which pretends to be real, valuable, ubiquitous, and even solemn, like book knowledge tended to be "given by God" himself. We tend to accept now that "cognitive" load is mainly "load," just like "knowledge economy" is mainly "economy." Sweller's theory on cognitive load stems from early ergonomic research that saw systems completely out of balance: too much control for a minor aspect in life. This book launches several chapters that converge in scope. Empathy and later Csikszentmihalyi's notion of "flow" brings us to the acceptance that humans are not just "users," "customers," or "recipients"; the human actor is "actor," "conductor," and creator. The

person who is in a media program is far away from a victim; he or she is a God(ess) who does not spend a second on deciding on how to give attention or leave a certain app; (s)he is just immersed in the sea of recommendations ("likes") that thousands did a few minutes ago. So, what is in the "cognitive load"? It is no longer an individual concern, as the Social Web did 95% of the filtering by human intelligence just before. In this sense, there is not a huge difference between the filtering process based upon authority and reputation, except that Web applications let thousands of persons play a role in this filtering process instead of two or four in the traditional editorial one.

4. Gamification: can we accept is as antidotum for those who suffer from marginal jobs, like the salespersons in exquisite warehouses, playing games to kill the confrontation of being obsolete if no customers are entering of buying? Of course, this painful societal fact of appointing persons to cynical jobs is the serious part of gaming. The game itself is joy, or at least divertissement. The other side of gamification is that sectors like escaping from taxpaying provokes a mutual game between the public opinion and the growing power of authorities to data-mine the citizen and "bringing them for justice." Dimensions like privacy and security will gain more and more momentum, until we admit that the story rather than "the data" are decisive in moral condemnation.
5. Publishing: The mere act of publishing has been overlooked in media campaigns; it is the last step before the public gets access to a creation. Shouting on a market square during a mass manifestation can be seen as publication. As long as the surrounding protesters see no conflict with their own message they will assimilate your words and maybe even copy and broadcast your words. Being in the wrong side of the football stadium may cause great problems as shouting few seconds earlier of later may soon lead to be excommunicated or worse. Indeed, the role of publisher stays important, not for the process of multiplication, but in the timing among seasons and the colour of the cover.
6. Neuroeconomics and rehabilitation: We may add the phenomenon of neuro-response plasticity; the fact that redundancy is still mentioned as the main role of gray matter, still demonstrates that we have not yet understood how essentially the cortex works autonomously and prepares during sleep. The disbelief that dancing in the school class and gaming before the surgeon enters the real patient gives better results demonstrates that we still carry the older belief that skills need to be compartmentalized (task analytical design) before we know how to train.
7. Education is still the abstraction from socialisation. As far as school education is concerned, this is one of the more pervasive attempts to locate the learning at the student, not at the teacher or the administrator. We do not need games in the schools; Students know and feel that there is already a game: to overcome complexity and know what the test will be about.
8. An enticing perspective is the chapter on "Learning Sociology in a Massively Multi-Student Online Learning Environment." Why do we need authors, editors, reviewers, and publishers if the best book on sociology is the blog with real life events and reflections. What authority do students need more? We may see a growing revolt from curricular and assessment designers who have problems to model such rich learning environments as its goals and eminent understanding may shift all the time; If it cannot be assessed, it is no use to learn it.

In summary, this book has opened the new arena for what I would call "critical media reflection." It goes beyond the scope designers and users; it is all about confronting society with fixations that cause too much frustration and overhead. Please dive into this book and seek contact with the authors

Piet Kommers
University of Twente, The Netherlands

Piet Kommers *is an Associate Professor at the University of Twente, The Netherlands. His specialty is social media for communication and organisation. As Conference Co-Chair of the IADIS Multi-Conference, he initiated the conferences of Web-Based Communities and Social Media, E-Society, Mobile Learning, and International Higher Education. He is Professor at the UNESCO Institute for Eastern European Studies in Educational Technology and Adjunct Professor at Curtin University in Perth, Australia.*

Preface

Those in the community of gaming theory and practice have pigeonholed gamification as a term. It is generally considered that it means applying gaming principles to non-gaming environments. Gaming theorists like Richard Bartle, wedded to his dated MUD model of gamers, speak out against gamification with passion, setting out why it never works and how videogames stand in their own right. Those in the area of Serious Games, often try to make themselves appear different to those in the area of gamification. In reality, Serious Games should be called serious gamification, if one sees Serious Games as a practical application of gaining principles to mundane or otherwise serious activities training + gamification. Serious Games are important, yet they are essentially gamified systems with a practical purpose that has been gamified. That is why they are in this book, which presents chapters on issues affected by gamification. This is evident in the first chapter of the second section (chapter 6), which shows how designing Serious Games for people with disabilities is basically dependent on the video game theories that people like Richard Bartle try to claim as their own and distinct from the topic.

On this basis, in addition to its more traditional meaning, one might further extend the definition of gamification to be the intentional application of gaming principles to an environment in such a way that participation within it is more enjoyable so that greater gratification and less discomfort is achieved for the greater good of those that form part of that environment. It is this definition that has led to the specific concentration of chapters for this book.

GAMIFICATION IN THE PUBLIC SPHERE

Shefali Virkar looks that the roles of gaming in shaping tax administration in Bangalore City in India (chapter 3). To those wedded to the idea of gamification being about video games in environments where video gaming is not used, this chapter may appear to be out-of-place. However, as a former politician on a council with tax varying powers, a former claimant of out-of-work welfare benefits, and a payer of fiscal taxes, I am of the view that the tax and benefits system is an area that needs to be tapped into according to the public policy of whoever is in power. It is an almost certain rule that in any economic system people will seek to maximize economic gain and minimize economic loss. Gamification can be applied to the tax and benefits system to encourage positive behaviors. This includes increasing the wealth and social mobility of those without much whilst preventing those who have too much from hoarding that which they do not need. Janice Anderson (chapter 12) made some interesting theoretical conceptualizations in her chapter, "Games and the Development of Students' Civic Engagement and Ecological Stewardship." She found that games or virtual environments that seek to engage students in a scaffolding way will need to develop intelligent tutors who prompt students to think about the application of scientific concepts learned during game play to issues in their own communities. Thus, Serious Games will have to look beyond their practical application, in a narrow sense, to make greater use of gamification beyond the computer.

GAMIFICATION AND ONLINE BEHAVIOR MANAGEMENT

Section 1 focuses on behavioral issues and the links between this and the activism associated with online environments. In the first chapter of this section, I consider the legal issues that those who run Websites that allow for folk gamification – where user-generated content such as social media can have a gaming element to it that can affect the liabilities for those that run a Website. The third chapter on tax administration in India was discussed earlier, and the chapter by Loren Falkenberg and Oleksiy Osiyevskyy (chapter 4) looks at the competing gaming and other strategies in decision making of diverse stakeholders in the publishing industry. The chapter communicates many of the challenges of contemporary corporate culture upon which gamification could play an important role.. Being the director of a print and publishing firm, with my first university qualification being in Design and Print, I felt that this chapter communicated many of the challenges of contemporary corporate culture upon which gamification could play an important role. The chapter provides a model for managing competing interests, which can be generalized outside the publishing arena. The fifth chapter, which is the final in section 1, is co-authored by Mark Goode and me. The chapter completes the section nicely by presenting neuro-economic models for understanding and influencing behavior, such as in the environments discussed in the earlier chapters. It shows that an important aspect of any system that seeks to influence the behavior of others is considering the dialogue and narratives of those that use it – to speak the user's language. Also considered in this chapter are the role of antecedents in online environments and their affects on Website strategies. My chapter, which is the first in this section, looks at how owners of Websites can open themselves up to legal claims if they do not moderate the effect of folk gamification such as pseudo-activism, which is where users encourage others to take part in an activity—such as "sharing" or "liking"—for their own benefit, such as a high number of shares of likes.

Stephanie B. Linek, Birgit Marte, and Dietrich Albert's chapter, "Background Music in Educational Games: Motivational Appeal and Cognitive Impact" (chapter 16), is an important chapter for understanding how to influence behavior. Although this chapter is in the education and simulation section, the chapter was not able to find either positive or negative effects of background music on learning. However, the chapter shows that background music can be considered as a motivating design element of educational games without negative side effects on learning.

GAMIFICATION FOR E-LEARNING ENHANCEMENT

E-Learning was a significant theme coming out of the chapters in this book, but selecting four of them in this section will help show their important contributions. Stephen Tang and Martin Hanneghan's chapter, "Designing Educational Games: A Pedagogical Approach" (chapter 11), provides a brief overview of educational games and game-based learning before highlighting theories of learning that is relevant to educational games. The chapter produces a number of guidelines for designing gamified e-learning systems. Like many chapters in the book, including my own, this chapter considers the concept of flow and the importance this plays in user engagement. Joel Foreman and Thomasina Borkman (chapter 13) provide a very important insight into the work on a Massively Multistudent Online Learning Environment for teaching a sociology curriculum. Forman and Borkman's long established work in the area of Multi-User Virtual Learning Environments (MUVLE), some of which is considered in this chapter, makes hyped concepts like Massively Open Online Courses (MOOCs) by Stephen Downs and George Siemens seem like they are trying to teach granny to suck eggs.

Holly Tootell and Alison Freeman present a very important contribution in terms of the application of gamification approaches to early childhood education (chapter 14). The chapter shows how Serious Games, by making use of technologies such as the iPad, could be effective at making compulsory education more fun in ways traditionally not associated with this application of gamification. Göknur Kaplan Akilli provides a brief theoretical framework and a fresh starting point for practitioners in the field who are interested in educational use of games and simulations and their integration into learning environments (chapter 17). This chapter shows how new instructional design and development models are needed to help designers create game-like learning environments that can armor students for the future and build powerful learning into their designs.

GAMIFICATION IN MULTIMEDIA COMMUNICATION

The second chapter in section 3, "From Chaos towards Sense: A Learner-Centric Narrative Virtual Learning Space" by Torsten Reiners, Lincoln C. Wood, and Jon Dron (chapter 15), takes an important look at this. They argue that taking a learner-centered approach in education environments can be more effective at getting an educator's messages across than other means. An important part of multimedia communication, especially where human-computer interaction principles are applied, is the consideration of factors like attention, perception, and memory. Wen-Hao David Huang and Sharon Tettegah's chapter, "Cognitive Load and Empathy in Serious Games: A Conceptual Framework" (chapter 2) is therefore an important contribution to the book. Affective computing is an emerging component of multimedia studies, where the effects of emotions of and between people using digital technologies. Their chapter shows that as users of gamified systems, like Serious Games, enjoy the entertaining, playful aspects of interactive games. They argue this includes an enthralling story, appealing characters, lush production values, a sense of social presence, making choices that affect the direction of the game, and assuming the role of a character and playing with a new personality or identity. State-of-the-art affective computing systems can using gaming principles to make the learning and use of emotions more interesting, as in Serious Games, using gamification principles.

CONCLUSION

Gamification is a simple concept of making non-gaming systems more engaging through applying gaming principles to them. Among its critics, it is a buzzword and concepts like video games will long outlast it. However, gamification is a term that is unlike connectvism, MOOCs, and e-learning 2.0, which are simply marketing gimmicks for long-established principles in education that have happened to make their way into electronic learning systems. Gamification existed long before the word was coined. As an economist, I have argued economics is 90% psychology and 11% mathematics! Nearly every economy or society has gaming principles applied to make life more interesting. Popular culture like Disney's 1964 film Mary Poppins make reference to mundane tasks like cleaning and finding fun in them in order to make them a game. One might ask whether cleaning with Mary Poppins as one's nanny would be the equivalent of a Serious Game. The application of gaming principles to any environment that people lack the motivation to use but it is essential for their success could be seen to be done by gamification and result in serious games.

On that basis, unlike historical figures like Richard A Bartle might think, gamification existed before he even picked up a joystick. Any advancement in technology leads to people using it in a new way as part of a gamification process. Whether this is through trying to influence trending on Twitter or by trying to maximize shares of an image on Facebook, humans will always find new ways to make non-gaming environments more interesting through gamification.

Jonathan Bishop
Centre for Research into Online Communities and E-Learning Sytems, European Parliament, Belgium

Section 1
Behaviour and Activism

Chapter 1
My Click is My Bond:
The Role of Contracts, Social Proof, and Gamification for Sysops to Reduce Pseudo-Activism and Internet Trolling

Jonathan Bishop
Centre for Research into Online Communities and E-Learning Sytems, European Parliament, Belgium

ABSTRACT

The growth in Internet use is not only placing pressure on service providers to maintain adequate bandwidth but also the people who run the Websites that operate through them. Called systems operators, or sysops, these people face a number of different obligations arising out of the use of their computer-mediated communication platforms. Most notable are contracts, which nearly all Websites have, and in the case of e-commerce sites in the European Union, there are contractual terms they must have. This chapter sets out to investigate how the role contract law can both help and hinder sysops and their users. Sysop powers are limited by sysop prerogative, which is everything they can do which has not been taken away by statute or given away by contract. The chapter finds that there are a number of special considerations for sysops in how they use contracts in order that they are not open to obligations through disabled or vulnerable users being abused by others.

INTRODUCTION

"And this is why we can't have nice things," is what it said on the website of Anonymous after it had been brought down by one of their dissident members (Yin, 2011). The group, known for their online acts of 'transgressive humour' often abused people online for their own entertainment. Groups like Anonymous have called themselves 'trolls' in order to legitimise their abusive behavior, with the term troll originally referring to people who try to entertain others through winding people

DOI: 10.4018/978-1-4666-5071-8.ch001

up. Others more subtly spread misinformation to a website's users in order to get kudos points such as share counts. Abusive behavior like this causes difficulties for website owners, but there are things they can do to mitigate this, using contract law.

It is commonplace in the present day for contracts to be completed either at a distance, such as on the Internet, or equally in person. In media and research organisations it is becoming commonplace to issue confidentiality contracts to new staff (Kalra, Gertz, Singleton, & Inskip, 2006). Since the Electronic Communications Act 2000 electronically signing contracts has become equally valid as signing them in person. Many legal cases around the ECA have been whether something constitutes a signature within the meaning of the Act. In J Pereira Fernandes SA v Mehta [2006] EWHC 813 (Ch) Judge Pelling ruled that if a party or a party's agent sending an e-mail types his or her or his or her principal's name to the extent required or permitted by existing case law in the body of an e-mail, then in his view that would be a sufficient signature for the purposes of Section 4 of the ECA.

BACKGROUND

It is sometimes the case that vulnerable and disabled people are at particular risk of entering into contracts with online communities believing that website requires standards of good behaviour, only to be targeted for their disability. This is despite it being argued by some that online medium can promote political equality by facilitating participation and engagement amongst those otherwise excluded from the public arena (Trevisan, 2010).

It is clear, however, that disabled people find the online medium offers easy and accessible communication, as well as greater command over aspects of their life (Bowker & Tuffin, 2002). Indeed, a poll found that disabled people go online and use e-mail twice as much as people without disabilities (Morrell, Mayhorn, & Bennett, 2002). For many disabled people the Internet offers a means where they can be judged on their merits in the messages they post and not be abused simply for having a medical condition (Pfeil & Zaphiris, 2009). However, whilst many disabled people will go online for help to deal with their impairments, all too often they will come across people all too eager to abuse them as offline if they disclose they have a disability. They are often put in a position where they rely on the owner of the website, known as a systems operator, or sysop, to intervene to help them. All too often no help is forthcoming, and the person with the disability is seen as a troublemaker and excluded from the website in the same way as they are excluded offline. This paper will investigate the extent and limitations of the power of sysops to treat disabled people less favourably than others in terms of contract, known as 'sysop prerogative.'

Pseudo-Activism and Its Effects

Pseudo-activism is a type of group-think, social loafing, or free-riding, where people join organisations based around activism, but their intentions are based more around a kind of empathism, where it is more important to be part of a group that believes, than be part of movement that does. In the South Wales Valleys for instance, if one is setting up a not-for-profit organisation, it is expected that it will be member based and not business-minded. Often in this locality people's worth is judged by how often they turn up to a meeting and not how much they actually contribute in real terms. The art of being seen to be doing something, while actually not doing anything is a type of pseudo-activism known as 'slacktivism', describing a form of "feel-good" activism which has little social or political effect (Morozov, 2011). An online equivalent of this is clicktivism, which is a form of pseudo-activism online which simply satisfies the need for affiliating with a group and for social interaction (Alexandrova, 2011)

Manipulation and Deception

It has been argued that if the precision of honest consumer opinions that firms manipulate is sufficiently high, firms of all types, as well as society, would be strictly better off if manipulation of online forums was not possible (Dellarocas, 2006). As one can see in Figure 1, campaigning groups with vested interests are likely to make wild claims to support their campaign, even if there is no truth or evidence to support their view. In the case of the 'Cannabis with Autism' campaign on the right, the group is seeking to justify the legalisation of cannabis because of unsupported claims it can cure conditions like cancer. It is is aimed at vulnerable persons looking for a cure for their illness, even if the efficacy of it for that condition is unproven. On the left hand side is an image claiming that if people share the images that Facebook will donate money to a cause, a perfect example of Clicktivism. The actual purpose of this image is 'hypersharing' – that is, it aims to get a higher number of people sharing it than others might have, by playing on people's sympathies and good-will, which they will happily offer to provide the 'social proof' that they are supportive of their friends who first share it.

Clearly, if a website is allowing the publication of this misinformation then they could be leaving themselves open for legal obligations. Someone who decides to take cannabis for cancer because of the post of the right of Figure 1, may die thinking it cures cancer when it only treats the side-effects of accepted treatments such as chemotherapy (Thomas et al., 2006). Death is something that a contract cannot be used to remove obligations for, meaning the risks of being made liable for this misleading content could be very real.

Aggression: Bleasure and Apprehension

The idea that computer-mediated environments that are violent and aggressive can cause violence and aggression in offline settings is unsupported (Williams & Skoric, 2005). But this does not mean that people are any less aggressive online than they would be if they were offline, just that the increase of aggression and threatening behaviour in one does not lead to greater aggression and threatening behaviour in the other. What is clear however is that is that both aggression and threatening behaviour can have a severe effect on people, in some cases to cause permanent psychological distress, especially if by unacceptable narcissistic individuation (Guillaumin, 1987). Or in other words, a psychological injury to someone without reasonable cause. Bleasure as a term exists

Figure 1. Examples of pseudo-activism: manipulative (left), deceptive (right)

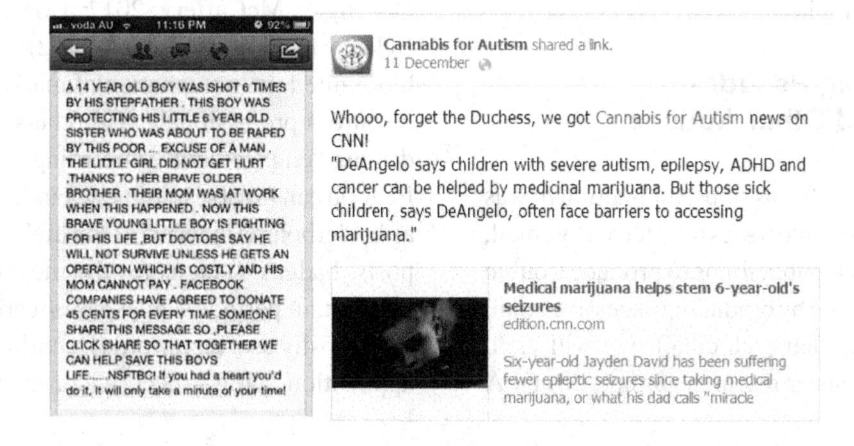

primarily in tort and contract law, deriving from the case of *King v Bristow Helicopters Ltd [2002]* 2002 Scot (D) 3/3, where it entered the English language from the interpretation of a French contract containing the word, 'blessure,' from which it is translated. The public law equivalent is likely to be 'apprehension,' which is an altered state of mind arising from some form of menacing, threatening or grossly offensive act according to Chambers v DPP [2012] EWHC 2157, when applied specifically to Internet trolling. Internet trolling can consist of online message communication over which flaming theory has no control due to the deliberate posting of inflammatory or controversial messages online (Gully, 2012). It has however been noted that online apprehension operates differently from traditional verbal notions of communication apprehension (Hunt, Atkin, & Krishnan, 2012).

It may be clear that a website telling someone their cancer can be cured by using cannabis is likely to end up with social impairment due to over-stimulation of the prefrontal cortex with dopamine (Bishop, 2011b; Bishop, 2012b; Thomas et al., 2006). Equally a vulnerable or disabled person might, depending on their condition, be more likely to fall for other pseudo-activism, thinking they are helping others especially if the number of people that have already used or shared the artefact is high (Burgoon, Stoner, Bonito, & Dunbar, 2003). This is known as social proof, as one is willing to agree to something if one's friends agree with it.

Counteracting Pseudo-Activism and Other Abuses

One of the earliest forms of pseudo-activism was chain emails. These messages transferred by email, often simply seek publicity or to produce trouble for a selected victim by producing excessive email traffic, including that each email sent will yield a donation for cancer research (Smith, 2011). A typical chain email starts with a hook to catch your interest (e.g. "Save a dying child!") followed by a threat (e.g. the child would die) and then finally a request (e.g. pass the email on), for which none are true (Espeland, 2003). Related are email petitions, which are some of the worst forms of pseudo-activism, as people get the sense of doing something when all it involves is forwarding on an email (Espeland, 2003).

Social Proof

As discussed above, social proof can have the adverse effect of friends taking advantage of other friends and trusted sources taking advantage of vulnerable persons, or in some cases disabled people, if they promised a wanted cure for instance. Having a friend forward an electronic message, such as e-mail, can be quite persuasive if it says what one wants to hear. In recent US elections, emails were circulating from the outset about Barrack Obama's life as a Muslim, and even though the emails were untrue because Obama is a Christian, many believed them (Clayton, 2010).

Gamfication

Gamification is the concept of applying game-design thinking through the use of game mechanics to drive game-like player behaviour to non-game applications to make them more fun and engaging (Deterding, Sicart, Nacke, O'Hara, & Dixon, 2011; Dorling & McCaffery, 2012; Downes-Le Guin, Baker, Mechling, & Ruylea, 2012). Gamification has ignited the imagination of marketers, human resources professionals, and others interested in driving "engagement" (Deterding, 2012). Gamification can involve as minor changes to a website as leaderboards in terms of the highest number of posts, badges, such as for the most kudospoints in a certain period. Whilst gamification has been seen mostly as a buzzword (Deterding, 2012), its application can go much wider than marketing, and can be used to enhance behaviour in online communities in order to reduce both lurking and flame trolling (Bishop, 2012a; Bishop, 2013). Gamification intends to capture the application of game design elements, such as high score lists, rewards, badges, among others, in non-gaming contexts (Cornillie, Lagatie, & Desmet, 2012). Indeed, a frequently used model for gamification is to equate an activity in the non-game context with points and have external rewards for reaching specified point thresholds (Nicholson, 2012).

As can be seen from Figure 2, some websites built around gamification can result in people making the effort to record their check-in to community venues. This has the advantage that disabled and vulnerable people, who might not want to make the effort to take part in their communities, could be motivated to do so through collecting points and badges. However, as can also be seen from Figure 2, when a particular person dominates a particular location, it may mean that less people will make the effort to check in, knowing it is impossible for them to compete.

Foursquare is a location-based social application that helps users to explore the world around them and share their experiences with friends (Sklar, Shaw, & Hogue, 2012). One intuition behind long-term users in Foursquare is that they conduct their usual activities within a similar time frame, with some random deviations (Melià-Seguí, Zhang, Bart, Price, & Brdiczka, 2012). It has been found that repetition of favourite activities has a moderate effect upon so-called Internet addiction, in line with the assertion of rational addiction theory (Chou & Ting, 2003). Through an analysis of treatment and convenience samples it has been found that instances of dependency are linked to many disabled groups, who may differ systematically from others (Sara et al., 2012). This places risk on sysops who use gamification on their website, whom may be liable for any harm to disabled people should they develop addition-like symptoms.

Systems like Foursquare can both aid and reduce pseudo-activism through social proof. If one's friend has checked into a place often, one might be pretty sure that it is a safe place to go. But equally, as it is currently easy to add any location to the platform, it can be a false one. This means that a vulnerable person could possibly be easily duped into thinking the claims of a pseudo-activist about a particular location are true, when they are not. On such a site a sysop would have to state in the contract terms that indicate the website should not be taken as providing accurate information, and that the users should seek third-party advice before using it. As Foursquare uses badges to promote certain behaviours, then it, or websites like it, might want to make clear that they do not endorse any premises or warrant for their suitability, and that accessing them are at users' own risk. It could also be stated that collecting badges, or climbing leader-boards, should be done responsibly and should not be considered as alternatives to a lifestyle one's physician recommends.

Figure 2. An example of a website using gamification – Foursquare

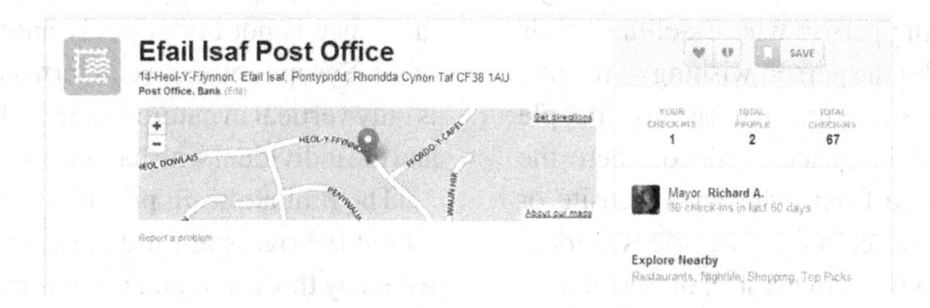

Sysop Prerogative

Sysop prerogative is the right that an owner of an online community has to take whatever actions they wish in order that the community functions the way they wish (Bishop, 2011a). In common law countries it is understood that someone can do whatever they want unless they give that right way or it is taken away from them by a superior authority, such as Parliament, in the case of the United Kingdom (UK). It can therefore be expected that the systems operator (sysop) who owns an online community is free to operate their community under any rules they want that has not been taken away from them by statute. After all the community is surely their property which they can use and dispose of in any way they choose? However, with each right or freedom comes certain duties and obligations, something which is enshrined in the European Convention on Human Rights. This convention, like the Treaty on the Functioning of the European Union, places obligations on nation states to guarantee certain rights and not to interfere with the enjoyment of these rights except where they conflict with other rights or duties. This means that sysop prerogative may be both guaranteed and limited by European Human Rights Law.

A contract is a legally enforceable agreement that gives rise to obligations for the parties involved (Koffman & Macdonald, 2007). In its most basic form, it is the agreement between two or more parties to exchange one thing for another. In the United Kingdom contracts require several things in order to be valid. These are; offer, acceptance, consideration and the intention to make legal relations. Unusually, an offer is not in legal terms made by the person who is selling goods or services, but by the person wishing to receive them. This may seem unusual because people usually pay whatever price is stated where the goods are accessed from, whether a website or traditional retail store. In fact, the prices marked up are only an 'invitation to treat.' This means that the prices shown are an invitation for the buyer to purchase the good or service at that price. They may, legally speaking, offer a different price for those goods, but the seller is under no obligation to sell it at that price, or even the original price they offered. In this context the price and goods are called 'consideration', which means they are the things the contract is considering for transfer between the parties. If the seller decides to accept the offer from the buyer then the condition of 'acceptance' is made. The remaining feature of a contract, the 'intention to create legal relations' is not often spoken of, but is an important part in deciding whether a contract can be legally enforced through the courts.

In terms of sysop prerogative the first of the elements of a contract are easily understandable. A system gives its users the opportunity to post content in exchange for them hosting it and making it available to other users. The user offers to post, the sysop may accept that offer to post, and the consideration is the provision of content and publication of that content. Where things are less certain however is in the case of the 'intention to create legal relations.'

It is a well established principle in contract law that an agreement cannot impose any additional obligations nor give any enforceable rights to a person who is not part of a contract (Collins, 2003). This has many advantages for sysops in terms of the extendibility of sysop prerogative, but is does not provide as a good a deal for users who have entered into agreements with a website. In terms of multi-user websites however, it means that even where a user abuses the rights of another user and both those users are agreeing to the terms and conditions prohibiting that conduct the one user's grievance is not horizontally enforceable on the other. Sysop prerogative in terms of contract law is only vertical in nature where it affects the sysop and the individual who has the contract with them.

The limitations in privity of contracts, which inhibit the rights of third parties who may be affected by those agreements was in part dealt with through the Contracts (Rights of Third Parties) Act 1999. This meant that where a third party was explicitly mentioned in a contract that they would have the rights to enforce it. As normal terms and conditions on websites do not refer to users by name, then it is unlikely users will have any further rights from this Act, as they are not mentioned directly within it.

Another important consideration in contracts that affect sysop prerogative are terms that are either explicit or implied. Explicit terms are those that the parties expressly articulate in the contract (Loots & Charrett, 2009). Implied terms are those that are added to, or place limits on, expressly stated terms (Cohen, 2000). These affect the rights and obligations of both sysops and users.

The explicit terms in contracts between users and sysops are often the terms and conditions for the website and sometimes separate 'house rules.' In common law jurisdictions where these contracts do not explicitly prohibit a certain act, the sysop can have the right to do what they want – unless there is an explicit term that says otherwise. Explicit terms can come from legislation in its various forms. One significant piece of legislation that affects those sysops who offer e-commerce services is the Distance Selling Directive [97/7/EC], which was transposed into the law of the United Kingdom through the Consumer Protection Distance Selling Regulations 2000. This legislation not only imposes implied terms on the sysop, but expects them to put these terms into their contracts, such as terms and conditions, as if they were expressed terms.

Related to expressed terms are implied terms, which are terms that can derive from statute or common law which while not expressed in the contract have effect as if they did. There are times when a court can order an implied term into a contract, such as where an issue arises upon which the parties have not made an express provision (Mead, Sagar, & Bampton, 2009). Sysops can try to limit some of their liability for not performance of their contract – such as allowing one user to abuse other users – by using exemption clauses. These exclude or restrict such liability and take the form of either exclusion clauses or limitation clauses (Bainbridge, 2007). However, even by inserting such terms, exemption clauses are considered inherently dangerous and detrimental to the weak party, so that their effects are usually levelled down by the courts (Girot, 2001). More importantly nowadays, exemption clauses are also controlled by statute. The Unfair Contract Terms Act 1977 limits the extent to which liability can be excluded or limited for breach of contract, or for negligence, or under the terms implied by the Sales of Goods Act 1979 and other legislation containing similar provisions (David I. Bainbridge, 2004).

Existence, Performance and Jurisdiction

The advent of computer technology and Internet commerce has probably resulted in an increase in the number of contracts formed across state and national boundaries, with consequent conflict-of-laws issues (Schwabach, 2006). This produces the problem of determining whether contracts exist, among others. While European Union Directives have harmonised contract law across countries that are part of it there are still differences in each jurisdiction relating to contracts (Minke, 2013).

Because of this differing legal set-up it can cause problems for sysops. When the sysop's website or information system accepts a fee for providing a service it is often very clear that a contract exists, but websites where information is provided for free have no contractual obligations to the user (McKenzie, 2001). Much of contract law in the United States applies across States via the Uniform Contract Code, or UCC, but the complications that exist in the EU also apply the US where contract law still differs in different states (Minke, 2013).

Two issues one can consider together in relation to contracts between users and sysops is breach and frustration. Many sysops will act as 'banhammers', which is the name of a mythical object used to ban people from online communities (Rufer-Bach, 2009). A breach of contract is the non-performance of a legal duty when due under a contract (Burton, 1980). The frustration of a contract is induced by the action of one of the parties to a contract resulting in that party is not thereby relieved from its contractual obligations (Blackshield, 2006).

A sysop could be considered to have frustrated the contract when through inaction they cause a user to suffer abuse, such as from Internet trolling. This may leave them open to legal action from users. Therefore sysops limit sysop prerogative by placing terms in contracts where they require users to agree not to be aggressive towards other users. If they fail to regulate the abuse of users then they are failing in their contract, meaning their rights are not enforcable. Equally if the sysop bans a user from their website when that user did not breach their side of the contract, such as if that user has a difference of opinion, then the sysop could be in breach of contract.

One option for either party during a dispute over breach of contract is to seek equitable remedies. Equitable remedies are court-ordered remedies that require the breaching party to do more than pay compensation, requiring a party to carry out a specific performance (Stark, 2002). Restitutionary remedies are also available for the reversal of an unjust enrichment by the defendant at the expense of the claimant, but does not strictly require a breach of contract occur (Oughton & Davis, 1996). This legal principle could pose problems for sysops if their terms and conditions contain behavioural codes or the website is described as 'non-political,' for instance. In the case of the former, the sysop could be liable for restitutionary remedies if they fail to enforce the behavioural codes, as this could be considered misrepresentation. Also, many independent websites claim to be 'non-political' and turn out to have politicised message boards, where users get accused and where restitutionary remedies may be applicable. An example of such a website is in Figure 3.

These independent online communities place their sysops in difficult situations when their websites are used by vulnerable or disabled people. One key challenge in the area of disablist hate crimes attaches to wider constructions of disabled people as 'vulnerable' (Roulstone, Thomas, & Balderston, 2011; Sin, Hedges, Cook, Mguni, & Comber, 2011), and thus they are more likely to be targeted by flame trollers.

Anonymity and Confidentiality

There is a general presumption online that anonymity is something that should be protected. Indeed, sysops are often not very willing to provide this information, arguing that the trust of their users to remain anonymous is the main pillar supporting

Figure 3. The Ponty Town website claims to be 'non-political' (left) but has a politics forum (right)

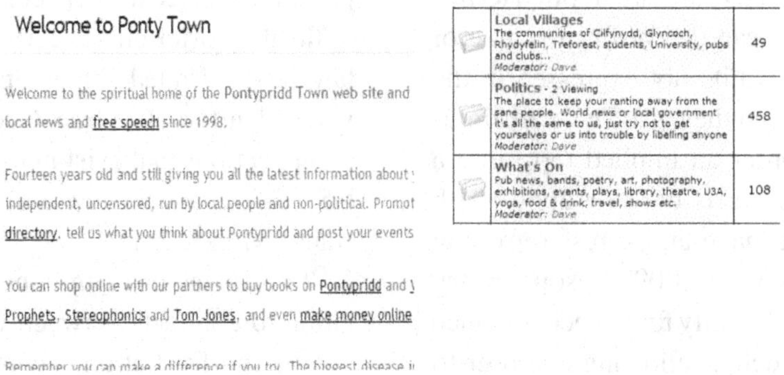

their contract (Roosendaal, 2007). When anonymity is no longer guaranteed, the number of clients immediately will decrease, it is claimed (ibid).

Until recently it was assumed that anonymity and the use of pseudonyms were synonymous. With the launch of 4chan and the rise of a collective identity of 'Anonymous', this is starting to change. Within this movement it is becoming common for Internet users not to sign their contributions with their real name or a similarly identifiable pseudonym, but to post along with others under the username 'Anonymous,' as presented in Figure 4.

Some have challenged whether anonymity online is protected in contractual terms online as it is in traditional journalism. Lambert (2008) argues that there are important distinctions between the protections applicable to a reporter's source arising out of a contract promising anonymity to the source, compared to where a blogger would typically lack such protections. One conclusion is certainly the continuing and vital importance of journalists being able to work with sources on an agreed contract of anonymity, a principle which lies at the heart of whistle-blowing in open societies (Hargreaves, 2005). It is not uncommon, however, for pseudonyms to be freely used online where there is not a genuine anonymity.

A problem where pseudonyms are used is that it increases the likelihood of flame trolling as the poster has a persistent identity, yet is able to hide their true identity. Clearly where a sysop has a contract with a user this could open them up to claims if they do not enforce it. One way around this for sysops is to have ''binding in honour only' clauses.' These are where the parties agree not to create a contract, even if it looks pretty clear that one exists (Keenan & Smith, 2007). The enforcability of these have been questioned however, to date the courts have decided that binding in honor only clauses and contracts are not legally binding (Jones, 1995).

Avatars and Embodiment of Trust

Millions of people are having meaningful social, economic, and medical interactions on a daily basis via avatars in online communities (Segovia,

Figure 4. An example of message board posts that are genuinely anonymous

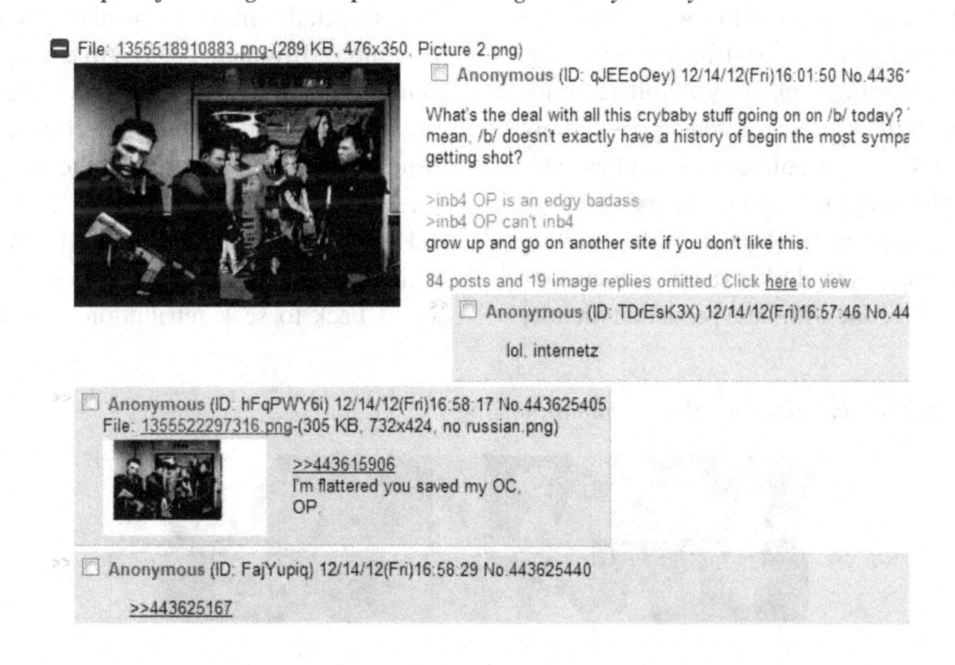

Bailenson, & Monin, 2009). An increasing number of studies highlight the fact that the embodiment of avatars in online communities enhances the development of interpersonal skills and online achievements can be successfully applied in offline settings (Desiato, 2009). Where the trust towards one users increases that of another, it can be seen to be 'social proof.' Such easy of building trust places vulnerable people at risk from people using avatars to build social proof among others in order to deceive them. It has been postulated that even though avatars are often used in virtual worlds people are able to gather assets, trade, communicate with the other players' avatars, make contracts, and so on. Others argue, however, that the contracts that govern virtual worlds in which avatars are used do not provide adequate protection for even the most basic property interests (Ludwig, 2011).

It is clear that if it is possible to make contracts in virtual worlds where avatars exist and the interaction is fluid then considerations need to be taken into account to protect people with disabilities or vulnerable people. Defences exist to the formation of contracts, including duress, undue influence, fraud or misrepresentation, mistake, and unconscionability (Ricks, 2011). These are many things that occur online with vulnerable people are particularly at risk from experiencing.

Figure 1 shows a morphology of troller avatars, which reflects the structure of stereotypes of those users who take part in online communities. This 12-character model is based on an extension of the 10-character model in (Bishop, 2008). Starting from the top left is the lurker (i.e. Loner avatar). A lurker is someone without a persistent identity in the online community. This may be reflected in them having the default avatar, as in this case, which could also include a white silhouette on a grey background. Lurkers are particularly vulnerable to abuse from the more experienced members. Even when they start posting and thus become 'novices' (Kim, 2000), there is still the risk they could go back to not posting if they are treated unfavourably. Unconscionability might apply in this case if the terms and conditions require the lurker, who may be a new member, to act in an appropriate way, yet this is unachievable to the way the community is biased against them. On that basis if a lurker is banned (i.e. ban-hammered) then they could have a claim against the sysop. Often those who target lurkers are elders (the image underneath the lurker in Figure 5. Known as 'trolling for newbies,' they will set little traps that they know newer members will fall for. In a clichéd way this would be an elder in a forum for Apple users to praise Microsoft, knowing that the novice Mac Fanatic will be irritated resulting in a barrage of abusive messages (called a 'flame war'), which the established members enjoy, but the novice does not.

The basis of both duress and undue influence is that one of the parties gave no real consent to the contract, this making it voidable, but not void (Salzedo & Brunner, 1995). An example of this could be that a person joins an online community and agrees to a behaviour code in order to respond to abuse by them, yet the sysop had no intention of letting them having a 'right to reply,' and instead makes things more difficult for them. Whilst it could be argued that an E-Venger who comes back to seek retribution in online com-

Figure 5. Avatars that exist in online communities

munities may be in breach of computer misuse laws for accessing a website they have been banned from, it might be considered proportionate to right a wrong if that wrong is to correct misinformation. A sysop that does not allow the right to reply and acts as a ban-hammer may not only be opening themselves up to claims of breach of contract but also claims for defamation.

A party who has entered into a contract in reliance on a false statement of fact by the other party will have an action for misrepresentation (Salzedo & Brunner, 1995). This could include a website which has a strong behaviour code, yet the sysop had no intention of enforcing it. In terms of misrepresentation it is important to show 'causation.' Fraudulent misrepresentation occurs where the person would not have entered in to the contract if they had known that a certain representation that was claimed to be part of the contract when actually was not (Chen-Wishart, 2012). Where the terms of conditions represent certain rights or codes of conduct, then if a sysop does not provide these and the user only joined the site because they existed, then the sysop is liable for fraudulent misrepresentation. In most instances of signing up to a website this is not the case, as most users would not even read the terms and conditions when signing up. In such a situation, the sysop may be liable for negligent misrepresentation. The crux of an action for negligent misrepresentation is that the defendant carelessly distributed information to those who could reasonably be expected to rely upon such information to their detriment (McMahon, 2012). This could include long terms and conditions which no one would likely read when they sign-up to a website. It could also include terms that one would not expect to be in the contract, such as transferring one's intellectual property as opposed to licensing it. Whilst negligent misrepresentation is considered by some to be considered part of tort law, others consider it to be more attributable to contract law where the obligation usually arises from (Gergen, 2011).

A unilateral mistake occurs when only one party makes a mistake regarding the contract, whereas when both parties to a contract make the same mistake, a mutual mistake occurs (Ashcroft & Ashcroft, 2007). This is most obvious in ecommerce websites where the sysop has made a mistake in terms of making the invitation to treat price less than it actually is. In the United Kingdom it is a criminal offence to buy something from an e-commerce website knowing it is not being sold at the right price. Under contract law in the UK it is the buyer that makes the 'offer' and not the seller, who gives the 'acceptance' to an offer. On this basis, a user that takes advantage of a mistake in relation to the invitation to treat not only carries out a criminal act, but also commits a mistake. The fact that the criminal offence exists, makes determining whether the breach is a unilateral or mutual mistake easier. Unconscionability is a judicial weapon against unfair contracts, which is based on the idea that a contract may be unenforceable because it is shockingly unfair (Emanuel, 2010). A website which gives the consumer far less rights that they might expect from a website is a good example. It might not be expected for a user to read all of the terms and conditions, but there are certain things which might be considered unfair, if they had read it.

All these factors place a concern on the Internet user that they could be conned if they bought from the wrong website. This is why the websites that build brands are the embodiment of trust among consumers, but the very nature of the Internet puts an unprecedented strain on that tie that binds more than offline (Upshaw & Taylor, 2000). Where undertakings see their main purpose in the use of the Internet compared to other platforms as the embodiment of trust on the Internet, it has been argued it is important to investigate offline forms of regulation-by-law that can fit in the online world and in order to help the trust establishment (Ažderska & Jerman Blažič, 2011).

Future Research Directions

Through seeking to understand the nature of contract law in relation to online communities, particular in relation to sysop prerogative and gamification, this chapter has gone some way to providing systems operators with guidance on the risks involved in running online communities. Empirical studies, such as an ethnography, need to be conducted in order to see the effect contracts such as terms and conditions have on different users. It may be that toughly worded contracts will put some people off, whilst unenforced others will put others off. It is clear that contract law will have an increasing role to play in online communities, particularly in protecting users from Internet trolling.

CONCLUSION

The saturation of the Internet and social networking social services is proving to be problematic not only for the Internet Service Providers trying to optimise bandwidth, but also the many websites which are hosted in cyberspace. The managers of these, called systems operators, or sysops face challenges in terms of navigating a legal landscape where for simply providing a computed-mediated communications platform they can be liable for all the actions of their users. Some sysops, however, will exploit the immediacy of cyberspace by using certain unfair terms and processes, which in some cases makes a contract unenforcable. Contracts, such as terms and conditions, or 'house rules' can offer a way for sysops to get out of obligations not due to their fault, so their users can avoid manipulation, aggression, and deception, particularly if they are from vulnerable or disabled groups. Through also using 'gamification,' which is the application of gaming principles to non-gaming environments.

The chapter finds that in order to maximise the use of gamification, sysops need to consider several issues. Namely; existence, performance and jurisdiction of contracts; user anonymity and confidentiality; and avatars and the embodiment of trust. The existence and performance of contracts agreed between sysops and users can restrict sysop prerogative in a number of ways. Sysops may say in their terms and conditions that certain abusive behaviours are not welcomed, but often will do nothing to enforce these policies. Issues such as anonymity, usually manifested through pseudonyms that are not anonymous, can increase flame trolling and this could leave sysops open to claims. The use of avatars as a means for users to create a persistent identity can be abused through the trust that users get which are embodied through them. Such means can lead to sysops being open to misrepresentation.

REFERENCES

Alexandrova, E. (2011). Metamorphoses of civil society and politics: From ganko's café to facebook. In G. Lozanov, & O. Spassov (Eds.), *Media and politics*. Konrad Adenauer Stiftung.

Ashcroft, J. D., & Ashcroft, J. E. (2007). *Law for business*. South-Western Pub.

Ažderska, T., & Jerman Blažič, B. (2011). A novel systemic taxonomy of trust in the online environment. *Towards a Service-Based Internet,* 122-133.

Bainbridge, D. I. (2007). *Introduction to information technology law*. Upper Saddle River, NJ: Prentice Hall.

Bainbridge. (2004). *Introduction to computer law*. New York: Longman.

Bishop, J. (2008). Increasing capital revenue in social networking communities: Building social and economic relationships through avatars and characters. In C. Romm-Livermore, & K. Setzekorn (Eds.), *Social networking communities and eDating services: Concepts and implications*. Hershey, PA: IGI Global. doi:10.4018/978-1-60566-104-9.ch004

Bishop, J. (2011a). All's well that ends well: A comparative analysis of the constitutional and administrative frameworks of cyberspace and the united kingdom. In A. Dudley-Sponaugle, & J. Braman (Eds.), *Investigating cyber law and cyber ethics: Issues, impacts and practices*. Hershey, PA: IGI Global. doi:10.4018/978-1-61350-132-0.ch012

Bishop, J. (2011b). *The role of the prefrontal cortex in social orientation construction: A pilot study.* Paper Presented to the BPS Welsh Conference on Wellbeing. Wrexham.

Bishop, J. (2012a). The psychology of trolling and lurking: The role of defriending and gamification for increasing participation in online communities using seductive narratives. In H. Li (Ed.), *Virtual community participation and motivation: Cross-disciplinary theories* (pp. 160–176). Hershey, PA: IGI Global. doi:10.4018/978-1-4666-0312-7.ch010

Bishop, J. (2012b). Taming the chatroom bob: The role of brain-computer interfaces that manipulate prefrontal cortex optimization for increasing participation of victims of traumatic sex and other abuse online. In *Proceedings of the 13th International Conference on Bioinformatics and Computational Biology (BIOCOMP'12)*. BIOCOMP.

Bishop, J. (2013). The psychology of trolling and lurking: The role of defriending and gamification for increasing participation in online communities using seductive narratives. In J. Bishop (Ed.), *Examining the concepts, issues and implications of internet trolling* (pp. 106–123). Hershey, PA: IGI Global. doi:10.4018/978-1-4666-2803-8.ch009

Blackshield, A. (2006). Constitutional issues affecting public private partnerships. *UNSWLJ*, *29*, 302.

Bowker, N., & Tuffin, K. (2002). Disability discourses for online identities. *Disability & Society*, *17*(3), 327–344. doi:10.1080/09687590220139883

Burgoon, J. K., Stoner, G., Bonito, J. A., & Dunbar, N. E. (2003). Trust and deception in mediated communication. In *Proceedings of the 36th Annual Hawaii International Conference*. IEEE.

Burton, S. J. (1980). Breach of contract and the common law duty to perform in good faith. *Harvard Law Review*, *94*, 369. doi:10.2307/1340584

Chen-Wishart, M. (2012). *Contract law*. Oxford, UK: Oxford University Press. doi:10.1093/he/9780199644841.001.0001

Chou, T. J., & Ting, C. C. (2003). The role of flow experience in cyber-game addiction. *Cyberpsychology & Behavior*, *6*(6), 663–675. doi:10.1089/109493103322725469 PMID:14756934

Clayton, D. M. (2010). *The presidential campaign of Barack Obama: A critical analysis of a racially transcendent strategy*. San Francisco: Taylor & Francis.

Cohen, G. M. (2000). Implied terms and interpretation in contract law. Encyclopedia of Law and Economics, 3, 78-99.

Collins, H. (2003). *The law of contract*. London: Butterworths.

Cornillie, F., Lagatie, R., & Desmet, P. (2012). Language learners' use of automatically generated corrective feedback in a gamified and task-based tutorial CALL program: A pilot study. Paper presented at the EuroCALL 2012: CALL: Using, Learning, Knowing. London, UK.

Dellarocas, C. (2006). Strategic manipulation of internet opinion forums: Implications for consumers and firms. *Management Science*, *52*(10), 1577–1593. doi:10.1287/mnsc.1060.0567

Desiato, C. (2009). *The conditions of permeability: How shared cyberworlds turn into laboratories of possible worlds*. Paper presented at the International Conference on CyberWorlds. New York, NY.

Deterding, S. (2012). Gamification: Designing for motivation. *Interaction*, *19*(4), 14–17. doi:10.1145/2212877.2212883

Deterding, S., Sicart, M., Nacke, L., O'Hara, K., & Dixon, D. (2011). Gamification: Using game-design elements in non-gaming contexts. In *Proceedings of the 2011 Annual Conference on Human Factors in Computing Systems*. New York, NY: IEEE.

Dorling, A., & McCaffery, F. (2012). The gamification of SPICE. In *Software Process Improvement and Capability Determination* (pp. 295–301). Academic Press. doi:10.1007/978-3-642-30439-2_35

Downes-Le Guin, T., Baker, R., Mechling, J., & Ruylea, E. (2012). Myths and realities of respondent engagement in online surveys. *International Journal of Market Research*, *54*(5), 1–21. doi:10.2501/IJMR-54-5-613-633

Emanuel, S. (2010). *Emanuel law outlines: Contracts*. Aspen Law & Business.

Espeland, P. (2003). *Life lists for teens: Tips, steps, hints, and how-tos for growing up, getting along, learning, and having fun*. Minneapolis, MN: Free Spirit Publishing.

Gergen, M. P. (2011). *Negligent misrepresentation as contract*. Berkeley, CA: University of California.

Girot, C. (2001). *User protection in IT contracts: A comparative study of the protection of the user against defective performance in information technology*. Boston: Kluwer Law International.

Guillaumin, J. (1987). Entre blessure et cicatrice: Le destin du négatif dans la psychanalyse. Ed.s Champ Vallon.

Gully, A. (2012). It's only a flaming game: A case study of arabic computer-mediated communication. *British Journal of Middle Eastern Studies*, *39*(1), 1–18. doi:10.1080/13530194.2012.659440

Hargreaves, I. (2005). The ethics of journalism: A summing-up for lord hutton. *Ethics, Law, and Society*, *1*, 153.

Hunt, D., Atkin, D., & Krishnan, A. (2012). The influence of computer-mediated communication apprehension on motives for facebook use. *Journal of Broadcasting & Electronic Media*, *56*(2), 187–202. doi:10.1080/08838151.2012.678717

Jones, M. A. (1995). Liability for psychiatric Illness—More principle, less subtlety? *Web Journal of Current Legal Issues Yearbook*, *258*, 259.

Kalra, D., Gertz, R., Singleton, P., & Inskip, H. M. (2006). Confidentiality of personal health information used for research. *British Medical Journal*, *333*(7560), 196. doi:10.1136/bmj.333.7560.196 PMID:16858053

Keenan, D. J., & Smith, K. (2007). *Smith and keenan's english law*. Harlow, UK: Longman Publishing Group.

Kim, A. J. (2000). *Community building on the web: Secret strategies for successful online communities*. Berkeley, CA: Peachpit Press.

Koffman, L., & Macdonald, E. (2007). *The law of contract*. Oxford, UK: OUP.

Lambert, K. A. (2008). Online identities unmasked. *Litig.News, 34,* 10.

Loots, P. C., & Charrett, D. (2009). *Practical guide to engineering and construction contracts.* CCH Australia Limited.

Ludwig, J. L. (2011). Protections for virtual property: A modern restitutionary approach: Why would anyone pay 50,000 for a virtual property. *Loy. LA Ent.L.Rev., 32,* 1.

McKenzie, A. (2001). Liability for information provision. In A. Scammell (Ed.), *Handbook of information management* (8th ed.). ASLIB Information Managment.

McMahon, M. J. (2012). Relationship between premium finance agency and insurance company is not sufficient to sustain a cause of action for negligent misrepresentation. *St. John's Law Review, 56*(2), 9.

Mead, L., Sagar, D., & Bampton, K. (2009). *Cima official learning system fundamentals of ethics, corporate governance and business law.* Cima Pub.

Melià-Seguí, J., Zhang, R., Bart, E., Price, B., & Brdiczka, O. (2012). Activity duration analysis for context-aware services using foursquare check-ins. In *Proceedings of the 2012 International Workshop on Self-Aware Internet of Things,* (pp. 13-18). IEEE.

Minke, A. G. (2013). *Conducting transatlantic business: Basic legal distinctions in the US and Europe.* London: Bookboon.

Morozov, E. (2011). *The net delusion: How not to liberate the world.* New York: Penguin.

Morrell, R. W., Mayhorn, C. B., & Bennett, J. (2002). Older adults online in the internet century. *Older Adults, Health Information, and the World Wide Web,* 43-57.

Nicholson, S. (2012). *A user-centered theoretical framework for meaningful gamification.* Paper presented at the Games+Learning+Society 8.0. New York, NY.

Oughton, D. W., & Davis, M. (1996). *Sourcebook on contract law.* Cavendish.

Pfeil, U., & Zaphiris, P. (2009). *Theories and methods for studying online communities for people with disabilities and older people.* Academic Press. doi:10.1201/9781420064995-c42

Ricks, V. (2011). *Assent is not an element of contract formation (No. SSRN 1898824).* Social Science Research Network.

Roosendaal, A. (2007). Elimination of anonymity in regard to liability for unlawful acts on the internet. *International Journal of Technoloy Transfer and Commercialisation, 6*(2), 184–195. doi:10.1504/IJTTC.2007.017805

Roulstone, A., Thomas, P., & Balderston, S. (2011). Between hate and vulnerability: Unpacking the British criminal justice system's construction of disablist hate crime. *Disability & Society, 26*(3), 351–364. doi:10.1080/09687599.2011.560418

Rufer-Bach, K. (2009). *The second life grid: The official guide to communication, collaboration, and community engagement.* Sybex.

Salzedo, S., & Brunner, P. (1995). *Briefcase on contract law.* Routledge-Cavendish.

Sara, G., Burgess, P., Harris, M., Malhi, G., Whiteford, H., & Hall, W. (2012). Stimulant use disorders: Characteristics and comorbidity in an australian population sample. *The Australian and New Zealand Journal of Psychiatry.* doi:10.1177/0004867412461057 PMID:22990432

Schwabach, A. (2006). *Internet and the law: Technology, society, and compromises.* Abc-clio.

Segovia, K. Y., Bailenson, J. N., & Monin, B. (2009). Morality in tele-immersive environments. In *Proceedings of the 2nd International Conference on Immersive Telecommunications*. IEEE.

Sin, C. H., Hedges, A., Cook, C., Mguni, N., & Comber, N. (2011). Adult protection and effective action in tackling violence and hostility against disabled people: Some tensions and challenges. *Journal of Adult Protection*, *13*(2), 63–75.

Sklar, M., Shaw, B., & Hogue, A. (2012). Recommending interesting events in real-time with foursquare check-ins. In *Proceedings of the Sixth ACM Conference on Recommender Systems,* (pp. 311-312). ACM.

Smith, R. E. (2011). *Elementary information security*. Jones & Bartlett Learning.

Stark, T. (2002). *Negotiating and drafting contract boilerplate*. Incisive Media, LLC.

Thomas, J., Andrysiak, T., Fairbanks, L., Goodnight, J., Sarna, G., & Jamison, K. (2006). Cannabis and cancer chemotherapy: A comparison of oral delta-9-thc and prochlorperazine. *Cancer*, *50*(4), 636–645. doi:10.1002/1097-0142(19820815)50:4<636::AID-CNCR2820500404>3.0.CO;2-4

Trevisan, F. (2010). *More barriers? Yes please: Strategies of control, co-optation and hijacking of online disability campaigns in scotland*. Glasgow, UK: Social Science Research Network. doi:10.2139/ssrn.1667427

Upshaw, L. B., & Taylor, E. L. (2000). *The masterbrand mandate: The management strategy that unifies companies and multiplies value*. Hoboken, NJ: Wiley.

Williams, D., & Skoric, M. (2005). Internet fantasy violence: A test of aggression in an online game. *Communication Monographs*, *72*(2), 217–233. doi:10.1080/03637750500111781

Yin, S. (2011). *Anonymous v anonymous: This is why we can't have nice things*. PM Magazine.

KEY TERMS AND DEFINITIONS

Gamification: Gamification is the systematic application of gaming principles from one environment into an environment where they are not typically used.

Internet Trolling: Internet trolling is the practice of posting messages to the Internet which are either provocative or offensive.

Pseudo-Activism: Pseudo-activism is the phenomenon where Internet users will post content to the Internet which says that if its message isn't acted on that some adverse consequences will occur. In its earliest for it was chain emails and more recently has been the posting of images on Facebook asking people to like or share them.

Social Proof: Social proof is the phenomenon where something is perceived to be desirable or valid because a friend has suggested that it is.

Systems Operator (or Sysop): A sysop is the owner or manager of a website who has the power to create and remove content, users and otherwise manage the online community they are in charge of.

Chapter 2
Cognitive Load and Empathy in Serious Games:
A Conceptual Framework

Wen-Hao David Huang
University of Illinois – Urbana-Champaign, USA

Sharon Tettegah
University of Illinois – Urbana-Champaign, USA

ABSTRACT

The design of serious games does not always address players' empathy in relation to their cognitive capacity within a demanding game environment. Consequently players with inherent limitations, such as limited working memory, might feel emotionally drained when the level of empathy required by a game hinders their ability to cognitively attain the desired learning outcome. Because of the increasing attention being given to serious games that aim to develop players' empathy along with their cognitive competencies, such as Darfur is Dying (Ruiz et al., 2006), there is a need to investigate the empirical relationship between players' cognitive load and empathy development capacity during serious game play. Therefore this chapter examines cognitive load theory and empirical work on empathy development to propose a conceptual framework to inform the research and design of serious games that have empathy as part of the learning outcomes. Future research should focus on implementation and empirical validation of the proposed framework.

INTRODUCTION

The concept of serious games and their processes was first noted by Abt in 1970. In this seminal work, Abt argued that serious games should require players to make consecutive decisions in order to achieve predetermined game objectives. Players' actions, in turn, are bound by rules and constraints while competing with others on various challenges (Abt, 1970). To these characteristics, Gredler (1994) and Suits (1978) add the aspects of voluntary participation of players

DOI: 10.4018/978-1-4666-5071-8.ch002

and entertainment. Prensky (2001) expanded the notion of where and how serious games could be played, arguing for their use in education and training by designing games for computer-based environments. Many of today's serious games are digital games delivered via computers or video game consoles for instructional purposes. Both types of games have gained a substantial level of attention in recent years (Huang & Johnson, 2008). Hence, their broad educational applications across organizations and disciplines have been widely recognized by scholars and industries (Federation of American Scientists, 2006; Serious Games Initiative, 2010). These games are also capable of emulating and rendering scenarios with high fidelity, which gradually diminishes the boundary between serious games and simulations (Raybourn, 2007).

Despite serious games' emphasis on education and training in digital formats, their core game components remain unchanged from other games. Crawford (1982) identified four independent but interconnected game components: representation, interaction, conflict, and safety. The representation of the game system consists of all participating agents (e.g., players, system interface, game rules, game objectives), which enables intended interactions. Conflict is the means and/or the end of interactions that requires players to resolve complicated situations. The safety component encourages players to experience the outcome of their game-playing actions without any real harm. Amory (2007) further suggested that in order to understand the effect of serious games on learning, we should also include *Game Space* (play, exploration, authenticity, tacit knowledge, etc.), *Visualization Space* (critical thinking, storylines, relevance, goals, etc.), *Elements Space* (fun, emotive, graphics, sounds, technology, etc.), *Problem Space* (communication, literacy level, memory, etc.), and *Social Space* (communication tools and social network analysis). In addition to their multi-component architecture, serious games also encompass numerous characteristics that enable them to develop players' holistic and complex skills. It is suggested that challenges, fantasy, competition, multimedia representation, role-playing, and goal-oriented actions, to name a few, may enhance the learning experience in serious games (Huang & Johnson, 2008).

In addition to providing players with information about current issues and topics such as health, environment, and human rights, serious games have also become a major medium to train and teach skills such as social etiquette and prosocial behavior. The focus of this chapter is serious games for change, with a specific emphasis on games for cultural change. In games that are developed for cultural change, one common design strategy is getting the player to feel sympathy and/or empathy for the characters in the game. Most games for change simulate real physical casualties so that the player develops an awareness of a situation where war and genocide may be central to everyday life. While other educational serious games may focus on teaching a specific concept or subject (e.g., algebra), games for cultural change center on a different concept, one that concentrates on behavioral or attitudinal changes where the purpose is to raise awareness and evoke empathic concern. Empathy becomes one of the primary outcomes of the game.

Given serious games' complexity, it is commonly understood that players will engage in intense cognitive and emotional processing (Gray, Braver, & Raichle, 2002 ; Gunter, Kenny, & Vick, 2008). Such game-playing experiences would very likely overload players' limited cognitive processing capacity if gameplay and its resulting cognitive load were not carefully managed (Ang, Zaphiris, & Mahmood, 2007). Therefore, players' cognitive load levels should be considered when designing serious games (Huang & Johnson, 2008; Low, Jin, & Sweller, this volume). However, while serious game designers are beginning to consider issues of cognitive load, little attention has been paid to affective/emotional interactions with cognitive load. This may be particularly relevant for games for change that focus on cultural awareness. Since serious game tasks often require players to make multiple attempts before accomplishing the objectives, players' full participation and control in the game may solicit emotional fluctuation in response to the "ups and downs" (i.e., frustration, anxiety) of the game-playing process. In games for change, a primary objective is to get players to develop compassion and put themselves in the position of the person (i.e., characters) or persons at risk. Players might feel encouraged and joyful after accomplishing the game task or winning the competition, which could leave them with more cognitive resources, or having more cognitive resources might allow them to be more successful in the challenges, which in turn makes them feel encouraged and joyful. On the other hand, the feeling of frustration and annoyance might be induced by multiple failures in pursuing particular game objectives, and fewer cognitive resources may result in more failures. Consequently players' emotional responses and cognitive gain and loss might influence each other (Grodal, 2000, p. 208). Therefore, the level of empathy a player has for a character in the game may influence the player's cognitive load, and vice versa.

We further argue that cognitive and affective empathy play critical roles associated with engagement within games such as *Darfur is Dying*. We raise questions in this chapter about the relationship between cognitive load and empathy in serious games designed to induce and promote empathy. Some questions we ask are these: If empathy is a critical component in the learning outcome of serious games, can it be managed with a purposeful design? Are there cognitive load implications for empathy (or other affective constructs, for that matter)? In other words, given learners' limited cognitive information-processing capacity, could the existence of empathy improve or impede a learner's mental model development in serious games? Since current literature has not investigated the relationship between cognitive load and empathic aspects of learning with serious games, this chapter proposes a conceptual framework for empirical investigation of the relationship between cognitive load and empathy development. The following questions will guide the conceptual framework development in this chapter: What might be the relationship between cognitive load and empathy, and how can we manage both cognitive load and empathic concern in serious games? To answer these questions, we will first provide a brief conceptual overview of empathy, including a discussion of empathic dispositions associated with factors that influence the social perspective-taking that is necessary for empathic learning. Next, we will discuss cognitive load as a theoretical framework in a complex learning environment such as serious games. We will then briefly describe the potential interaction of empathy and cognitive load, including some examples taken from the game *Darfur is Dying* (Ruitz et al., 2006). Finally, we will propose a conceptual framework for future investigations of cognitive load and empathy in serious games.

BACKGROUND

Empathy

There have been several definitions of empathy associated with perspective-taking experiences. The following have been most prominent in the literature: cognitive empathy, affective empathy, and multidimensional empathy. The primary means of designing for empathic outcomes lie in asking the player/users to infer the mental states and experiences of the characters in the game. The mental states of the player are provoked by the virtual experiences of the characters in the game, which hopefully promote cognitive and affective empathy on the part of the player.

Cognitive empathy refers to the ability of the player to infer something about the mental state of others, whereas affective empathy refers to the ability of the player to infer the emotional state of another person. (Davis 1994; Hoffman, 2000). The third kind of empathy, multidimensional, is conceptualized as encompassing both cognitive and affective aspects and includes personal distress, empathic concern, fantasy, and perspective-taking. Each of the aforementioned dimensions is presumed in the storyline of *Darfur is Dying*, and therefore, we focus our discussion of empathy in this and other persuasive games as multidimensional empathy.

Cognitive empathy is more than role taking. "Role taking refers to the process in which one individual attempts to imagine the world of another" (Davis, 1994, p. 17). Cognitive empathy refers only to the act of constructing for oneself another person's mental state (Davis, 1994). Hogan (1969) defined cognitive empathy as "the intellectual or imaginative apprehension of another's condition without actually experiencing that person's feelings" (p. 308). In cognitive empathy, an individual must have a cognitive sense of others; otherwise, empathy cannot occur through direct association. Direct association involves the cognitive process of developing new mental models in order to better relate to others' thinking patterns. Insufficient cognitive information processing capacity might impede such a mental modeling process, thus leading to a lack of cognitive empathy. In the context of serious games, this is very likely to occur since players are constantly situated in a cognitively demanding environment. Managing cognitive load in the serious game-playing process, therefore, becomes crucial so that players acquire the desired empathic concerns for the characters in the game.

Empathy is distinguished from sympathy. Sympathy involves "a heightened awareness of the suffering of another person as something to be alleviated" (Wispé, 1987, p. 318). Empathy involves a set of attributes that children develop and acquire through experiences with others who exhibit behaviors such as perspective-taking, an ability to put themselves in another place, and a feeling of believing that what is happening to another is also happening to the individual (Tettegah & Neville, 2007). Empathy is considered a moral emotion, which is defined as the capacity one person has to feel and relate her/himself into the consciousness of another person (Wispé, 1987). It is the ability to step into someone else's shoes. Empathy involves a cognitive awareness of other's internal states, emotions, thoughts, feelings, and ways of perceiving and behaving in the world. As such, it involves perspective-taking, is multidimensional, and includes affective and cognitive aspects (Davis, 1994; Hoffman, 2000; Tettegah, 2007; Tettegah & Anderson 2007).

Cognitive Load

The gap between information structures presented in the instructional material and human cognitive architecture must be bridged so that learners can use their working memory more efficiently. In other words, learners should invest less mental effort to accomplish the learning task if the instructional material closely aligns with how learners cognitively process information. The level of mental effort investment during the learning process is defined as "cognitive load" (Paas, Tuovinen, Tabbers, & van Gerven, 2003). The purpose of cognitive load theory (CLT) (Chandler & Sweller, 1991), which has established a sound theoretical foundation to connect cognitive research on human learning with instructional design and development (van Merriënboer, Clark, & de Croock, 2002), is to be that bridge.

In the context of CLT, "learning" involves acquisition – the process of how learners construct schema and store them in long-term memory – and automation – how learners perform certain tasks *without* accessing working memory – of schema. Information required for the performance of a task is retrieved directly from the long-term memory (Paas et al., 2003). Since both attributes require little working memory capacity yet are still critical to meaningful learning, successful acquisition and automation of schema will lead to more efficient use of working memory for a desired performance (Mayer, 2001).

As suggested by CLT, there are three types of cognitive load which, combined, compose the total cognitive load: intrinsic, extraneous, and germane. For learning to occur, the total cognitive load can never exceed a learner's working memory capacity. The total extraneous and germane cognitive load, combined, is assumed to be equal to the total cognitive load minus the intrinsic cognitive load. Since the intrinsic cognitive load is fixed (i.e., the load cannot be manipulated by instructional design), instructional design's main purpose is to *reduce* the extraneous while *increasing* the germane cognitive load (van Gerven, Paas, van Merriënboer, & Schmidt, 2006).

Intrinsic cognitive load is associated with the element interactivity – the degree to which information can be understood alone without the involvement of other elements – inherent to the instructional material itself. Information with high element interactivity is difficult to understand because it usually depends on the involvement of other information units in order to see the full interaction. Therefore, instructional material with high element interactivity is assumed to induce a higher intrinsic cognitive load, since the instruction requires more working memory for information processing (Paas et al., 2003). Intrinsic cognitive load is considered to be independent of instructional manipulations because the manipulation only involves the amount of information a learner needs to hold in working memory without decreasing the inherent element interactivity (Pollock, Chandler, & Sweller, 2002). The extraneous cognitive load and germane cognitive load, in contrast, *can* be manipulated by instructional design (Brüken, Plass, & Leutner, 2003).

Extraneous cognitive load, also known as *ineffective cognitive load,* as it only involves the process of searching for information within working memory as opposed to the process of constructing schemas in long-term memory, can be influenced by the way information is presented and the amount of working memory required for given learning tasks (Paas et al., 2003). Considered a necessary cost of processing information, yet not related to the understanding of that information, extraneous, or ineffective, cognitive load must be reduced by instructional design. (Brüken et al., 2003). One method found to be successful in reducing extraneous cognitive load is the use of well-designed instructional multimedia components (Khalil, Paas, Johnson, & Payer, 2005; Mayer & Moreno, 2003). For the design of instructional materials, Cobb (1997) suggested the use of multimedia components (nonverbal and nontextual) as cognitive capacity external to learners' working memory, to facilitate cognitive efficiency and information processing. In theory, learners should spend less cognitive effort to understand the given information.

In contrast to the desired low degree of the extraneous cognitive load, instructional materials should be designed to *increase* the germane cognitive load. Also known as effective *cognitive load*, the germane cognitive load is described as the effort learners invest in order to facilitate the process of constructing schema and automation (Paas et al., 2003). The higher the germane cognitive load, the deeper the learning, since, by the design of the instructional material, learners are compelled to reexamine every new piece of information (de Crook, van Merriënboer, & Paas, 1998). Although the overall goal of manipulating cognitive load with instructional design is to *decrease* the level of ineffective cognitive load and to *increase* the effective cognitive load that promotes deeper learning, CLT suggests that the combination of extraneous and germane cognitive load should remain relatively constant after removing the fixed intrinsic cognitive load (Paas et al., 2003). The decrease of extraneous cognitive load should lead to the increase of germane cognitive load, or vice versa (Paas et al., 2003; van Gerven et al., 2006).

COGNITIVE LOAD AND EMPATHY IN SERIOUS GAMES

Cognition and emotion are regarded as two interrelated aspects of human functioning (Immordino-Yang, & Damasio, 2008, p. 192). Immordino-Yang & Damasio, (2008) present a discussion of the neurological relationship between cognition and emotion. They argue that emotional thought encompasses processes of learning, memory, and decision making, in both social and nonsocial contexts (p. 193). Prior research on emotions and cognitive load in particular also indicates that there is an effect of emotions on learning and mental effort investment (Um, Song, & Plass, 2007). Emotions entail the perception of an emotionally competent trigger, a situation either real or imagined that has the power to induce an emotion, as well as a chain of physiological events that will enable changes in the body & mind (Immordino-Yang & Damasio, 2008, p. 192). Clearly, cognitive load is apt to be affected by the presence or absence of empathy. Emotions such as empathy help to direct our ability to reason in learning.

Learning with serious games is a complex process. Learners might engage in in-depth cognitive information processing while also experiencing intense emotional fluctuations. For instance, *Darfur is Dying* (http://www.darfurisdying.com/) (Ruitz, et al., 2006), an online game based on the genocide in Sudan, aims to evoke empathy in the player. Although players only interact with the game using a computer keyboard, the game designers hope its effect on learners' empathy level towards Darfur will be long-lasting. The game developers for *Darfur is Dying* state,

Darfur Is Dying is a narrative-based simulation where the user, from the perspective of the displaced Darfurian, negotiates forces that threaten the survival of his or her refugee camp. It offers a faint glimpse of what it's like for the more than 2.5 millions who have been internally displaced by the crisis in Sudan (http://www.darfurisdying.com/aboutgame.html).

The goals of the developers and instructional designers of *Darfur is Dying* include raising awareness so that the player/user shares fear, empathy, and other emotions associated with victims of war. *Darfur is Dying* was developed with goals to educate, provide support and inspire. The influence of the characters on the player can cause emotional responses that are positive, negative, or both. In addition to the basic cognitive information processing required in order to understand the rules of the gameplay, the interactions might play a significant role in affording high or low empathic experiences.

In gameplay of *Darfur is Dying* or other social games, learning outcomes are often associated with specific characters in the game and the things that happen to them. For example, in *Darfur is Dying*, the characters need to risk their lives to protect their village. An ideal outcome is for players to feel empathic about the character's experiences portrayed in the games. In serious games that focus on behavioral and attitudinal changes, it is necessary for the player to connect and take the perspective of the character. The developers are hoping that the connection is at such a deep level empathy the player may experience will transfer to empathy for the actual Darfurian's real-life experiences. Hence, a goal is for the player to seek further cognitive processing of the situation with hopes that those further actions will lead to more effort to save the characters (victims) in future actions in the game. We argue that if the player does not have empathy for the characters in the game, then it is quite likely the game designers have failed to attain their primary goal of getting the player to work hard to save the characters and their village.

Empathy is particularly important in gameplay that involves characters. *Darfur is Dying* is designed to make players feel empathy for, or take on the role of, the characters who are living in a dangerous environment. In this sense, the game designers intend to engage users in social perspective-taking. The designers expect for the player to "infer the mental state of others (their thoughts, motives, intentions)" (Davis, 1994, p.49). Characters in *Darfur is Dying* face snipers, thieves, and insurgents on a daily basis while carrying out simple tasks such as foraging for water. One expects players might be motivated to learn more about Darfur and/or develop empathy through social perspective-taking with the characters, and possibly with the genocide of the people in Darfur. Social perspective-taking in games such as *Darfur is Dying* is primary, and the game designers assume that the player will be able to associate with and take on the role of the characters in the game. Role or perspective-taking occurs when one imagines how the victim or character feels or how one would feel in the victim's situation (Hoffman, 2000). In this sense, the designers are attempting to induce empathy with the hope that the emotional response to the character's situation will encourage the user to play the game.

In light of the aforementioned observations on cognitive load and empathy development in serious games like *Darfur is Dying*, it is clear that the play–learn process not only provides ample cognitive stimulation but also fosters the complex development of empathy. Both outcomes require a substantial and concurrent amount of cognitive processing capacity from the player. As a result, if we design games without acknowledging the drain that empathy formulation may place on cognitive resources in addition to other cognitive drains associated with interface, message design, etc., we may unwittingly exhaust the cognitive capacity of our learners in games in general. This has particular relevance, obviously, to games with empathy as the intended outcome. We must investigate whether or not the management of effective cognitive load and empathy will lead us to better understand a character's role in gameplay and its effects on the cognitive load of individuals who are engaged in gameplay.

MANAGING COGNITIVE LOAD AND EMPATHY

You have real feelings for imaginary events, which—even as you laugh or weep—you know to be fictitious . . . cognitive evaluations that engender emotions are sufficiently crude that they contain no reality check. (Johnson-Laird & Oatley, 2000, p. 465)

This proposed framework focuses on how researchers can manipulate different types of cognitive load during the play–learn process in order to investigate the relation of cognitive load and empathy in serious games. Because a serious game is a closed system, researchers should be able to control variables in a systematic manner via purposeful design. That is, the investigative variables can be integrated at the beginning of the design process with corresponding independent and dependent variables. Because the serious game environment contains multiple layers and dimensions that influence players' cognitive and empathy development, initial empirical investigations should be conducted in well-controlled laboratory settings. For example, competitiveness as one prominent characteristic of a serious game (Huang & Johnson, 2008) could be one independent variable, while players' cognitive level and empathy could be the dependent variable. Or, using models that focus on the measurement of empathy associated with the visual representation of specific characters may affect cognitive load and empathy.

This framework recommends three interrelated research perspectives for systematically designing investigative treatments in serious games: environments, characters, and activities. Each perspective can have a significant impact on players' final learning outcomes, and also affords possibilities for researchers to isolate variables or factors to efficiently investigate the play–learn process. Because all perspectives are required simultaneously to provide a comprehensive play–learn experience, the interactions among all three perspectives must also be observed.

Because the ultimate goal is to enhance learning, all investigative treatments should focus on how they might help players reduce the ineffective cognitive load (extraneous cognitive load) while increasing the effective cognitive load (germane cognitive load) through empathy evoked by the characters in the game. By minimizing the ineffective cognitive load, players not only will have more cognitive capacity to develop desired mental models for the cognitive tasks (e.g., knowledge gain, skill development), but this minimization also allows cognitive space for the necessary development of empathy.

Environments in Serious Games

This perspective encompasses the information accessibility, scenario representation, and game event delivery in serious games. If managed improperly, any component alone would induce players' cognitive overload and leave insufficient cognitive capacity to enable the presence or absence of empathy for the characters in the game. Accessibility of information in serious games mainly concerns the design of how players interact with the information. The actions and information should facilitate the process for players to identify and retrieve relevant information in order to move forward in the game and not be stifled by high levels of cognitive load induced by a complex interface. Player control, for instance, often requires players to use game console controllers or personal computer input devices (keyboard and mouse) without consideration of how players might need time to get acquainted with the interface. Representation of scenarios in serious games usually requires heavy utilization of multimedia (e.g., audio, video, and animation) to create intense and immersive scenarios. But overly incorporated multimedia would overload players' limited cognitive capacity. In terms of the delivery of game events, instead of pouring everything at the player at once, the delivery should focus on how often and in which manner game events should be made available to players to enable cognitively efficient information processing. Research in this perspective bears rich opportunities since all three components can be examined individually while monitoring their interaction effect on each other, which would also connect practical design issues of serious games with empirical research grounded in CLT.

Characters in Serious Games

This perspective emphasizes how players might empathize with, identify with, and relate themselves to the characters in the serious game. This perspective is particularly crucial for empathy development because players will dispense less mental effort in taking others' perspectives if they can easily identify with the character in the game. *Full Spectrum Warrior* (Institute for Creative Technologies, 2009), for example, is a serious game designed for training urban combat tactics. Characters in this game are designed based on real human experiences in refugee camps and combat positions in light infantry combat units in order to enable players (i.e., in-service armed forces personnel) to identify with the characters. Another approach to this research perspective is shifting the game character design to the game players. Once players have full control over the appearance of their characters, they might be expected to identify more with their characters and, therefore, be more likely to develop empathy.

Activities in Serious Games

This perspective aims to investigate the interaction between the player and the serious game environment and characters. Activities in serious games should require that players constantly retrieve relevant mental models from their long-term memory and repurpose them in various problem-solving tasks. In games that include empathy as an outcome, social interactions between players also need to be included in serious game activities since most tasks, in reality, demand collaborative efforts, and they may also facilitate the development of empathy between the player, his or her character, and individuals in the physical world. Activities irrelevant to cognitive as well as empathic gain, however, should be avoided. Most information-searching activities in serious games might fall into this less desired category.

FUTURE RESEARCH DIRECTIONS

Current research in serious games has not investigated the relationship between cognitive load and empathy. In games such as *Darfur is Dying*, characters depict actual situations that occur in real life. The objectives are to induce empathy for the characters in the game that will lead to cultural awareness and for a cultural group's true experiences in the Sudan. In other words, empathy becomes the main learning outcome as compared to learning about a specific content. Game designers in this case might have assumed that players already have a disposition for empathy for the characters in the game or, for that matter, have the ability to be empathic. The players might have some understanding about Darfur before participating in the game and therefore demand less cognitive capacity to process their empathic responses. It is also possible, however, that the game was designed without the aforementioned assumption. One intended outcome involves the development of the player's empathy for the character(s) in the game; however, the assumption here is that the individual has a similar experience that can help the player to connect through direct association with the victim characters in *Darfur is Dying*. With this approach, the characters within the game itself become the main source to provide critical information to evoke empathy in the player. Players must acquire an understanding of the situation while developing empathy simultaneously in the game. Cognitive overload is likely to occur due to the dual-tasking if the design of game environment, characters, and activities lacks empirical ground.

In order to resolve issues associated with design, research must investigate the relationship between the management of cognitive load via grounded design and players' presence or absence of empathy as a learning outcome. Other research has begun to document the relationship between learning, emotions, and empathy (Bransford, Brown, & Cocking, 2000; Diamond & Hopson, 1998; Sousa, 2005), but cognitive load and empathy in serious games has not been systematically investigated. The research should begin with correlational design to explore preliminary relationships between design variables (e.g., ease of interaction/intuitiveness of the interface, the degree of multimedia utilization in scenario representation, players' control in creating their own game characters) and the intended learning outcome (i.e., empathy level). Cognitive load then could be used to gauge players' cognitive capacity efficiency during the game-playing process. With that exploratory finding in place, we can employ experimental design to investigate relationships between specific variables. For example, researchers grounded in CLT can manipulate the representation of game character information (audio versus textual) and investigate its impact on the player's empathy development.

Designers and researchers should examine the presence and absence of empathy in players to determine whether or not players have the ability to empathize with characters in general. The working memory of the player can be examined using empathy as a dependent variable. Player attitudes and perceptions about the characters can be examined through the following methods: Players could design their own personal avatars for the game to measure the level of ownership related to perspective-taking if a character shares similar physical attributes as the player. If game designers and developers seek social perspective-taking and the presence of empathy in serious games, they must consider the gender and appearance of the avatar. The consideration may be obtained through user/player control over the development of the avatar and connections with universal facial expressions, which can promote players empathic identification with the characters in the game (Hoffman, 2000). Furthermore, more flexibility can be given to the player so that the player can develop his or her own survival scenario after playing *Darfur is Dying* or similar games. A better understanding of players' moral internalization and motivations will also assist developers in ensuring the transference of empathy to physical world individuals instead of remaining isolated to the game.

Game developers, psychologists, neuroscientists, and designers should investigate possibilities to ensure players are not making negative causal attributions to characters in the game and transferring those negative causal attributions to Darfurians who experience everyday survival and victimization in the Sudan (Hoffman, 2000). Hoffman (2000) stated, "Training in multiple empathizing, which is not a natural thing to do, may capitalize on rather than be defeated by the natural human proclivity to empathize more with kin than strangers" (p. 298).

The relationship between cognitive load and empathy was laid out earlier in this chapter with a specific focus on games for change. As we continue to develop games with a focus on perspective-taking and empathy we have to consider the association between cognitive load and the absence or presence of empathy. We cannot assume that empathy is a given but must begin with investigating whether or not the player has the ability to empathize and to manage the interaction between cognitive load and empathy. Empathy is a very complex moral emotion that requires the ability to step into another's place cognitively. Game designers must realize that there must be a connection to prior experiences, or the long-term memory, for the player to feel connected to the character's plight. Otherwise, the experience becomes a game, not a serious game.

CONCLUSION

Players enjoy the entertaining, playful aspects of interactive games, which could include an enthralling story; appealing characters; lush production values; a sense of social presence; making choices that affect the direction of the game; assuming the role of a character and playing with a new personality or identity; the extreme emotions that come with failure and success; and the pleasure of interacting with other characters and players. These experiences can heighten players' emotional responses to an interactive game and motive their effort to learn. (Lieberman, 2006, p. 381)

This chapter proposes several factors that inform how serious game designers can incorporate and manage elements to support the development of empathy while considering players' limited cognitive capacity. The first is cognitive load induced by the game environment, characters, and activities; the second is empathy and social perspective-taking resulting from the gameplay; and the last is the potential relationship between cognitive load and empathy and the impact of that relationship on learning in serious games. We argue that the ability to be empathic in serious games should affect the performance and learning of the player. In the meantime, players must have sufficient cognitive capacity allowance to develop their empathy. Thus the design of serious social games must consider the equilibrium between cognitive loads that engage players in the learning process and the cognitive allowance that supports empathy development.

REFERENCES

Abt, C. (1970). *Serious games*. New York: Viking Press.

Amory, A. (2007). Game object model version II: A theoretical framework for educational game development. *Educational Technology Research and Development*, *55*, 55–77. doi:10.1007/s11423-006-9001-x

Ang, C. S., Zaphiris, P., & Mahmood, S. (2007). A model of cognitive loads in massively multiplayer online role playing games. *Interacting with Computers*, *19*, 167–179. doi:10.1016/j.intcom.2006.08.006

Bransford, J., Brown, A., & Cocking, R. (2000). *How people learn: Brain, mind, and experience & school*. Washington, DC: National Academy Press.

Brünken, R., Plass, J., & Leutner, D. (2003). Direct measurement of cognitive load in multimedia learning. *Educational Psychologist*, *38*, 53–61. doi:10.1207/S15326985EP3801_7

Chandler, P., & Sweller, J. (1991). Cognitive load theory and the format of instruction. *Cognition and Instruction*, *8*, 293–332. doi:10.1207/s1532690xci0804_2

Cobb, T. (1997). Cognitive efficiency: Toward a revised theory of media. *Educational Technology Research and Development*, *45*, 1042–1062. doi:10.1007/BF02299681

Crawford, C. (1982). *The art of computer game design*. Retrieved from http://www.vancouver.wsu.edu/fac/peabody/game-book/Coverpage.html

Davis, M. H. (1994). *Empathy: A social psychological approach*. Dubuque, IA: Brown and Benchmark Publishers.

de Crook, M. B. M., van Merriënboer, J. J. G., & Paas, F. G. W. C. (1998). High versus low contextual interference in simulation-based training of troubleshooting skills: Effects on transfer performance and invested mental effort. *Computers in Human Behavior*, *14*, 249–267. doi:10.1016/S0747-5632(98)00005-3

Diamond, M., & Hopson, J. (1998). *Magic trees of the mind*. New York: Penguin.

Federation of American Scientists. (2006). Harnessing the power of video games for learning [report]. *Summit on Educational Games*. Federation of American Scientists, Washington, DC. Retrieved from http://fas.org/gamesummit/Resources/Summit%20on%20Educational%20Games.pdf

Gagné, R. M., & Driscoll, D. (1988). *Essentials of learning for instruction*. Englewood Cliffs, NJ: Prentice-Hall.

Gray, J. R., Braver, T. S., & Raichle, M. E. (2002, March 19). Integration of emotion and cognition in the lateral prefrontal cortex. *Proceedings of the National Academy of Sciences of the United States of America, 99*(6), 4115–4020. PubMed doi:10.1073/pnas.062381899

Gredler, M. (1994). *Designing and evaluating games and simulations: A process approach*. Houston, TX: Gulf Publishing Company.

Grodal, T. (2000). Video games and the pleasures of control. In D. Zillmann, & P. Vorderer (Eds.), *Media entertainment: The psychology of its appeal*. Mahwah, NJ: Lawrence Erlbaum Associates.

Gunter, G. A., Kenny, R. F., & Vick, E. H. (2008). Taking educational games seriously: Using the RETAIN model to design endogenous fantasy into standalone educational games. *Educational Technology Research and Development*, *56*, 511–537. doi:10.1007/s11423-007-9073-2

Hoffman, M. (2000). *Empathy & moral development: Implications for caring and justice*. New York: Cambridge University Press. doi:10.1017/CBO9780511805851

Hogan, R. (1969). Development of an empathy scale. *Journal of Consulting and Clinical Psychology, 33*, 307–316. PubMed doi:10.1037/h0027580

Huang, W. D., & Johnson, J. (2009). Let's get serious about E-games: A design research approach towards emergence perspective. In B. Cope, & M. Kalantzis (Eds.), *Ubiquitous learning*. Champaign, IL: University of Illinois Press.

Huang, W. D., & Johnson, T. (2008). Instructional game design using Cognitive Load Theory. In R. Ferdig (Ed.), *Handbook of research on effective electronic gaming in education*. Hershey, PA: Information Science Reference. doi:10.4018/978-1-59904-808-6.ch066

Immordino-Yang, M. H., & Damasio, A. (2008). *We feel, therefore we learn: The relevance of affective and social neuroscience to education. In the brain and learning. The Jossey-Bass reader*. San Francisco: Jossey-Bass.

Institute for Creative Technologies. (2009). *Full spectrum warrior*. Retrieved from http://ict.usc.edu/projects/full_spectrum_warrior

Johnson-Laird, P. N., & Oatley, K. (2000). Cognitive & social construction in emotions. In M. Lewis, & J. M. Haviland-Jones (Eds.), *Handbook of emotions* (2nd ed.). New York: The Guildford Press.

Khalil, M., Paas, F., Johnson, T. E., & Payer, A. (2005). Design of interactive and dynamic anatomical visualizations: The implication of cognitive load theory. *Anatomical Record. Part B, New Anatomist, 286B*, 15–20. doi:10.1002/ar.b.20078 PMID:16177992

Lieberman, D. A. (2006). Can we learn from playing interactive games? In P. Vorderer, & J. Bryant (Eds.), *Playing video games: Motives, responses, & consequences*. Mahwah, NJ: Lawrence Erlbaum Associates.

Low, R., Jin, P., & Sweller, J. (2010). Learner's cognitive load when using education technology. In R. Van Eck (Ed.), *Interdisciplinary models and tools for serious games: Emerging concepts and future directions*. Hershey, PA: IGI.

Mayer, R. E. (2001). *Multimedia learning*. New York: Cambridge University Press. doi:10.1017/CBO9781139164603

Mayer, R. E., & Moreno, R. (2003). Nine ways to reduce cognitive load in multimedia learning. *Educational Psychologist, 38*, 43–52. doi:10.1207/S15326985EP3801_6

Paas, F., Tuovinen, J. E., Tabbers, H., & van Gerven, P. W. M. (2003). Cognitive load measurement as a means to advance cognitive load theory. *Educational Psychologist, 38*, 63–71. doi:10.1207/S15326985EP3801_8

Pollock, E., Chandler, P., & Sweller, J. (2002). Assimilating complex information. *Learning and Instruction, 12*, 61–86. doi:10.1016/S0959-4752(01)00016-0

Prensky, M. (2001). *Digital game-based learning*. New York: McGraw-Hill Publishing Company.

Raybourn, E. M. (2007). Applying simulation experience design methods to creating serious game-based adaptive training systems. *Interacting with Computers, 19*, 206–214. doi:10.1016/j.intcom.2006.08.001

Ruiz, S., York, A., Truong, H., Tarr, A., Keating, N., Stein, M., et al. (2006). *Darfur is Dying*. Thesis Project. Retrieved from http://interactive.usc.edu/projects/games/20070125-darfur_is_.php

Serious Games Initiative. (2010). Retrieved on January 14, 2010, from http://www.seriousgames.org/about2.html

Sousa, D. (2005). *How the brain learns*. Thousand Oaks, CA: Corwin Press.

Suits, B. (1978). *The grasshopper: Games, life, and utopia*. Ontario, CA: University of Toronto Press.

Tettegah, S. (2007). Pre-service teachers, victim empathy, and problem solving using animated narrative vignettes. *Technology, Instruction. Cognition and Learning*, *5*, 41–68.

Tettegah, S., & Anderson, C. (2007). Pre-service teachers' empathy and cognitions: Statistical analysis of text data by graphical models. *Contemporary Educational Psychology*, *32*, 48–82. doi:10.1016/j.cedpsych.2006.10.010

Tettegah, S., & Neville, H. (2007). Empathy among Black youth: Simulating race-related aggression in the classroom. *Scientia Paedagogica Experimentalis, XLIV*, *1*, 33–48.

Um, E., Song, H., & Plass, J. L. (2007). *The effect of positive emotions on multimedia learning*. Paper presented at the World Conference on Educational Multimedia, Hypermedia & Telecommunications (ED-MEDIA 2007) in Vancouver, Canada, June 25–29, 2007.

van Gerven, P. W. M., Paas, F., van Merriënboer, J. J. G., & Schmidt, H. G. (2006). Modality and variability as factors in training the elderly. *Applied Cognitive Psychology*, *20*, 311–320. doi:10.1002/acp.1247

van Merriënboer, J. J. G., Clark, R. E., & de Croock, M. B. M. (2002). Blueprints for complex learning: The 4C/ID-model. *Educational Technology Research and Development*, *50*, 39–64. doi:10.1007/BF02504993

Wispé, L. (1987). History of the concept of empathy. In N. Eisenberg, & J. Strayer (Eds.), *Empathy and its development*. Cambridge, UK: Cambridge University Press.

KEY TERMS AND DEFINITIONS

Cognitive Load Theory: The he process of how learners construct schema and store them in long-term memory – and automation – how learners perform certain tasks without accessing working memory – of schema.

Darfur is Dying: An application for social perspective-taking in games in which game designers assume that the player will be able to associate with and take on the role of the characters in the game.

Empathy: The ability to experience the same emotional state as another within a social situation or through prior experience of those emotions.

Extraneous Cognitive Load: Also known as ineffective cognitive load, Extraneous cognitive load involves the process of searching for information within working memory as opposed to the process of constructing schemas in long-term memory.

Player: A social actor who engages with a gaming environment for a particular purpose, such as to receive an education.

Serious Games: E-Learning or computer based training environments that have been transformed through gamification to make them more engaging and effective for learning.

Sympathy: Sympathy involves a heightened awareness of the suffering of another person as something to be alleviated through dialogue.

APPENDIX: ADDITIONAL READING

"Must-Reads" for This Topic

Aldrich, C. (2004). *Simulations and the future of learning*. San Francisco: Pfeiffer.

Blakemore, S.-J., & Frith, U. (2005). *The learning brain: Lessons for education*. Oxford, UK: Blackwell.

Clark, R., Nguyen, F., & Sweller, J. (2006). *Efficiency in learning: Evidence-based guidelines to manage cognitive load*. San Francisco, CA: Pfeiffer.

Restak, R. (2006). *The naked brain: How the emerging neuro-society is changing how we live, work, and love*. New York: Harmony Books.

Salen, K., & Zimmerman, E. (2004). *Rules of play: Game design fundamentals*. Cambridge, MA: The MIT Press.

Schroeder, R. & Axelsson, A.-S. (Eds.). (2006). *Avatars at work and play*. The Netherlands: Springer.

Vorderer, P., & Bryant, J. (Eds.). (2006). *Playing video games: Motives, responses, & consequences*. Hillsdale, NJ: Lawrence Erlbaum Associates.

Top Texts for Interdisciplinary Studies of Serious Games

Michael, D., & Chen, S. (2005). *Serious games: Games that educate, train & inform*. Washington, DC: Thomson Course Technology PTR.

Clark, R., Nguyen, F., & Sweller, J. (2006). *Efficiency in learning: Evidence-based guidelines to manage cognitive load*. San Francisco, CA: Pfeiffer.

This work was previously published in Gaming and Cognition: Theories and Practice from the Learning Sciences, edited by Richard Van Eck, pp. 137-151, copyright 2010 by Information Science Reference (an imprint of IGI Global).

Chapter 3
What's in a Game?
The Politics of Shaping Property Tax Administration in Bangalore City, India

Shefali Virkar
University of Oxford, UK

ABSTRACT

Much as been written about e-government within a growing stream of literature on ICT for development, generating countervailing perspectives where optimistic, technocratic approaches are countered by far more sceptical standpoints on technological innovation. This body of work is, however, not without its limitations: a large proportion is anecdotal in its style and overly deterministic in its logic, with far less being empirical, and there is a tendency for models offered up by scholarly research to neglect the actual attitudes, choices, and behaviour of the wide array of actors involved in the implementation and use of new technology in real organisations. Drawing on the theoretical perspectives of the Ecology of Games framework and the Design-Actuality Gap model, this chapter focuses on the conception and implementation of an electronic property tax collection system in Bangalore (India) between 1998 and 2008. The work contributes to not just an understanding of the role of ICTs in public administrative reform, but also towards an emerging body of research that is critical of managerial rationalism for an organization as a whole, and which is sensitive to an ecology of actors, choices, and motivations within the organisation.

INTRODUCTION

Over the course of the last two decades, globalisation and information technology have been rapidly dismantling traditional barriers to trade, travel and communication, fuelling great promise for progress towards greater global equity and prosperity. Attracted by the 'hype and hope' of Information and Communication Technologies (ICTs), development actors across the world have adopted computer-based systems and related ICTs for use in government as a means reforming the

DOI: 10.4018/978-1-4666-5071-8.ch003

inefficiencies in public service provision. Whilst a number of these electronic government or 'e-government' projects have achieved significant results, evidence from the field indicates that despite the reported success stories, the rate of project failure remains particularly high.

Much as been written about e-government within a growing stream of literature on ICT for development, generating countervailing perspectives where optimistic, technocratic approaches are countered by far more sceptical standpoints on technological innovation. However, in trying to analyse both their potential and real value, there has been a tendency for scholars to see e-government applications as isolated technical artefacts, analysed solely as a collection of hardware and software. Far less work is based on empirical field research, and models put forward by scholars and practitioners alike often neglect the actual attitudes, choices and behaviour of the wide array of actors involved in the implementation and use of new technology in real organisations as well as the way in which the application shapes and is shaped by existing social, organisational and environmental contexts.

This chapter seeks to unravel the social dynamics shaping e-government projects used to reform public sector institutions. The value of such an approach is based on a review of existing development literature, which tends to be overly systems-rational in its approach. As a consequence, the literature does not recognise the degree to which project failure (*viz.* the general inability of the project design to meet stated goals and resolve both predicted and emerging problems) is symptomatic of a broader, much more complex set of interrelated inequalities, unresolved problems and lopsided power-relationships both within the adopting organisation and in the surrounding environmental context.

The case study from which this paper is drawn, focused on a project aimed at digitising property tax records and administrative processes within the Revenue Department of the Greater Bangalore City Municipal Corporation. In recognising the need to turn property tax into a viable revenue instrument that delivers high tax yields without compromising on citizen acceptance, the Bangalore City Corporation has sought to improve its property tax administration system through the introduction of a computerised database and the use of digital mapping techniques to track compliance and check evasion.

BACKGROUND

Simultaneous with the shift towards a more inclusive process of participation in political decision-making and public sector reform has been an increased interest in the new digital Information and Communication Technologies (ICTs) and the ways in which they may be used to effectively complement and reform existing political processes. Developments in communication technologies have historically resulted in changes in the way in which governments function, often challenging them to find new ways in which to communicate and interact with their citizens, and ICTs today are seen to possess the potential to change institutions as well as the mechanisms of service delivery, bringing about a fundamental change in the way government operates and a transformation in the dynamic between government and its citizens (Misra, 2005).

e-Governance thus does not merely involve the insertion of computers and computer operators into an organisation, instead it involves the creation of systems wherein electronic Internet-enabled technologies are integrated with administrative processes, human resources, and the desire of public sector employees to dispense services and information to people fast and accurately. The concept thus consists of two distinct but intertwined dimensions– political and technical aspects relating to the improvement of public sector management capacity and citizen participation (Bhatnagar, 2003). Conceptually, e-Governance may be divided into **e-Democracy**, defined by the express intent to increase the participation of citizens in decision-making through the use of digital media, and **e-Government**, the use of Information and Communication Technologies by government departments and agencies to improve internal functioning and public service provision (Virkar, 2011a).

Over the last 10 years, a number of scholars and international organisations have defined e-government in an attempt to capture its true nature and scope. A selection of key definitions is highlighted in Box 1. Almost all definitions of e-government indicate three critical transformational areas in which ICTs have an impact (Ndou, 2004), illustrating that e-government is not just about the Internet and the use of Internet- and web-based systems with government and citizen interfaces; instead it includes office automation, internal management, the management of information and expert systems, and the design, and adoption of such technologies into the workplace (Margetts, 2006).

- **The Internal Arena:** wher E Information and Communication Technologies are used to enhance the efficiency and effectiveness of internal government functions and processes by intermediating between employees, public managers, departments, and agencies.
- **The External Arena:** Where ICTs open up new possibilities for governments to be more transparent to citizens and businesses by providing multiple channels that allow them improved access to a greater range of government information.
- **The Relational Sphere:** Where ICT adoption has the potential to bring about fundamental changes in the relationships between government employees and their managers, citizens and the state, and between nation states.

Thus, although the term e-government is primarily used to refer to the usage of ICTs to improve administrative efficiency, it arguably produces other effects that would give rise to increased transparency and accountability, reflect on the relationship between government and citizens, and help build new spaces for citizens to participate in their overall development (Gascó, 2003). Broadly speaking, the concept may be divided into two distinct areas: (1) **e-Administration**, which refers to the improvement of government processes and to the streamlining of the internal workings of the public sector using ICT-based information systems, and (2) **e-Services**, which

Box 1. Definitions of e-Government

Tapscott (1996): "eGovernment is an Internet-worked government which links new technology with legal systems internally and in turn links such government information infrastructure externally with everything digital and with everybody – the tax payer, suppliers, business customers, voters and every other institution in the society."
Fraga (2002): "Government is the transformation of public sector internal and external relationships through net-enabled operations, IT and communications, in order to improve: Government service delivery; Constituency participation; Society."
Commonwealth Centre for E-Governance (2002): "E-government constitutes the way public sector institutions use technology to apply public administration principles and conduct the business of government. This is government using new tools to enhance the delivery of existing services."
World Bank (2010): "*E-Government* refers to the use by government agencies of information technologies (such as Wide Area Networks, the Internet, and mobile computing) that have the ability to transform relations with citizens, businesses, and other arms of government. These technologies can serve a variety of different ends: better delivery of government services to citizens, improved interactions with business and industry, citizen empowerment through access to information, or more efficient government management. The resulting benefits can be less corruption, increased transparency, greater convenience, revenue growth, and/or cost reductions."
(Source: Commonwealth Centre for E-Governance (2002), Ndou(2004), World Bank(2010))

refers to the improved delivery of public services to citizens through ICT-based platforms.

The adoption of technology often involves interactions to reform the way governments, their agencies, and individual political actors work, share information, and deliver services to internal and external clients by harnessing the power of digital Information and Communication Technologies – primarily computers and networks – for use in the public sector to deliver information and services to citizens and businesses (Bhatnagar, 2003). However, whilst online e-government service initiatives have become common in many countries, and in a variety of contexts, such applications are characteristically built with a primary focus on administration-citizen interaction, rather than on explicitly supporting plans for strategic organisational development.

Although considerable attention has been focused on how e-government can help public bodies improve their services, there are relatively few studies which focus the long-term sustainability of e-government initiatives, particularly in the developing world. In contrast, this project focused on in this study seeks to illustrate that the potential for improved government-citizen interaction through e-government and public sector reform could be realised not only through developing the 'virtual front office' but also through their effect on back-office organisation and culture.

CENTRAL RESEARCH QUESTIONS

The main goal of this chapter is thus to approach the issues thrown up by the organisational and institutional transformations that occur in public administration from a multidisciplinary perspective and, through the use of a case study, attempt to bring a new perspective to bear on the following questions:

1. Does the introduction and implementation of Information and Communication Technologies within developing world bureaucracies have an impact on the internal dynamics of the group and relationships between actors operating within these organisations?
2. If, so what are the types of interactions that may arise between the actors concerned and how do these impact project outcome?

Whilst a single case study cannot provide closed-end answer to these questions, it can suggest ways of addressing them that could be applicable to a wider variety of cases.

UNDERSTANDING ACTOR BEHAVIOUR

The central issue that needs to be understood whilst studying the implementation of ICTs through an analysis of actor interactions is thus: *Why do people do what they do?* One approach to understanding behaviour is to look at the rationality of individual actors, rather than the system as a whole. This is largely because political actors are driven by a combination of organisational and institutional roles and duties and calculated self -interest, with political interaction being organised around the construction and interpretation of meaning as well as the making of choices (Virkar, 2011b).

Political actors, in general, have a complex set of goals including power, income, prestige, security, convenience, loyalty (to an idea, an institution or the nation), pride in work well done, and a desire to serve the public interest (as the individual actor conceives it). According to Downs (1964) actors range from being purely self-interested ('*climbers*' or '*conservers*' motivated entirely by goals which benefit themselves and their status quo rather than their organizations or the society at large) to having mixed motives ('*zealots*', '*advocates*' and '*statesmen*' motivated by goals which combine self interest and altruistic loyalty with larger values).

Introducing e-government initiatives into public bodies is a tricky game to play, as computerisation alters the work-load, work profile and content of the average public sector employee; impacting accountability, reducing the opportunities for exercising discretion, making performance more visible and flattening the hierarchy (Bhatnagar, 2004), and often forcing the need for retraining and retooling and sometimes creating redundancy. Many projects tend to face internal resistance from staff – particularly from the middle to lower levels of the civil service – with moves made to reengineer processes and effect back-end computerisation having a profound effect on the way civil servants perform their duties and perceive their jobs. Very often in developing countries, it is the fear of the unknown that drives this resistance, especially if the introduction of new technology results in a change of procedures and the need for new skills. Further, in corrupt service delivery departments, there may be pressure to slow down or delay the introduction of technology-led reforms due to the impending loss of additional income

PROPERTY TAX: DEFINITION AND SCOPE

Property tax may be defined as a recurrent tax on real property (land and/or improvements) in urban areas (Dillinger, 1988). Just like other taxes, it may be considered as 'a compulsory transfer of money…from private individuals, institutions or groups to the government…[as] one of the principal means by which a government finances its expenditure' (Bannock et.al, 1987). Further, Rosengard (1998) defines Property Tax as '…[an] *ad valorem* ("according to the value" tax, as opposed to a unit tax), *in rem* ("against the object" tax as opposed to a personal or *in personam* tax) levied on the ownership, occupation or development of land and/or buildings. Property taxes usually are assessed annually upon the capital value of a property, or upon proxies for capital value such as presumed or actual rental income. Taxes not confined to immovable property, such as net wealth taxes and general capital gains taxes, are not commonly classified as property taxes."

Property tax is appealing to local governments in developing countries for a number of reasons. First and possibly most importantly, it is a potential revenue generator, particularly given the high-income elasticity of property ownership in developing countries. It is a relatively stable source of income, and it is easy to implement slight adjustments and incremental rate changes. The tax is generally equitable and progressive for residential properties. It is hard to avoid legally due to the high visibility and relative immobility of property, with asset immobility also conferring a high degree of economic efficiency on the tax. It is clearly enforceable, particularly through the seizure and liquidation of property. The tax has the potential to enhance the local government agency's responsiveness to local priorities, particularly when used to finance local goods.

Changing the mindsets of key actors in games related to property tax administration is often central to successfully reforming the method of valuation. On the one hand, public reluctance to adopt new methods is a major obstacle to changes in assessment when undertaking property tax reform, often arising from a certain degree of unawareness and an aversion to shouldering an increased tax burden. In addition, those taxpayers who seek to use existing flaws in the system to their own advantage – generally to partially or completely evade taxes – may also be reluctant to accept a better, more foolproof system. At the same time, the attitudes of government officials also need to be dealt with during the reform process as any change in procedure is bound to bring about modifications in existing systems, with resistance arising when staff are confronted with the need to develop new skills and where well-entrenched power structures and old mental models are challenged (Virkar, 2011b) The degree to which both the public and the bureaucracy are willing to adapt to reforms is thus important, as these attitudes often shape the political response to changes in the administrative set-up of tax regime and to reassessments of the tax rate.

ANALYTICAL FRAMEWORK: THE ECOLOGY OF GAMES

From the turn of the century to the present, there has been a progressive movement away from the view that governance is the outcome of rational calculation to achieve specific goals by a unitary governmental actor, and in that context metaphors based on games have been extremely useful in developing new ways to think about the policy process. A look through the literature reveals that although political games have been described by scholars within differing contexts ranging from electoral politics to administrative functioning, there exists no comprehensive description of the public organization as a system of these various interactions.

The use of Game Theory and most other game metaphors (although differing widely in their orientation) have had, according to scholars, one major limitation for clarifying policy processes: they focus squarely on a single arena or field of action; be it a school, a county, a legislature, etc. Yet, by their very nature, policy making and implementation cut across these separate arenas, in both their development and impact (Firestone, 1989). In e-government projects for instance, systems built by both public and private enterprises for use by government employees and citizens across different political constituencies must be enforced by legislative acts created and interpreted by national branches of government. In addition, actors at different levels of the policy system encounter divergent problems posed by the system in question and their actions are influenced by varied motives. What is needed, therefore, is a framework that goes beyond single games in order to focus on how games 'mesh or miss' each other to influence governance and policy decisions. One of the few efforts to look at this interaction and interdependence was Norton Long's (1958) discussion of "The Local Community as an Ecology of Games."

During the period immediately following World War II, Institutionalist approaches came into their own in the late 1980s under the guise of New Institutionalism as a result of a growing number of scholars attempting to describe and understand in concrete terms the political world around them (Peters, 2000). Contrary to both Behavioural Theory and Rational Choice Theory, New Institutionalists considered observable behaviour to occur and be understood solely within the context of institutions, leading to the creation and development of two new branches of theory, namely Rational Choice Institutionalism and Behavioural Institutionalism (Immergut, 1998).

The basic assumption is that not only may social phenomena be explained as the outcome of interactions amongst intentional actors – individual, collective, or corporate – but that these interactions are structured and outcomes are shaped by the characteristics of the institutional settings within which they occur (Scharpf, 1997). As the basic argument of rational choice approaches is that utility maximisation can and will remain the primary motivation of individuals, rational choice approaches to institutions all presume the same egoistic behavioural characteristics found in similar approaches to other aspects of political behaviour (Peters, 2000). However, the institutional variants of the approach focus attention on the importance of *institutions* as mechanisms for channelling and constraining individual behaviour.

The Ecology of Games framework, as first laid out by Long in the late 1950s offers a New Institutionalist perspective on organisational and institutional analysis. As with most theories of New Institutionalism, it recognises that political institutions are not simple echoes of social forces, and that routines, rules and forms within organisations and institutions evolve through historically interdependent processes that do not reliably and quickly reach equilibrium (March & Olsen, 1989). Long developed the idea of the ecology of games as a way of reconciling existing debates about who governed local communities as he believed they had significant flaws.

The crucial insight in Long's theory however, was not the idea of games *per se* which, as has been discussed earlier, was already well developed, but his linking of that notion to the metaphor of an ecology (Firestone, 1989). Ecology as a concept relates to the interrelationships of species in their environment, allowing for numerous relationships amongst entities, and has been used to understand the relationships among individuals and more complex social systems. This speaks of a singular interdependence between different actors within a given territory. Although there may be other relationships as well, what is significantly missing is a single, rational, coordinating presence.

Games themselves are social constructs that vary over time and across social contexts (Crozier & Friedberg, 1980). Similar types of games might recur within similar social settings, but all games tend to be uniquely situated in place and time, and any typology of games that might emerge across a cumulative body of studies is likely to remain quite abstract. Despite this, Dutton (1992) has identified several key attributes all games may share: first, every game has a set of goals, purposes, or objectives, with some games having multiple aims. Second, a game has a set of prizes, which may vary widely from profit to authority to recognition, and are distinct from the objectives of the players. Third, games have rules that govern the strategies or moves open to players depending on the organisational or institutional settings within which they are played. Rules need not be public or fair (depending on whether public or private interests are involved), may change over time, and may or may not need consensus to be accepted. Finally, a game has a set of players, defined by the fact that they interact – compete or cooperate – with another in pursuing the game's objectives.

For Long, territories (or fields of play) were defined quite literally by being local communities. Moved from the community context to the world of e-government design, adoption and implementation, territories may be diverse – from the inner circle of the project design team, through to the adopting organisation, the nation and finally the international policy arena – but the idea of each stage being a political community or a collection of actors whose actions have political implications is still very much applicable. The ecology of games metaphor thus provides us with a useful way to think about how the various players interact in making and carrying out administration and developing policy.

ASSESSING PROJECT OUTCOME: A DISCUSSION OF THE DESIGN-ACTUALITY GAP MODEL

Like all political interactions, the behaviour of actors related to the design and uptake of e-government projects is circumscribed by the organisations and institutions within which they are played out, and by the range of actors taken from the individuals and groups directly and indirectly involved with the process of governance. The outcome of an e-government project therefore does not depend on a single project entity alone, and instead depends on the interaction between different actors in the process and the nature of the relationships between them. Gaps in project design and implementation can in reality be seen as expressions of differences arising from the interaction between different (often conflicting) actor moves and strategies, determined to a large extent by actor perceptions, and played out within the context of set circumstances.

In order to assess the extent to which the case study in question has succeeded or failed, this research project will first attempt to locate it within Heeks' seminal three-fold categorisation. By examining numerous case studies related to ICTs and e-government failure in developing countries, Heeks (2002) identified three dominant categories of reported outcome: *total failure*, *partial failure*, and *success*. Though not theoretically exhaustive (they do not, for instance take into account the mutation of outcomes over time), these categories are nonetheless valuable and comprise the first step of a framework within which a project might be evaluated.

Heeks (2003) concluded that the major factor determining project outcome was the degree of mismatch between the current realities of a situation ('where are we now') and the models, conceptions, and assumptions built into a project's design (the 'where the e-government project wants to get us'). From this perspective, e-government success and failure depends largely on the size of this 'design-actuality' gap: the larger gap, the greater the risk of e-government failure, the smaller the gap, the greater the chance of success. He also identified three so-called 'archetypes of failure', situations when a large design-actuality gap – and, hence, failure – is more likely to emerge. These may be classified as Hard-Soft Gaps, Public-Private Gaps and Country Context Gaps (Dada, 2006), and are summarised below:

Hard-Soft Gaps

Hard-soft gaps refer to the difference between the actual, rational design of the technology (hard) and the actuality of the social context – people, culture, politics, etc. – within which the system operates (soft). These sorts of gaps are commonly cited in examples of e-government failure in developing countries, where 'soft' human issues that are not initially taken into account whilst designing a project result in undesirable effects after implementation. Many scholars, such as Stanforth (2006), see technology as just one of a number of heterogeneous socio-technical elements that must be considered and managed during the design and implementation of a successful e-government project, whilst Madon (2004) has discussed different sets of case studies which have revealed that numerous factors that have allowed individuals in developing countries to access ICTs (and which depend on resources, skill-levels, values, beliefs, and motivations, etc.) are often ignored. It may thus be inferred that a lack of training, skills, and change management efforts would all affect rates of failure, as it is these factors that would bridge the gap between the technology itself and the context within which it exits.

Hard-soft gaps thus may be seen as the outcome of interactions played out primarily at the level of the project itself, between individuals and agencies involved with the design and acceptance of the technology. For instance, decisions taken by senior officials relating to issues of change management and skill levels might be motivated by the desire of the top brass to curtail and keep in check the power of their junior employees and to maintain control over their territories. Similarly a clash between powerful rivals on a project planning committee could result in either half-baked compromise decisions or strong decisions that are not followed through, leading to chaos at the implementation stage that has repercussions on more junior staff.

Private-Public Gaps

The next archetype put forward by Heeks (2003) is that of private-public gaps, which refers to the difference between organisations in the private and public sectors, and the mismatch that results when technology meant for private organisations is used in the public sector without being adapted to suit the role and aims of the adopting public organisation. A common problem is again the lack of highly skilled professionals in the public sector, resulting primarily from uncompetitive rates of pay in that sector as compared to the private sector (Ciborra & Navarra, 2005). The design of e-government projects is consequently outsourced to the private sector, resulting in a clash of values, objectives, culture, and large design-actuality gaps.

Public-private gaps often arise out of games played at the level of the adopting government agency, generally between the agency and its private sector counterparts, although it is not uncommon to find interactions between public and private individuals on project committees having an impact on the outcome of a project as well.

Country Context Gaps

The final archetype of failure defined by Heeks (2003) is the country context gap, or the gap that arises when a system designed for one country is transferred into the reality of another. This is particularly true for systems transferred between developed and developing countries, where designs for one may clash with the actualities within the other. Country context gaps are, according to Dada (2006) closely related to hard-soft gaps as they arise from, amongst other things, differences in technological infrastructure, skill sets, education levels, and working cultures.

Country-context gaps emerge chiefly as a result of games played by national, provincial and international actors operating across borders. For instance, decisions to adopt or promote a certain management style or value system, buy or sell a particular technology from a particular organisation or country, or collaborate with particular government agencies in different parts of the world all stem from games of international trade, aid, and diplomacy.

In conclusion, Heeks' model is particularly useful given the large investments made by developing country governments in e-government systems and the large opportunity costs associated with implementation, as it encourages project planners to take a focused, holistic view of problem solving; making them consider concurrently the technology at hand, the current circumstances, the impact of actors' motivations and actions, and possible vested interests. It may be used both as a predictive tool anticipating potential failings and heading them off at the initial stages, as well as being used to diagnose problems during the execution of the project. The framework is thus a means of evaluating outcome and problem solving strategies at all stages during the development of a project, and not just to examine what went wrong in hindsight.

RESEARCH METHODOLOGY

The ultimate aim of this research project was to contribute to the development of a conceptual framework that was relevant to policy discussions of e-government within an Indian, and hopefully a broader developing world context. In order to augment theoretical discussions of administrative reform in a digitised world, this research used a case study to explore the central research issues. Within the case study, a mixed methods approach was selected in order to inform and strengthen the understanding of the relationships between the actors, inputs, and project outputs. The aim of such a study would be to evolve ideas, which may be generalised across similar situations, and would involve the following steps:

- In-depth review of existing theoretical perspectives and literature surrounding corruption and tax evasion, ICTs and public administration, and property tax reform.
- Qualitative analysis of official documents
- Collection and analysis of quantitative data relevant to the case
- Developing case studies through in-depth personal interviews
- Data analysis and interpretation
- Preparation of conclusions and their validation
- Recommendations for the future

The use of mixed-method case study research is becoming increasingly popular in the social sciences, recognised as a successful approach for investigating contemporary phenomena in a real-life context when the boundaries between phenomenon and context are not evident and where multiple sources of evidence present themselves (Yin, 2003). It is thus a particularly apt way of studying the nature and impact of actor actions and motivations on e-government project outcome, where the aim is not simply to judge whether the project at hand represents a success or failure, but to understand the qualities that have made it so. In such a case quantitative data alone is thus not a sufficient measure of impact.

DIGITISING PROPERTY TAX RECORDS IN BANGALORE, INDIA: EXAMINING ACTOR ATTITUDES AND PERCEPTIONS

Against the background of technological innovation in Karnataka state, project planners from the Greater Bangalore City Corporation (BBMP) felt that the manual system of property tax administration was archaic, opaque, and inefficient. All the members of the core project group believed that property tax collections under the manual system had suffered from poor record keeping and bad information management practices, slow processing times, and overcomplicated assessment and payment procedures. These had, in turn, created frustration amongst taxpayers and resulted in low levels of compliance. The computerised property tax system was thus borne out of a need to reform the manual system of property tax administration in Bangalore and improve tax revenues and compliance through the improvement of back-office efficiency, the simplification of tax collection, and the reduction of money lost through malpractice through the effective detection and deterrence of tax evasion – spurred on by the need to enhance power, authority and reputations.

Interviews with tax officials revealed that most felt that there had been serious problems with the manual system of tax administration. They claimed that the biggest hurdles to the efficient administration of tax that they encountered prior to the introduction of the computerised database were poor and haphazard recordkeeping and large amounts of paperwork that needed to be done manually. Information was scattered and the process of calculating tax due, administering collections and checking up on defaulters was extremely unsystematic. While, as expected, none of the revenue officials interviewed mentioned government employee corruption as being serious problem, many interviewees spoke of the difficulties they faced in identifying and catching tax evaders. Most officials interviewed felt that the introduction of technology had greatly impacted old work processes and had helped alleviate the difficulties they faced under the manual system. They believed that the centralisation of data, the ease with which citizens could access their tax information, and the setting up of tax collection points across the city had all helped in bringing more properties into the tax net and contributed significantly towards improving tax payer compliance. All the officials interviewed felt that their interactions with the public had significantly decreased since the introduction of the computerised system, and a little over half them believed their overall relationship with citizens had improved as a result.

However, while acknowledging that the use of digitised records, computer printouts and online databases had had a positive impact on their work, some interviewees were quick to point out that technology had been used simply to automate existing processes, and that old infrastructural problems (such as poor electricity supply and old computers) and problems related to a lack of skills and training on the computerised system had not been resolved.

Only a small percentage of revenue officials reported that they had been consulted during the design stage of the project. Further, there appeared to be no mechanism in place to solicit user feedback once the initial system had been developed. Almost all the officials interviewed said they felt disconnected from system. Most professed a high degree of unfamiliarity with the system, and were completely unaware of its key features. For instance, only one tax official mentioned the introduction of GIS mapping techniques as being useful to his work and that of his staff, a worrying fact given that the core project team had placed much store by the GIS maps as a tool to track property tax payments and identify defaulters. These are not good signs, as effective system implementation requires employees to fully accept and adopt the technology in the belief that it will do them some well-defined good.

Further, none of the officials interviewed knew how to operate even its most basic features. With no scheme in place to give them any formal training on the system, all the interviewees reported to be completely dependent on a private computer operator to feed in, change and retrieve electronic property tax data. This, this researcher feels, created a new problem within revenue offices and limited the effectiveness of the system, as it resulted in a shift in the balance of power within the workplace to the disadvantage of revenue officials and consequently hardened their attitude towards computerisation. Senior officers, once enthusiastic about the system, spoke about the frustration they felt at being unable to fulfil their supervisory role and at being put at the mercy of a junior employee. Junior tax officials, already slightly sceptical of the system, feared that their skill levels would put them at a disadvantage within the office and could eventually result in redundancy.

Opinions were divided about whether or not computerisation of the system that had led to improved tax yields. Most tax officials felt that while the introduction of the computerised system had positively impacted tax collections to some extent, there were many other reasons as to why tax yields had improved. For others, the introduction of the Self Assessment Scheme as a means of shifting the responsibility of tax payments onto the shoulders of the citizens and reducing the workload of revenue staff was almost as (if not more) important as the introduction of technology into the workplace. It may be concluded from the interviews that general citizen apathy towards property tax is to a large extent a consequence of poor public awareness about the benefits of paying property tax, a lack of enforcement measures and a general dislike of cumbersome processes – problems which cannot be solved through the introduction of technology alone.

IDENTIFYING GAMES THAT IMPACT THE PLANNING AND UPTAKE OF ICTS IN DEVELOPING WORLD BUREAUCRACIES

An examination of the interviews and other data collected during field research reveals that the eventual outcome of the revenue department project can be interpreted as the consequence of a number of players making moves within a number of separate but interrelated games related to the project's design, implementation and adoption. At least six kinds of games have influenced the effect the system has had on tax administration in Bangalore city. They include expertise games, power and influence games, policy games, turf struggles, games of persuasion and business games.

From the games identified during the course of research, a four-fold taxonomy has been developed which classifies analyses games depending on the level of actor interactions on the basis of the field of play, the key actors involved, the main objective(s) of the game under study and the nature and/or spirit in which the game has been played. The four categories, which are derived from this author's research, are elaborated below:

1. **Arena or Field of Play:** Actor interactions may be classified according to the arena within which they are played out. In other words, this classification – which has its roots in initial work done by Vedel (1989)

and Dutton (1992) – focuses on the reach and influence of actors within a given context, and the impact of their actions (both direct and indirect) on project outcomes.

a. **Project-Specific Games:** are generally played by individuals and groups of actors directly involved with the case under study. Such interactions usually occur during the planning and execution of a project and impact.
b. **Organisation-Specific Games:** are played out within the department or organisation within which the case study is based, involving not only actors directly concerned with the case study but also others within the institution whose moves come to bear influence on the project at hand.
c. **City or Regional Level Games:** include those interactions between actors whose power or reach extends to the level of the city or region within which the project is based, and who are playing power games for relatively high stakes. The goals, moves and strategies chosen by actors at this level may or may not have a direct link to the case study, however they come to bear either a direct or indirect influence on its eventual outcome.
d. **National Level Games:** involve players who have their eye on attaining some sort of national prestige or who are influenced by other actors or discourses operating at the national level. Here again, actors may or may not be directly attached to the project or organisation under study.
e. **International Level Games:** are played chiefly by actors or groups of actors possessing international clout and/or aspirations. Games played at this level usually do not have a direct bearing on the project under study, however, actors might indirectly influence outcomes by attempting to gain power/prestige through adhering to popular trends, binding project planners to third-party conditonalities or merely by subscribing to certain schools of thought.

2. **Key Actors Involved:** Games may also be classified according to the key actors involved in each interaction studied. This axis thus aims to study interactions within the context of the key players – who they are and who they interact with.

a. **Interactions Internal to the Project Planning/Core Group:** Includes any games being played exclusively between constituent elements of the project planning committee or the core group responsible for the design and execution of the project under study.
b. **Core Project Group vs. Other Members of Implementing Department:** Cover games played between members of the core project committee and other individuals and/or groups within the implementing department who are otherwise not directly involved on the project at hand.
c. **Games within the Implementing Organisation:** Are played out between groups and individual actors who are members of the implementing organisation. Such interactions may or may not be directly related to the ICT4D project, but their outcome would have an impact on its eventual success or failure.
d. **Department/Organisation vs. External Players:** Cover interactions between the implementing department/organisation acting in a unified, institutional capacity and other external players such as the media, citizens and civil society organisations.

e. **Games Played by External Actors:** Which have little or no direct connection to the current project, but which nonetheless have a significant impact on its eventual outcome.

3. **Actor Goals:** A third way of classifying actor interactions is based on the goals that different actor groups seek to attain by engaging with other players. Actors within each game are bound to have multiple goals that motivate them to act in certain ways, and thus it is important when applying this classification to identify the primary motivating factor behind each move.
   a. **Games of Power and Prestige:** Involve moves to enable actors to gain or shore up their individual power and prestige or those of their group.
   b. **Games to Maintain Status Quo:** Are those interactions whereby players seek to maintain the status quo. These games are generally played when actors perceive a threat to their current position or status, and thus act to preserve their current standing in the hierarchy.
   c. **Games to Achieve Change:** Are those interactions that attempt to change a current situation or process within a department or organisation, primarily through the attainment of project goals and objectives.
   d. **Games to Achieve Political and Policy Aims:** Are those moves and strategies played by actors to achieve certain political or policy aims which may or may not have a direct relationship or bearing on the project under study.
   e. **Games to Further Ideology and/or Discourse:** Comprise chiefly those games played by actors who are generally driven by a particular ideology or discourse and wish to use their political influence to impose their ideas on either the implementing organisation or on the project planners themselves.
4. **Nature of Game Play:** The final axis against which games may be classified analyses the nature of the political dynamic between the key actors within which the project was conceived and implemented. In other words, this axis differentiates between positive and negative actors and the impact of their actions on their sphere of influence.
   a. **Constructive Game Play:** Includes altruistic and other positive moves, where competition is seen to be constructive and controlled/restrained rivalry brings about positive results. Such games are therefore win-win situations, and include all those moves that have a positive impact on the adoption of new technologies within a development context.
   b. **Destructive Game Play:** Involves fierce rivalries and negative competition, resulting in zero-sum games where actors act purposefully to win at the cost of their so-called 'opponents', thereby creating a negative project environment and often resulting in a large wastage of time and resources.

WORKPLACE ORGANISATION, STRUCTURE AND POLITICAL INSTITUTIONS:

Exploring Adaptation to Change in the Age Of the Internet

The discussion above reveals that at the heart of a design-actuality gap usually lies a power struggle brought about through a deep-seated mistrust between different actor groups. In particular, the case study demonstrates that gaps arise because those with the power and authority to take design or implementation decisions are usually unwilling to allow any initiative to go ahead that would give the other actor group(s) in the game more autonomy over the system.

The identification of the actors related to property tax administration in India at its most basic, and the discussion of the games they play during the process of tax administration and reform, highlighted the fact that if property tax is administered and reformed almost exclusively by a local government authority using conventional policy and fiscal tools, then the arena within which games are played out remains highly localised with the number of actors restricted and their moves limited. However, as the case study illustrates, the introduction of ICTs into the reform process will not only add more actors to the mix but also introduce different levels of interaction and open up the playing field to a larger number of moves and decisions, as the use of technology in development is connected to much larger national and international policy discourses.

As the analysis has shown, certain key games with local impacts get played out in different arenas between actors influenced by not only local but also national and international factors. Design-actuality gaps open up and give way to unfavourable project outcomes if designers and top managers assume that localised outcomes result only from direct local influences, discounting the impact of other factors external to the project at hand. In the light of such an evaluation, what impact, it may be asked, has the computerised system had on the process of organisational reform, and what effect has this in turn had on institutional change?

From the discussion and analysis of the system presented above and in previous sections of this chapter, it may be concluded that despite the presence of self-interested and competitive game-play during the development of the Revenue Department system, co-operation during the introduction and adoption stages has resulted in positive steps being taken towards organisational reform and institutional change. In terms of changes at the level of the organisation, the digitisation of the tax registers and the automation of processes have increased the efficiency of the Department by speeding up tax administration processes, reducing mistakes, and lowering workloads.

Institutionally, the project has also had some success. By allowing citizens to access their records and pay at their convenience (either online or at kiosks), the corporation has adopted a radical citizen-centric approach to tax administration which has not only shifted the balance of power in the government-citizen relationship in favour of the citizens, but has, at the same time, made citizens more responsible. This might be considered to be a real departure from traditional notions of Indian bureaucracy where power is concentrated in the hands of the bureaucrats, and citizens can afford to be passive actors in administration meta-games. However, as some of the key organisational and institutional variables have not yet been put in place, and there is no mechanism by which problems may be identified and bridged during the implementation process, the long-term direction of these changes is still uncertain. Of particular concern are the attitudes of revenue employees towards the changes in work processes and the shifts in hierarchies.

IMPLICATIONS AND FUTURE RESEARCH DIRECTIONS

The Ecology of Games provides a theoretical framework for discussing the strength and interplay of groups and interests shaping e-government projects. But the question must be asked: what added value has the use of this theory brought to the study of e-government and ICT4D projects? The answer to this question has been brought out through the discussion of the case study in previous chapters and its most salient points may be summarised below.

Firstly, the notion of an Ecology of Games offers a framework for thinking about an extremely complex set of interactions, identifying and highlighting the roles played by those who shape and are shaped by the rules of the game, and the impact that each player has on a project's ultimate outcome. It particularly emphasises the potential for unanticipated, unplanned developments on project success; raising doubts on more conventional, information-systems views of e-government that see project implementation as governed by a more controlled, isolated, predictable system of action.

The framework also focuses on 'symbolic politics', what Dutton (1992) in his discussion of the theory has described as the role ideas play in political change. Whilst democratic politics and the formulation of policy is in part a contest over ideas about how to define and achieve the common good, empirical scholars of politics have been remarkably resistant to giving ideas a central explanatory place in their accounts (Mehta, 2010). The Ecology of Games amends this, emphasising not only the way in which the development of ideas shapes political interactions, but also highlighting the emergence and role of new bearers and interpreters of those ideas (like the media) as key players in the ultimate success or failure of a project.

The Ecology of Games has, according to Dutton (1992) yet another advantage as an approach to research, in that it helps identify cross-pressures facing key players often involved in more than one game; recognising that e-government project development is not, contrary to conventional frameworks, a self-contained system of action. Instead, as illustrated by the case study, the framework recognises that projects are being formulated and implemented in parallel with other policies. Many players in one policy area are playing simultaneously in others. The outcome of the political process in one arena often shapes play within another. In doing so, it provides a more nuanced interpretation of the broader system of action in which the development of an e-government project develops, emphasising again the role of unplanned, unanticipated interactions between various interests, and the formation of unconventional interests and alliances as a result of shared goals. Nothing is new about many of these interactions, they often repeat themselves through the ages in a variety of situations, yet conventional theories tend to ignore or underplay them.

Using the Ecology of Games perspective does, however, mean that a number of limitations and difficulties need to be acknowledged. The first is that it is essentially a 'sensitizing' concept, a background theory that offers a certain way of seeing, organizing, and understanding complex reality (Dutton, 1992). Whilst this is not necessarily a weakness, it does imply not only a limited usefulness for quantitative or formal mathematical approaches, but also a large degree of interpretive flexibility. Consequently, different researchers applying the Ecology of Games to the same situation are likely to perceive different ecologies, games, actors, and interactions. Any one interpretation can thus be challenged by others, or by any researcher who can critically assess the depiction of a specific ecology.

A second, related criticism it that whilst it provides a point of view and indicates a set of methods to conduct case studies, the Ecology of Games theory can only give an indication of the likely nature of the dynamics shaping outcomes. Based on human behaviour motivated by a particular set of influences, it is thus limited as a predictive theory, in the sense that it will not be able to predict the concrete outcomes necessary for both micro- and more generalised macro-level decision-making. A final problem with the framework is that it may lead to an extremely complex mapping of social reality (Dutton, Schneider, & Vedel, 2011). Its innate flexibility can lead a researcher to read deeper and deeper into what might be, in reality, only a few large meta-games. And finally, partly as a result of this tendency towards increased complexity, it becomes necessary to arbitrarily limit the depth of any analysis lest it become too unwieldy. Such an arbitrary truncation feeds back into the discussion surrounding the theory's interpretive flexibility and its value as a predictive tool, as different studies of the same of object would be likely to result in different analyses and (especially if used as a policy tool) different policy decisions.

These disadvantages, however, might be overcome; as seen in recent work combining the Ecology of Games with other sociological perspectives such as Network Theory and Social Constructivism (Cornwell, Curry, & Schwirian, 2003). It may be concluded that the central strength of the Ecology of Games perspective compared with other theoretical frameworks is this: without taking away from the central issues at hand or diverting attention from the central field of action, the framework focuses on a variety of phenomena – personality, values, historical circumstance, environment – that are all to often peripheral to the central action of conventional theories but in truth form the central core of the policy process and are often key forces behind organisational and institutional change. In doing so, the combination helps provide researchers with a nuanced understanding of how actor dynamics impact political and policy outcomes.

CONCLUSION

Rapidly evolving economic and social contexts mean that political institutions and the people who constitute them cannot afford to get bogged down in traditional work practices or be impervious or resistant to change themselves. Whilst this does not necessarily mean a wholesale rejection of what has gone before, it does mean that there needs to be a constant assessment and reassessment of workplace values and current practices, eliminating those which result in behaviours that are detrimental to the functioning of the organisation and encouraging those that promote positive interactions. Organisations and institutions, particularly those which form the political core of a society, cannot afford to be seen to have been left behind, as the people within those institutions are generally looked to as political trendsetters and role models in addition to being responsible for societal welfare.

The discussion brought out in this chapter reveals that at the heart of a political game usually lies a power struggle brought about through a deep-seated mistrust between different actor groups. In particular, the case study put forward demonstrates that gaps arise because those with the power and authority to take design or implementation decisions are usually unwilling to allow any initiative to go ahead that would give the other actor group(s) in the game more autonomy over the system. Further, certain key games with local impacts get played out in different arenas between actors influenced by not only local but also national and international factors. Problems arise if designers and top managers assume that localised outcomes result only from direct local influences, discounting the impact of other factors external to the project at hand.

Added to this, there is a tendency for power elites to lose touch with ground realities when devising projects for their organisations as well as for their citizens, especially when planners comprise the higher echelons of government and operate within a top-down command-and-control system of management. There is also a danger that high-level project planners will, in looking at macro-outcomes, ignore outliers and how these may precipitate unexpected turns of events. This holds particularly true when existing patterns of communication and information exchange fail to be flexible or unable to adapt to changing situations.

REFERENCES

Asquith, A. (1998). Non-elite employees: Perceptions of organizational change in english local government. *International Journal of Public Sector Management*, *11*(4), 262–280. doi:10.1108/09513559810225825

Avgerou, C., & Walsham. (2000). Introduction: IT in developing countries. In C. Avgerou & G. Walsham (Eds.), *Information technology in context: Studies from the perspective of developing countries*. Aldershot, UK: Ashgate.

Bahl, R. W., & Linn. (1992). *Urban public finance in developing countries*. New York: Oxford University Press.

Bannock, G. Baxter, & Davis. (1987). The Penguin dictionary of economics (4th ed.). London: Penguin Books.

Basu, S. (2004). E-government and developing countries: An overview. *International Review of Law Computers & Technology*, *18*(1), 109–132. doi:10.1080/13600860410001674779

Beinhocker, E. D. (2006). *The origin of wealth: Evolution, complexity and the radical remaking of economics*. Boston: Harvard Business School Press.

Bellamy, C., & Taylor. (1994). Introduction: Exploiting IT in public administration – Towards the information polity? *Public Administration*, *72*, 1–12. doi:10.1111/j.1467-9299.1994.tb00996.x

Bellamy, C. (2000). The politics of public information systems. In G. D. Garson (Ed.), *Handbook of public information systems*. New York: Marcel Dekker Inc.

Bhatnagar, S. (2004). *E-government: From vision to implementation*. Thousand Oaks, CA: Sage Publications.

Bhatnagar, S. (2005). *E-government: Opportunities and challenges*. Retrieved from http://siteresources.worldbank.org/INTEDEVELOPMENT/Resources/559323-1114798035525/ 1055531-1114798256329/ 10555556- 1114798371392/Bhatnagar1.ppt

Bruhat Bangalore Mahanagara Palike. (2000). Property tax self-assessment scheme: Golden jubilee year 2000. Mahanagara Palike Council Resolution No. 194/99-2000, Bangalore.

Bruhat Bangalore Mahanagara Palike. (2007). *Assessment and calculation of property tax under the capital value system (new SAS), 2007- 2008*. Unpublished.

Casely, J. (2004, March 13). Public sector reform and corruption: CARD facade in andhra pradesh. *Economic and Political Weekly*, 1151–1156.

Chadwick, A., & Howard. (Eds.). (2009). *The handbook of internet politics*. London: Routledge.

Ciborra, C. (2005). Interpreting e-government and development: Efficiency, transparency or governance at a distance? *Information Technology & People*, *18*(3), 260–279. doi:10.1108/09593840510615879

Ciborra, C., & Navarra. (2005). Good governance, development theory and aid policy: Risks and challenges of e-government in Jordan. *Information Technology for Development*, *11*(2), 141–159. doi:10.1002/itdj.20008

Cornwell, B., Curry, & Schwirian. (2003). Revisiting norton long's ecology of games: A network approach. *City & Community*, *2*(2), 121–142. doi:10.1111/1540-6040.00044

Crozier, M., & Friedberg. (1980). *Actors and systems*. Chicago: University of Chicago Press.

Dada, D. (2006). The failure of e-government in developing countries: A literature review. *The Electronic Journal on Information Systems in Developing Countries*, *26*(7), 1–10.

De, R. (2007). Antecedents of corruption and the role of e-government systems in developing countries. In *Proceedings of Ongoing Research*. Retrieved from http://www.iimb.ernet.in/~rahulde/CorruptionPaperEgov07_RDe.pdf

Dillinger, W. (1988). *Urban property taxation in developing countries*. Retrieved from http://ideas.repec.org/p/wbk/wbrwps/41.html

Downs, A. (1964). *Inside bureaucracy*. Boston: Little Brown.

Dunleavy, P. Margetts, Bastow, & Tinkler. (2006). Digital era governance: IT corporations, the state and e-government. Oxford, UK: Oxford University Press.

Dutton, W. H. (1992). The ecology of games shaping telecommunications policy. *Communication Theory, 2*(4), 303–324. doi:10.1111/j.1468-2885.1992.tb00046.x

Dutton, W. H. (1999). *Society on the line: Information politics in the digital age*. Oxford, UK: Oxford University Press.

Dutton, W. H. Schneider, & Vedel. (2011). Large technical systems as ecologies of games: Cases from telecommunications to the internet. In J. Bauer, A. Lang, & V. Schneider (Eds.), Innovation policy and governance in high-tech industries: The complexity of coordination. Berlin: Springer.

Fehr, E., & Gachter. (1998). Reciprocity and economics: The economic implications of homo reciprocans. *European Economic Review, 42*(3), 845–859. doi:10.1016/S0014-2921(97)00131-1

Fehr, E., & Gachter. (2002). Altruistic punishment in humans. *Nature, 415*, 137–145. doi:10.1038/415137a PMID:11805825

Fink, C., & Kenny. (2003). W(h)ither the digital divide? *Info: The Journal of Policy. Regulation and Strategy for Telecommunications, 5*(6), 15–24. doi:10.1108/14636690310507180

Firestone, W. A. (1989). Educational policy as an ecology of games. *Educational Researcher, 18*(7), 18–24.

Gintis, H. (2000). *Game theory evolving*. Princeton, NJ: Princeton University Press.

Gintis, H. (2006). *The economy as a complex adaptive system - A review of Eric D. Beinhocker, the origins of wealth: Evolution, complexity, and the radical remaking of economics*. MacArthur Research Foundation. Retrieved from http://www.umass.edu/preferen/Class%20Material/Readings%20in %20Market%20Dynamics/ Complexity%20Economics.pdf

Gopal Jaya, N. (2006). Introduction. In N. Gopal Jaya, A. Prakash, & P. K. Sharma (Eds.), *Local governance in India: Decentralisation and beyond*. New Delhi: Oxford University Press.

Hacker, K. L., & van Dijk. (2000). Introduction: What is digital democracy?. In K. L. Hacker & J. van Dijk (Eds.), *Digital democracy: Issues of theory and practice*. London: Sage Publications.

Hall, P. A. & Taylor. (1996). Political science and the three new institutionalisms. *MPIFG Discussion Paper 96/9*.

Hechter, M., & Kanazawa. (1997). Sociological rational choice theory. *Annual Review of Sociology, 23*(1), 191–214. doi:10.1146/annurev.soc.23.1.191

Heeks, R. (2003). Most eGovernment-for-development projects fail: How can the risks be reduced? *i-Government Working Paper Series,* Paper No. 14, IDPM.

Heeks, R. (2005). eGovernment as a carrier of context. *Journal of Public Policy, 25*(1), 51–74. doi:10.1017/S0143814X05000206

Heeks, R. (2006). *Implementing and managing eGovernment – An international text*. New Delhi: Vistar Publications.

Isaac-Henry, K. (1997). Development and change in the public sector. In K. Isaac-Henry, C. Painter, & C. Barnes (Eds.), *Management in the public sector: Challenge and change*. London: International Thomson Business Press.

Jha, G. (1983). Area basis of valuation of property tax: An evaluation. In A. Datta (Ed.), *Property taxation in India*. New Delhi: Centre for Urban Indian Studies – The Indian Institute of Public Administration.

Jha, S.N., & Mathur. (1999). *Decentralization and local politics*. New Delhi: Sage Publications.

Jick, T. D. (1979). Mixing qualitative and quantitative methods: Triangulation in action. *Administrative Science Quarterly*, *24*(4), 602–611. doi:10.2307/2392366

Johnson, R. N., & Libecap. (1994). *The federal civil service system and the problem of bureaucracy*. Chicago: University of Chicago Press.

Kenman, H. (1996). Konkordanzdemokratie und korporatismus aus der perspektive eines rationalen institutionalismus. *Politische Vierteljahresschrift*, *37*, 494–515.

Laver, M., & Schofield. (1990). *Multiparty government: The politics of coalition in Europe*. Oxford, UK: Oxford University Press.

Lewis, A. (1982). *The psychology of taxation*. Oxford, UK: Martin Robertson & Company.

Lieten, G. K., & Srivatsava. (1999). *Unequal partners: Power relations, devolution and development in uttar Pradesh*. New Delhi: Sage Publications.

Long, N. E. (1958). The local community as an ecology of games. *American Journal of Sociology*, *64*(3), 251–261. doi:10.1086/222468

Madon, S. (1993). Introducing administrative reform through the application of computer-based information systems: A case study in India. *Public Administration and Development*, *13*, 37–48. doi:10.1002/pad.4230130104

Madon, S. (1997). Information-based global economy and socio-economic development: The case of Bangalore. *The Information Society*, *13*, 227–243. doi:10.1080/019722497129115

Madon, S., & Bhatnagar. (2000). Institutional decentralised information systems for local level planning: Comparing approaches across two states in India. *Journal of Global Information Technology Management*, *3*(4), 45–59.

Madon, S. (2004). Evaluating the developmental impact of e-governance initiatives: An exploratory framework. *Electronic Journal of Information Systems in Developing Countries*, *20*(5), 1–13.

Madon, S., & Krishna, S., & Michael. (2010). Health information systems, decentralisation and democratic accountability. *Public Administration and Development*, *30*(4), 247–260. doi:10.1002/pad.571

Madon, S., & Sahay, S., & Sahay. (2004). Implementing property tax reforms in Bangalore: An actor-network perspective. *Information and Organization*, *14*, 269–295. doi:10.1016/j.infoandorg.2004.07.002

March, J.G., & Olsen. (1984). The new institutionalism: Organisational factors in political life. *The American Political Science Review*, *78*(3), 734–749. doi:10.2307/1961840

March, J. G., & Olsen. (1989). *Rediscovering institutions: The organisational basis of politics*. New York: The Free Press.

Margetts, H. (1998). *Information technology in government: Britain and America*. London: Routledge.

Margetts, H. (2006). Transparency and digital government. In C. Hood, & D. Heald (Eds.), *Transparency: The key to better governance?* London: The British Academy. doi:10.5871/bacad/9780197263839.003.0012

McCubbins, M. D., & Sullivan. (1987). *Congress: Structure and policy*. Cambridge, UK: Cambridge University Press.

Mehta, J. (2010). Ideas and politics: Towards a second generation. *Perspectives on Politics*. Retrieved from http://www.allacademic.com//meta_/p_mla_apa_research_citation/ 0/2/2/1/1/pages22111/p22111-1.php

Mintzberg, H. (1985). The organisation as political arena. *Journal of Management Studies*, *22*(2), 133–154. doi:10.1111/j.1467-6486.1985.tb00069.x

Misra, S. (2005). eGovernance: Responsive and transparent service delivery mechanism. In A. Singh (Ed.), Administrative reforms: Towards sustainable practices. New Delhi: Sage Publications.

Moon, M. J. (2002). The evolution of e-government among municipalities: Rhetoric or reality? *Public Administration Review*, *62*(4), 424–433. doi:10.1111/0033-3352.00196

National Institute of Urban Affairs. (2004). *Reforming the property tax system*. New Delhi: NIUA Press.

Ostrom, E. Gardner, & Walker. (1994). Rules, games and common-pool resources. Ann Arbor, MI: University of Michigan Press.

Peters, B. G. (2000). *Institutional theory in political science: The new institutionalism*. London: Continuum.

Ronaghan, S. A. (2002). Benchmarking e-government: A global perspective. New York: The United Nations Division for Public Economics and Public Administration (DPEPA) Report.

Rose, R. (2005). A global diffusion model of e-governance. *Journal of Public Policy*, *25*(1), 5–28. doi:10.1017/S0143814X05000279

Rose-Ackerman, S. (1978). *Corruption: A study in political economy*. New York: Academic Press.

Rosengard, J. K. (1998). *Property tax reform in developing countries*. Boston: Kluwer Academic Publications. doi:10.1007/978-1-4615-5667-1

Scharpf, F. W. (1997). *Games real actors play: Actor-centered institutionalism in policy research*. Oxford, UK: Westview Press.

Schech, S. (2005). Wired for change: The links between ICTs and development discourses. *Journal of International Development*, *14*(1), 13–23. doi:10.1002/jid.870

Schiller, H. I. (1981). *Who knows: Information in the age of the fortune 500*. Norwood, NJ: Ablex Publishing Corp.

Simon, H. A. (1955). A behavioural model of rational choice. *The Quarterly Journal of Economics*, *69*(1), 99–118. doi:10.2307/1884852

Sinha, K. P. (1981). *Property taxation in a developing economy*. New Delhi: Puja Publications.

Stanforth, C. (2006). *Analysing eGovernment in developing countries using actor-network theory*. iGovernment Working Paper Series – Paper no. 17.

Tanzi, V. (1987). Quantitative characteristics of the tax systems of developing countries. In D. Newbery, & N. Stern (Eds.), *The theory of taxation for developing countries*. New York: Oxford University Press.

Tsebelis, G. (1990). *Nested games: Rational choice in comparative politics*. Berkeley, CA: University of California Press.

Vedel, T. (1989). Télématique et configurations d'acteurs: Une perspective européenne. *Reseaux*, *7*(37), 9–28.

Virkar, S. (2011a). Exploring property tax administration reform through the use of information and communication technologies: A study of e-government in Karnataka, India. In J. Steyn, & S. Fahey (Eds.), *ICTs and sustainable solutions for global development: Theory, practice and the digital divide*. Hershey, PA: IGI Global.

Virkar, S. (2011b). *The politics of implementing e-government for development: The ecology of games shaping property tax administration in Bangalore City*. (Unpublished Doctoral Thesis). University of Oxford, Oxford, UK.

Wade, R. H. (1985). The market for public office: Why the Indian state is not better at development. *World Development*, *13*(4), 467–497. doi:10.1016/0305-750X(85)90052-X

World Bank - Global Information and Communication Technologies Department. (2002). *The networking revolution: Opportunities and challenges for developing countries*. Washington, DC: World Bank.

Yin, R. K. (2003). *Case study research: Design and methods*. London: Sage Publications.

KEY TERMS AND DEFINITIONS

Actor Goals and Motivations: The aims that key actors seek to attain and maintain from interacting with other players, encompassing both broader long-term achievements as well as more short- to medium-term rewards.

Actor(s): The individuals, groups or other entities whose interactions shape the direction and nature of a particular game being considered.

Game(s): Arena(s) of competition and cooperation structured by a set of rules and assumptions about how to act in order for actors to achieve a particular set of objectives.

Gamification: The application of gaming principles to non-gaming environments.

Moves: May be defined as actions, decisions and other plays made by key actors taken to arrive at key goals, usually if not always based on their strategy of choice.

Rules: The written or unwritten codes of conduct that shape actor moves and choices during a game.

Serious Gaming: The application of gaming principles to training environments.

Strategies: Include tactics, ruses, and ploys adopted by key actors during the course of a game to keep the balance of the engagement in their favour.

Chapter 4
Should We Publish That?
Managing Conflicting Stakeholder Expectations in the Publishing Industry

Loren Falkenberg
University of Calgary, Canada

Oleksiy Osiyevskyy
University of Calgary, Canada

ABSTRACT

As the responsibilities of modern business expand to multiple stakeholders, there is an increased need to understand how to manage conflicting normative expectations of different stakeholders. Corporate responsibilities to stakeholders are based on the need to minimize or correct harm from operations (respect negative injunctions) while contributing to the social welfare of communities the firm operates in (engage in positive duties). By comparing multiple decision scenarios in the traditional and online publishing industry, the chapter explores the tensions that arise between these imperatives. Based on these tensions, the chapter outlines a framework and a practical industry-independent heuristic decision making process, embracing normative expectations, the consequences to a company and to stakeholders, and potential mitigating actions. The proposed heuristic approach allows balancing the tensions among stakeholder expectations to ensure selection of the appropriate alternative. The discussion is finished by pointing out the usefulness and applicability of the proposed heuristics in other industries and settings of the contemporary business environment.

INTRODUCTION

Corporations that fail to respond to stakeholder pressures risk losing goodwill and damaging their public image and reputation (Julian, Ofori-Dankwa, & Justis, 2008). Numerous frameworks have been suggested for managing stakeholder relations (Agle, Mitchell & Sonnenfeld, 1999; Doh & Guay; 2006; Mitchell et.al, 1997; Peloza & Falkenberg, 2009), with the underlying assumption that a company should focus on the most threatening, urgent or visible stakeholders,

DOI: 10.4018/978-1-4666-5071-8.ch004

or where they have the greatest capability (Julian, Ofori-Dankwa & Justis, 2008). We suggest there are times when the focus of analysis should shift from stakeholder management to managing the tension between competing normative responsibilities: the negative injunction and positive duty imperatives.

Since the 1960s a dominant doctrine for prescribing business responsibilities has been manifested in the libertarian views of Milton Friedman: the goal of a corporation is to optimize its performance in terms of augmenting shareholders' wealth, with laws setting the boundaries for unacceptable actions. Contemporary understanding of these ideas – referred to as the "moral minimum" – require the managers to avoid, minimize and correct self-caused harm to stakeholders or communities (Wettstein, 2010), or respect ***negative injunctions*** (Simon, Powers, & Gunnemann, 1983).

Another view of responsibilities to stakeholders is based on a ***positive duty*** of making the world a better place. Mulligan (1993) notes "the moral mission of business is not fulfilled simply by doing what is required in order to survive in the social environment… a business must deserve to survive as a result of its honest choices and deliberate accomplishments" (p. 70). The assignment of "positive duties" to businesses has also been labeled "affirmative duties" or "positive obligations" by business ethicists (Simon et al, 1983; Wettstein, 2010), and integrated into traditional profit-oriented strategic management theories (e.g., Porter & Kramer, 2011). The extreme version of positive duties imperative stems from the premise of consequentialism, which treats all negative events a company fails to prevent as equal to those it causes directly, by this means making business morally responsible for all social problems it had nothing to do with in the first place (e.g., Scheffler, 2001).

Therefore, throughout the text we are referring to the business's moral minimum of not hurting stakeholders as negative injunctions; the contentious imperative of improving the society well-being will be labeled positive duty. Conflicting normative responsibilities occur in situations where expectations of different stakeholders are based on these two different premises – doing no harm (i.e., respecting negative injunctions) and contributing to the improved welfare of a community or stakeholder group (i.e., engaging in positive duties). For example, the editors of the Danish newspaper Jyllands-Posten were concerned about a growing trend of self-censorship and chose to publish cartoons critical of Muslim religious dogmas, and hurting some of Muslims by depicting the Prophet Muhammad. The editors anticipated a negative reaction from the stakeholder groups, but still chose to support the principles of free speech by publishing the cartoons. The cartoons led to an increased awareness of the important role social commentary has in supporting the free speech (i.e., positive duty), while they also led to demonstrations and loss of life and moral harm (i.e., violation of a negation injunction).

BACKGROUND

Little has been written about how to effectively manage competing expectations based on the normative responsibilities of negative injunctions and positive duties. This paper explores the tensions created by these two normative corporate responsibilities through a series of illustrative vignettes (i.e., descriptions of stakeholder expectations, reactions and consequences to stakeholders and the company) of controversial publications in the publishing industry. The paper begins with a review of normative responsibilities that arise from negative injunctions and positive duties, followed by a discussion of the specific responsibilities associated with the role of publishers in supporting freedom of speech. Next a series of vignettes that illustrate competing stakeholder expectations in the publishing industry, and the consequences that have occurred when publishers responded to negative injunctions and positive duties are reviewed. The paper ends with a heuristic decision making process for comparing the benefits and costs of respecting a negative injunction or engaging in a positive duty for both the company and stakeholders in a given situation. A detailed description of the qualitative research methodology followed to identify, develop and analyze the vignettes is provided in Appendix.

NORMATIVE RESPONSIBILITIES OF BUSINESS

Today, a dominant doctrine for prescribing business responsibilities is the "moral minimum" of Milton Friedman (2002): any private business's *raison d'être* is maximizing shareholders' wealth, with laws and legal requirements setting the boundaries for unacceptable actions. This position can be traced back to the original ideas of Adam Smith, who professed that self-interest, not the voluntary benevolent actions – corporate social responsibility beyond obeying the laws – should drive the commercial arena (Bragues, 2009). This orientation, serving the basis for legal systems of Western democratic countries, implies that managers should make decisions that limit unnecessary harm to stakeholders and communities, or respect ***negative injunctions***. Friedman's initial libertarian views were later expanded in management scholarship by resource dependency and stakeholder theories, which imply that multiple stakeholder demands should be prioritized according to their impact on shareholder wealth (Pfeffer & Salancik; 1978). Mitchell, Agle and Wood (1997) further refined this prioritization according to three stakeholder attributes: power, legitimacy and urgency. The "stakeholder view" has led to researchers exploring links between responsiveness to stakeholder demands and financial performance (e.g., Greenley & Foxall, 1997), reputation (Carter, 2006), and perceived legitimacy of demands (Fineman & Clarke, 1996). The expansion of corporate responsibilities from shareholders to stakeholders has led to two management principles: the principle of corporate rights, which guides managers to respect the legal and cultural rights of other actors, and the principle of corporate effects, which assigns responsibility for the negative effects of business activities on other actors (Dubbnick, 2004; Evan & Freeman, 1995).

The opposite view of corporate responsibility to society implies assigning of "positive duties" to businesses, explicitly requiring companies to improve the world, deserving by this means the right to exist, through deliberate accomplishments and honest choices (Mulligan, 1993). This view of proactive business's engagement in resolving the social problems is held by many business ethicists, environmental and social activists. The positive duties associated with business activities align with the expectations and responsibilities of professional groups. For example, although skilled scientists and surgeons are generously compensated for their work, no one assumes their professional duties are based on existing laws; rather they are expected to use professional judgment to limit harm and improve society/patients. In line with these responsibilities, it is inappropriate to assume that an entrepreneur's or a manager's only motivation is to make money. Moreover, this view is not always incompatible with firms' *raison d'être* of shareholder value maximization, and can be integrated into traditional profit-oriented strategic management frameworks, such as value-based theory of the firm (Becerra, 2009) or shared value framework (Porter & Kramer, 2011). Porter and Kramer (2011) recognize both profit and social imperatives, by pointing out that "capitalism can create economic value by creating societal value" (p.67), and that "capitalism is an unparalleled vehicle for meeting human needs, improving efficiency, creating jobs, and building wealth" (p.64). The attributes of entrepreneurship, innovation and efficiency found in a capitalist economy can also be applied to most of the humanity's problems;

thus managers have the skills and knowledge to expand the pie of pooled economic and social value, without the need to make a hard choice between the two (Porter & Kramer, 2011).

A substantial number of corporate leaders have integrated positive duties into their firm's strategic goals and management plans. The reasons given for integrating positive duties into the firm's strategic planning range from a purely pragmatic need to avoid negative publicity to "an intrinsic belief in the social contributions of business in a post-modern society" (Doh & Guay, 2006; p.54).

Scholars have recognized the limitations of focusing on negative injunctions as there is often a time lag between stakeholders incurring injury and recognition of the cause (i.e., business activities); and externalities (i.e., the unanticipated harm that hurts bystanders or unidentified stakeholders) are difficult to anticipate (Martin, 2008; Stone, 1975). In contrast, focusing on positive duties can lead to stakeholders within a given community engaging in power tactics to ensure their priorities visibly dominate over those of other stakeholders (Julian, Ofori-Dankwa & Justis, 2008; Mitchell, Agle & Wood, 1997). And, managers often lack the expertise to fully understand what is needed to improve social welfare within a given community (Friedman, 2002; Kahn, 2006).

Competing stakeholder expectations on how publishers should protect free speech provide visible examples of the tensions between negative injunctions and positive duties. Freedom of speech is a fundamental right to express ideas and access information that is recognized by international bodies, philosophers and most developed countries. Article 19 in the Universal Declaration of Human Rights states "Everyone has the right to freedom of opinion and expression, this right includes the freedom to hold opinions without interference and to seek, receive and impart information and ideas through any media". Free speech is critical to the development of prosperous and stable societies. For example, in the Islamic Golden Age, from the eighth to the 16th centuries, scholars collected and shared knowledge from all over the world. During this period, scholars followed Prophet Muhammad Ibn Abdullah's statement in the Quran "The scholar's ink is more sacred than the blood of martyr" (Falagas, Zarkadoulia & Samanis, 2006). Later, in Britain John Stuart Mill asserted that a nation's social and intellectual progress hinges, *inter alia*, on freedom of information (Mill, 1974). In more recent centuries the strongest economies have developed in countries where freedom of speech is a fundamental value (Zakaria, 2008).

Publishing companies continually balance multiple stakeholder expectations, including ensuring communication channels for the expression of ideas (e.g., pornography, political criticism), limiting injury to vulnerable individuals (e.g., hate speech, terrorist communications), and respecting community norms. Publishers are gatekeepers in the sense that they control access to data and ideas through production, copyrights and distribution channels. They act as mediators and moderators between the authors of ideas and the general public. As mediators, they serve as connecting links, distorting the information along the way; while as moderators they interact with the information, attenuating or amplifying different messages. As noted by one commentator, "Free speech is free only on a mountain top; all else is editing" (Jenkins, 2006).

The assignment of moral responsibilities to publishers has occurred since the invention of the printing press; as noted by Thomas Carlyle, a Scottish satirical writer in the Victorian era, "publishing represents both the Stock Exchange and the Cathedral, or more bluntly, money and morals" (Gedin, 2004). The traditional view of publisher responsibilities was based on the principle of "primary non nocere" or a negative injunction. This view was justified because publishers were not expected to publish every submission; rather, editorial decisions were based on profit or demand, as long as the publication did not harm stakeholders.

Currently, a dominant negative injunction is to avoid violating human rights. The UNCHR (2004) "Embedding Human Rights in Business Practices" (p. 19) identifies two conditions in which businesses are complicit in violating human rights (i.e., violating negative injunctions). The first condition is when a company authorizes, tolerates or knowingly ignores human rights abuses committed by an entity associated with it (i.e., permits), and the second is when company activities support or influence ongoing abuses (Brenkert, 2009). The application of these two conditions implies that publishers should knowingly avoid any harm that could be caused by their printed or electronic publications, and stop disseminating a publication if potential harm has been identified. However, it is not easy to predict stakeholder reactions, and the negative impact of a publication may not be known until the publication has been in the public domain for more than a year, and it is too late to withdraw it (Martin, 2008). Publishers cannot control the ultimate use or destination of publications or websites found through search engines; thus, it is not feasible for a publisher to pull its products/outputs from every community where there is a potential for a group of stakeholders to be hurt (Martin, 2008).

In terms of positive duties, publishers are expected to assume an active role in social and public issues; which involves supporting free speech and the discussion of controversial issues by providing channels in which to communicate and debate (Brenkert, 2009; Nunziato, 2009). However, positive duties are not always obvious because of competing stakeholder interests and the globalization of publications. Often publications desired by some stakeholder groups and justified on the basis of free speech can offend or harm other stakeholders.

The background research for this paper identified controversial decisions through media and academic databases searches for three types of controversial publications: pornography, hate speech and sedition. Vignettes of specific controversial publications were developed when there was sufficient information (i.e., multiple sources validating the key points) to contrast normative expectations, consequences to stakeholders and the publisher.

CONTROVERSIAL PUBLICATIONS

Pornography: Obscenity or Literary Merit

One of the largest sources of entertainment revenue is the demand for pornography in print or electronic forms; however, publishers continually balance revenue generation with potential stakeholder backlash. Pornographic novels were one of the first forms of publically available erotica, with the Internet now being the dominant source. Approximately 25% of all Internet users visit pornographic sites (D'Orlando, 2009). Although there is a huge variation in what is defined as pornographic, it is the most common form of censorship, with most countries having laws limiting the explicitness of sexual activity that can be published in books or accessed through Internet search engines (Zittrain & Palfrey, 2008).

A classic example of the ambiguous boundaries distinguishing erotica from pornography is the evolving acceptance of Lady Chatterley's Lover as a literary work. This novel was first printed in Italy in 1928, and banned in Great Britain and the United States. At the time the US Senate was moving to end the practice of US Customs censoring imported books, and one Senator who vigorously opposed ending the censorship claimed Lady Chatterley's Lover was an example of an obscene book that must not reach domestic audiences (*Time*, 1930). This novel was not printed in the United Kingdom until 1960. Once it was published in Britain, Penguin Books was prosecuted under the Obscene Publications Act of 1959. This Act made it possible for publishers to escape conviction if it could be shown that a potential pornographic work had literary merit. The jury acquitted Penguin Publishers and the 1961 second edition of this book, contained the following dedication: "For having published this book, Penguin Books were prosecuted under the Obscene Publications Act, 1959 at the Old Bailey in London from 20 October to 2 November, 1960. This edition is therefore dedicated to the twelve jurors, three women and nine men, who returned a verdict of 'Not Guilty' and thus made D.H. Lawrence's last novel available for the first time to the public in the United Kingdom." More recently critics have looked past the sexual antics of the game keeper and his lover for deeper symbolic meaning. In 2006 Doris Lessing wrote "DH Lawrence's landmark novel, created in the shadow of war as he was dying of tuberculosis, is an invocation to intimacy and one of the most powerful anti-war novels ever written" (Lessing, 2006).

Currently publishers attempt to maintain a fine line in avoiding antagonizing anti-pornography groups while not losing market share to competitors. Stakeholder backlash can be very strong, as occurred when Telus (a Canadian telecommunications company) offered an "adult content" service where customers could download nude photographs or videos on a pay-view basis. Telus was the first wireless carrier in North America to offer the service, although it was being offered in Europe (Austen, 2007; Galt, 2007). Thousands of customers downloaded the photographs, while hundreds of other stakeholders complained, and Telus pulled the service to stop the stakeholder reaction. A spokesman for Telus noted that there is significant market demand for pornography, but there is also a stigma attached to companies that provide sex-related content, as well as a market for ethically minded companies.

Although child pornography is outlawed in most countries (Maitra & McGowan, 2007; Zittrain &Palfrey, 2008), with significant consequences for perpetrators, development and viewing of child pornography has grown significantly with the development of the Internet. Prior to the Internet it was difficult for pedophiles to disseminate illegal material; purchasing and viewing of the material was more visible and easily caught, and pedophiles were isolated and unable to profit from massive distribution. Now, with ease of use and anonymity provided by Internet, the situation is radically different, and there is increasing pressure on publishers to report anyone posting pedophilia.

Police forces have successfully cooperated across international boundaries in trying to control this abusive activity. This led to a request for Google's Brazilian subsidiary to release to the government information on 3,000 users of its ORKUT social networking service who may have been involved in the spread of child pornography on its website. Initially, Google eliminated the users from the ORKUT groups, but refused to release the names of those allegedly involved in child pornography. However, later it reversed the decision and handed over its list of potential users of pedophilia (*The Wall Street Journal*, 2008).

Key stakeholder groups for publishers disseminating pornography are civil society groups who advocate for specific standards and monitor for publications that violate these standards. An issue for publishers is the lack of consensus among these civil society groups as standards are evolving and based on cultural, religious and moral beliefs (D'Orlando, 2009). The only unambiguous negative injunction is to avoid any pornographic material involving children. Publishers (i.e., Internet search engine companies) cooperate with both not-for-profits and industry associations that monitor for obscenity and child pornography. The goal of monitoring groups, such as Project Clean Feed and Internet Watch Foundation, is to fight the child pornography, as well as to prevent youth under 18 years from accessing pornography. When unacceptable websites are identified, they are placed on a "black list", which is sent to search engine companies and broad band providers with a request to remove the site or material (Ramachander, 2008).

Unfortunately, even these advocacy groups are not able to apply unambiguous standards to identify child pornography and obscenity. In 2008 the Internet Watch Foundation (IWF) categorized a decades old album cover of a nude young girl as child pornography. The album cover was in a site on Wikipedia; thus several British ISPs blocked access to Wikipedia (Morphy, 2008). After Wikipedia administrators met with the IWF, the not-for-profit organization invoked its Appeals Procedure and posted the following statement on its website "The IWF has given careful consideration to the issues involved in this case. The procedure is now complete and has confirmed that the image in question is potentially in breach of the Protection of Children Act 1978. However, the IWF Board has today (9 December 2008) considered these findings and the contextual issues involved in this specific case and, in light of the length of time the image has existed and its wide availability, the decision has been taken to remove this webpage from our list" (IWF, 2008).

In summary, publishers must continually balance the demand of pornography (i.e., positive duty to support free speech) with the potential harm created by extreme pornographic material (i.e., negative injunctions), the expectations of civil society groups (i.e., negative injunctions) and the evolution of societal standards (i.e., positive duty). There are obvious cases where the application of a negative injunction dominates, such as child pornography, even if this material is condoned in some countries. Some advocacy groups would argue that publishers could easily justify not publishing any pornographic material to protect cultural and religious norms, and vulnerable groups. A blanket negative injunction (i.e., do not contribute to any potential harm) is not appropriate because of a combination of factors, including the significant demand for the material, the ambiguity associated with terms like "indecent" and "patently offensive", and the potential to limit the dissemination of creative works that push boundaries, such as Lady Chatterly's Lover.

The potential consequences to stakeholders of publishers not respecting the negative injunctions associated with pornographic material are vulnerable groups (e.g., children) experiencing direct injury by having access to the material, violation of community or religious norms and loss of access to desired materials. The potential consequences to publishers of not respecting the negative injunctions are loss of social licence to operate in certain communities, and retaliation from offended stakeholders. The potential consequences to stakeholders of publishers engaging in a positive duty of supporting free speech is access to desired materials, and the ability to push community standards in the name of free speech. The potential consequences to publishers of supporting free speech are maintaining independent decision making, fulfilling duty to protect free speech and, again, loss of licence to operate in certain communities.

Hate Speech: Extreme Intolerance or Cultural Criticism

Generally, hate speech is defined as expressions of extreme intolerance or dislike; however, it is difficult to distinguish hatred from ordinary dislike or disagreement (Post, 2009). The prohibitions against hate speech were initially based on racial discrimination, but have shifted to protection of religious differences and sexual orientation. Across cultures there are significant differences in terms of the level of harm accepted to protect free speech. The United States accepts "hate speech" as part of the costs that are incurred under the protection of free speech (Heinze, 2009); while in Europe and other countries state regulations against hate speech are accepted as appropriate protection of vulnerable populations. In a recent European Gallup survey less than 10% of respondents in France, German and the United Kingdom believe racial slurs should be allowed under the principle of free speech (Rheault & Moghaed, 2008).

In the 1960s there was almost complete worldwide acceptance of hate speech bans based on international human rights treaties, with the significant exception of the United States which has not passed any regulations on hate speech. Article 4 of the International Convention on the Elimination of All Forms of Racial Discrimination states "… State parties condemn all propaganda and all organizations which are based on ideas or theories of superiority of one race or group of persons of one colour or ethnic origin, or which attempt to justify or promote racial hatred and discrimination in any form, and undertake to adopt immediate and positive measures designed to eradicate all incitement to, or acts of, such discrimination…". More recently, there has been a controversial move to include criticism of religious groups in the definition of hate speech. On March 26, 2009 the United Nations Human Rights Council adopted a resolution condemning "defamation of religion" which was originally proposed by the 56 nation Organization of the Islamic Conference (OIC), and put to the Human Rights Council by Pakistan. Prior to the resolution being passed, 180 secular, religious and media groups from around the world urged diplomats to reject the resolution which they said may be used in certain countries to silence and intimidate human rights activities, religious dissenters and other independent voices and ultimately restrict freedoms (MacInnis, 2009). Even with the international conventions the determination of what is defamation or hate speech is left to small groups of individuals, and although there are laws and regulations against hate speech, there is little consensus as to how to define or identify it.

Under the principle of free speech, neo-Nazi web sites and the sale of Mein Kempf are allowed in the United States. In contrast, the German state of Bavaria holds copyright for the book and its publication is banned until 2015; and in both Germany and France neo-Nazi websites are not allowed to be listed by search engines. The head offices of both Google and Yahoo! are located in the United States; and initially they operated under the US interpretation of free speech and did not filter the neo-Nazi sites in their European operations (Le Menestrel, Hunter & de Bettignies, 2002). However, after a series of court decisions, Google and Yahoo! opted to comply with these laws.

A perverse outcome of decisions based on the need to protect vulnerable groups (i.e., negative injunction) is the gradual but significant growth of censorship and suppression of civil liberties justified as protecting the vulnerable (Bernstein, 2003; Wente, 2009). Many critics complain the avoidance of offence is now regarded as more important than the abstract right to freedom of expression (Malik, 2009). Three examples of publishers experiencing extreme pressure not to publish books/articles on the basis of limiting hate speech are Salman Rushdie's Satanic Verses published in 1986, the Jyllands-Posten Muhammad cartoons in 2005, and Sherry Jones' Jewel of Medina in 2008. Satanic Verses won the Whitbread novel of the year award. The novel is based on the disputed story of the Prophet adding Qua'rn verses which he later revoked saying he had been deceived by Satan. Rushdie intended for the book to be a satire on Islam, and initially Muslims did not react. However an Islamist group in India developed a campaign against the book leading to the Ayatollah Khomeini's fatwa against Rushdie. Eventually the book was banned in 11 countries, and in Britain the anti-Rushdie campaign led to a march of 1000 Muslims and a burning of the book in Bradford (Malik, 2009).

An example of a book publisher unintentionally insulting the Islamic faith is Random House, publishing the "Jewel of Medina" by Sherry Jones. Although many readers did not perceive the novel as insulting to Muslims, many Muslims took offense to the book. After major backlash to the book executives at Random House stopped distributing it. At one time the publisher's website contained an explanation of the decision to stop distribution of the book, with the following paragraph: "We stand firmly by our responsibility to support our authors and the free discussion of ideas, even those that may be construed as offensive by some. However, a publisher must weigh that responsibility against the others that is also bears … the safety of the author, employees of Random House, booksellers…" Meanwhile, another publisher, Gibson House, decided to publish and distribute the book. However, their offices were subsequently bombed and this publisher dropped the book. It is now published by Beaufort books in the United States (Spencer, 2009).

On September 30, 2005 a Danish newspaper the Jyllands-Posten's printed 12 cartoons, most depicting the Islamic Prophet Muhammad (a major offence from Muslim point of view), "in response to several incidents of self-censorship in Europe caused by widening fears and feelings of intimidation in dealing with issues related to Islam" (Rose, 2006). The following comment by the newspaper's culture editor was included in the newspaper: "The modern, secular society is rejected by some Muslims. They demand a special position, insisting on special consideration of their own religious feelings. It is incompatible with contemporary democracy and freedom of speech, where one must be ready to put up with insults, mockery and ridicule. It is certainly not always attractive and nice to look at, and it does not mean that religious feelings should be made fun of at any price, … we are on our way to a slippery slope where no-one can tell how the self-censorship will end."

A small group of Danish Muslims organizations filed a legal claim that the cartoons were "publicly ridiculing or insulting the dogmas of worship"; however, after the Danish legal system did not agree with the claim, a Danish imam took the cartoons to the Middle East and met with other imams. In order to ensure the imams reacted, he supplemented the 12 cartoons with additional ones that were far more offensive. The reaction of the imams ignited riots and violence, including attacks on Danish embassies in Lebanon and Syria, death threats against the cartoonist and editors. Over 100 people died and 800 were injured in Muslim countries.

At the end of January 2006, major European newspapers reprinted the cartoons, leading to more protests; however, most major newspapers in Canada, the USA, and the United Kingdom did not reprint the actual pictures (Spiegelman, 2006). In an editorial to North Americans, the cultural editor of the Jyllands-Posten concluded that the cartoons supported a constructive debate in Denmark and Europe about freedom of expression, freedom of religion and respect for immigrants and people's beliefs (Rose, 2006).

In summary, there is a cultural gap in stakeholder expectations between the negative injunctions supporting banning of material advocating intolerance and the positive duty of allowing cultural criticism. The preceding examples illustrate publishers' responses to competing normative expectations, from supporting free speech as a positive duty (e.g., Jyllands-Posten), supporting local government regulations based on negative injunctions (e.g., posting of Nazi materials) to filling a positive duty of supporting the free discussion of ideas, and ultimately respecting a negative injunction to protect employees and booksellers (e.g., Random House).

Our interpretation of the editors' decision at Random House is the company initially intended to fulfill its positive duty (support of authors and free discussion of ideas), but found this option too costly, and ultimately respected a negative injunction to protect employees and booksellers. The decision of publishers to reprint the Jyllands-Posten cartoons illustrates different approaches to a "positive duty" to inform stakeholders of the controversy. While some publishers republished the actual cartoons to continue an informed discussion on political correctness, others adopted a combined "negative injunction and positive duty" orientation by publishing commentary without including the pictures to avoid inciting harm against their companies or other stakeholders.

The potential consequences to stakeholders of publishers respecting the negative injunctions associated with potential hate speech are the development of invisible communication barriers and reinforcement of the chilling effect. The potential consequences to publishers of respecting the negative injunctions are loss of social licence to operate in certain communities and independence in decision making. The potential consequences to stakeholders of publishers engaging in a positive duty of supporting free speech are direct and indirect injury to vulnerable groups, providing access (for some stakeholders) to desired but potentially offensive materials increasing stakeholder awareness of evolving problems, and development of erroneous/negative interpretations. The potential consequences to publishers of supporting free speech are maintaining independent decision making, fulfilling duty to protect free speech and loss of licence to operate in certain communities.

Sedition: Political Criticism or Terrorism

The ability to criticize and offend governments and institutional leaders is critical for a flourishing democracy (Faris & Villeneue, 2008; Zakaria, 2008), while governments sometimes attempt to censor sensitive information to protect their positions. The Economist (2007) reported a study by OpenNet Initiative which concluded that government censorship of the Internet had grown from a few countries in 2002 to over 26 nations in 2007. Many current governments want to both control what information is available on the Internet as well as read e-mails. For example, the government of Iran pressured Google and Yahoo! to slow down the digital signals; then in March 2010 Iran banned Google e-mail (i.e., gmail) and replaced it with a national e-mail service provided by Iran's telecom agency (Fathi, 2010). Part of the speculation for gmail being disrupted is that Google's encryption had prevented the government from reading e-mails. The American legal system recognizes the need for electronic free speech, as noted when Judge Dazell wrote in a legal decision (Reno vs ACLU) that the Internet "deserves the highest protection from governmental intrusion". However, American civil liberties and public interest groups still have to protest against government censorship of different forms through the legal system.

Not all government censorship is considered wrong. One form of government censorship accepted as legitimate is restricting access to the ideas or communications of violent anarchists or terrorists (Doyle, 2004; Zittrain & Palfrey, 2008). "Dark nets" have developed on the Internet where militants, insurgents and criminal elements organize, communicate and mobilize illegal and harmful activities (Diebert & Rohozinski, 2008). The European Union passed legislation that criminalizes the publication of bomb making instructions on the Internet (Charter & Richards, 2009). The need for governments to protect citizens from harmful publications is supported by one respected philosopher, who has argued only negative outcomes occur from easy access to bomb making instructions, as the only purpose for accessing these instructions is to cause harm. He states "every right, even free speech, needs to be balanced with others, especially the right to live" (Etzioni, 1997, p.66). However, there are costs to monitoring and censoring websites; enforcement is expensive and it legitimizes invasions of privacy. One commentator notes there is limited evidence, if any, of the number of homemade explosions going up in the United States since the introduction of the Web in the mid 1990s (Doyle, 2004).

A visible example of Internet censorship violating human rights involves the government of China requiring Yahoo! and Google to censor or modify information sent to Chinese citizens on their search engines. The Chinese government recognized the need to increase Internet access for its citizens and build the technical skills of its own information technology specialists, while wanting to control access to what information was available through search engines (Gorgan & Brett, 2006). For example, search results for Tiananmen Square on Chinese search engines showed tourism pictures, while Google.com search engines showed the massacre that occurred (Hamilton, Knousse & Hill, 2009). The Chinese government employs more than 30,000 Internet police to monitor and control the information on web sites (Pan, 2006).

Initially Google avoided government censorship by locating the servers off-shore; but the Internet messages were still filtered through China's Great Firewall (developed by Cisco), and this was causing significant delays in the search requests. Eventually Google and Yahoo! began losing market share (Brenkert, 2009). Faster responses would happen if Google located the servers in China, where the filtering would occur at the front end of the search. Moving the servers inside China violated Google's code of conduct (e.g., complying with local laws and regulations only if they do not harm the vulnerable groups) and led to potential backlash from stakeholders in the United States. However, ultimately Google, Yahoo!, and Microsoft opted to conform to the Chinese government censorship requirements in order to operate in China. Google executives justified the decision to impose self-censorship in China on the basis of three reasons. First, it could not operate in China without agreeing to the government censorship. Second, by operating in China, Google met the need for faster service and the search engine offered better results than any other competitor. Third, by operating in China communication channels with the government were still open, and users still had access to Google.com which was open and unfiltered (Brenkert, 2009). As the company's senior policy counsel Andrew McLaughlin put it, "While removing search results is inconsistent with Google's mission, providing no information (or a heavily degraded user experience that amounts to no information) is more inconsistent with our mission" (Oliver & Shinal, 2006).

However, Google did take mitigating steps, in that every time Google's filters removed information from a website, users received a warning that that the search results had been removed in accord with Chinese "laws, regulations, and policies" (Gorgan & Brett, 2006). The filtering continued until the end of 2009 when Google found the Chinese government had been hacking into its service and using it as a platform to hack into other private Internet systems.

A publisher dependent on its reputation for valid and reliable information is Wikipedia. Its purpose is to disseminate knowledge that can be verified through other sources (Wikipedia, n.d.), and the information is provided through individual contribution, which may be reviewed and changed by other individuals. However, Wikipedia is not an absolute tool of free speech, as it intentionally does not include original ideas or information that cannot be validated through different sources. Wikipedia refused to change content in sections like Mao Zedong, Tiananmen Square and Tibet and so was censored by the Chinese government (Wikipedia, 2011; Wu, 2006). After blocking the wiki site for almost a year, Chinese authorities in October allowed access to most of the online encyclopedia's English-language entries. One explanation for the change in access is that "Beijing realized it was better to have mainland Chinese citizens contributing to the user-created encyclopedia than to have entries written exclusively by those critical of the regime" (Wu, 2006, p.14).

Opposite to Wikipedia is the WikiLeaks website which was started in 2007 to provide a forum for individuals to anonymously publish previously classified or sensitive documents (Bernard, 2010). The first leaked documents were about Swiss banks, Scientology and Guanatanamo Bay. In early 2010 the site posted a classified video of a US airstrike in Baghdad, that led to questioning whether the site was a whistleblowing or information terrorism site (The Economist, 2010).

One commentator noted "Purloining secret documents is an old business but technology has hugely increased the ease and potential quantity of theft and the ubiquity of publication of stolen information" (The Economist, 2010). The content of any one memo published on WikiLeaks has so far not been very important; however, the memos have created threats to individuals who in "good-faith" provided information to different government representatives. One consequence of the open publication is the restricted ability of a government to collect much needed security information (Harshaw, 2010). For example, a cable about a firm involved in selling arms to Iran identifies the informant as a well-connected businessman, listing family connections, place of business, education and sport accomplishments, by this means not only putting the individual in danger, but also reducing any motivation for the man to ever provide information on critical matters again (The Economist, 2010).

We must acknowledge that despite the amount of critique, the position and legitimacy of Wikileaks remains contentious. Sometimes, the site remains the only source of important and sensitive information concerning the government activities the citizens have the right to know. The ability to provide crucial facts and evidence is particularly important for the citizens of counties with not well-established democracies and institutions of open society. Unfortunately, sometimes the pragmatic considerations of Western countries prevent them from making such data (obtained, for instance, through intelligence) available to the public that needs it. In such cases Wikileaks may fulfill the positive duty of information dissemination, while obviously violating the negative injunctions imperative by making sensitive information public. Arguably, sometimes the gains from the former can outweigh the losses from the latter.

Wikipedia and Wikileaks are both non-profit websites involved in editing and posting content from individual contributions; one based on expertise and the other based on access to private documents. The editors of Wikipedia, as part of a negative injunction, monitor for information that is stolen, plagiarized or intended to disclose private data. Stolen or secret information cannot be validated through open debate and publically accessed information, and can harm individuals. In contrast, the stated purpose of Wikileaks is a positive duty to facilitate free speech by providing communication channels for whistleblowers and access to private government documents (Harshaw, 2010); thus the published information is often "stolen from an owner". It could be argued that the editors of Wikileaks act out of a positive duty by increasing access to important documents; however, others have suggested that types of documents published have moved beyond a positive duty of facilitating free speech to information terrorism (Harshaw, 2010) or creating unacceptable harm to many individuals and free speech (i.e., violating important negative injunctions) because of the lack of editorial judgement.

Publishers are responding to negative injunctions by not disseminating information on topics such as bomb making, planning a suicide or allowing dark nets, and when they boycott operating in countries where censorship is expected. George Brenkert (2009, p. 461) argues that Google's acquiescence to the Chinese government violates a negative injunction, because "China has set up rules regarding access to information that are unjustified and that do violate people's right to information… (and) playing by those rules and engaging in the resulting system is to engage in unethical practices." Publishers engage in positive duties when they provide channels for valid and reliable information and publically oppose government censorship (e.g., Wikipedia), and combine positive and negative duties to visibly notify users of censorship (e.g., Google notifying users in China of government censors).

In some situations publishers can protect vulnerable stakeholders or communities from direct injury by limiting the dissemination of information that could support sedition or treasonous activities. However, if they support the principles of free speech and disseminate information that could be used for sedition they also increase stakeholder awareness of evolving problems. When publishers follow government censorship regulations that support authority positions, stakeholders can develop erroneous interpretations, be unaware of barriers to accessing information, and lose access to desired materials, while positions of abusive authority are reinforced. At the same time publishers could maintain some communication with the government and have the opportunity to work within the political system to push for change. However they would lose independence in decision making and a loss of social license with some stakeholder groups. If they engage in a positive duty of refusing to comply with government censorship requirements stakeholders may be more aware of their lack of access to desired information/communication channels.

Analysis of Publishers' Responses

In the preceding vignettes publishers adopted one of three alternatives in response to conflicting normative expectations: support the principles of free speech, support the government censorship or engage in editorial censorship (see Table 1).

In most cases where the intent is to support free speech the publisher is responding on the basis of a positive duty. Many philosophers and advocacy groups oppose any form of censorship in the belief a "functioning vibrant market place of ideas is a public forum where competing ideas and theories are presented in oral or published formats in the open for all to hear, read, ponder, dissect, discuss and evaluate" (Greco, 1996). The duty to protect free speech is important when a publisher is the only accessible source of accurate information. Supporting free speech has trade-offs because it can mean creating harm to vulnerable groups through perceived offensive speech (e.g., racist web sites advocating Nazism, or descriptions of obscene acts) and to significant direct costs in terms of lost reputation, violent stakeholder reactions, and lost opportunities from failing to support government censorship expectations. Column one in Table 1 summarizes the positive duties, consequences to stakeholders and the publisher, and mitigating actions associated with the support of free speech.

Publishers supporting government censorship are assumed to violate freedom of speech; however, governments enforce censorship expectations to either protect vulnerable groups (i.e., respect a negative injunction), or to protect individuals who are in a position of authority or power (i.e., violate a negative injunction). More acceptable forms of government censorship are the limitations placed on hate speech in Europe, and the restrictions on publications of information from "dark nets" as they are intended to protect vulnerable groups (Le Menestrel, Hunter & de Bettignies, 2002). Column two in Table 1 summarizes the justifications and consequences of supporting government censorship.

Editorial censorship occurs when an editor chooses not to publish information to avoid harming one or more stakeholders (see column three in Table 1). Editorial censorship is supported at the international level by the authors of the 2009 Human Rights Commission Special Report to Parliament, stating that "Words and ideas have power. That power, while overwhelmingly positive, can also be used to undermine democracy and freedom. One classic argument in favour of unrestricted freedom of expression posits that in the battle of ideas, good ideas will inevitably win out over bad ideas. While good ideas gain sway over bad ideas most of the time, history tells that this is not always the case". In parallel, the European free speech doctrine incorporates the belief and experience that certain ideas can destroy public order, and with it any semblance debate; therefore editorial censorship can be justified on ethical grounds(Le Menestrel, Hunter & de Bettignies, 2002).

Visible examples of editorial censorship occurred with Random House stopping publication of the "Jewel of Medina" because of stakeholder backlash, or Yahoo!'s "notice and take down" policy.

Table 1. Publisher alternatives and associated contextual factors

	Support Free Speech	Support Government Censorship	Editorial Censorship
Justification for Decision	• Need to foster cultural review and debate • Need to provide accurate information from experts • Need to increase access to controversial documents	• Cannot force change without cooperation and access to decision makers • Limiting access to potentially dangerous information	• Respect for cultural and religious norms • Protecting vulnerable from harmful information • Other sources of controversial information currently exist
Potential consequences to stakeholders	• Open and free speech in every community • Access to needed information and communication channels • Increased awareness of global dynamics or evolving problems • Misuse of distribution channels leading to community harm • Vulnerable groups experience direct injury (e.g., injustice)	• Reinforcement of authority positions • Protect vulnerable groups From direct injury • Restricted access to needed materials • Development of erroneous interpretations	• Restricted access to information and communication channels • Reinforced chilling effect • Invisible barriers to accessing information • Limit stakeholder retaliation • Protect vulnerable groups from direct injury
Potential consequences to publisher	• Loss of social license to operate in certain communities • Receive support from stakeholders who work to support free speech • Retaliation from offended stakeholders • Fulfill duty to protect free speech • Maintain independent decision making • Violate regulatory standards leading to fines and loss of legal license	• Maintainability to operate within specific countries • Loss of ability to work within specific country • Retaliation from offended stakeholders • Work within a political system to create need for change • Attraction attention to government censorship	• Protection of staff and assets from stakeholder retaliation • Protection of social license to operate within specific communities • Loss of independence in decision making
Mitigating Actions	• Work with authorities to identify criminal activity • Ensure all stakeholders have a voice • Attract attention to government censorship	• Ensure transparency of actions • Attract attention to government censorship	• Ensure transparency of decision criteria • Cooperate with competitors to reduce externalities

Although reacting to stakeholder demands may appear to be a positive duty (i.e., acting to ensure appropriate material is published), editorial censorship can be more insidious than government censorship because it is more often than not invisible, occurring in every country under every type of government. It can lead to a "chilling effect", in that restricting freedom of speech because of the threat of violence only provides an incentive for further threats from those who do not want to see their views criticized (Le Menestrel, Hunter & de Bettignies, 2002; Zittrain & Palfrey, 2008). The end consequence is a "politically correct society", where everyone is afraid of offending or being prosecuted. Thus publishers walk a narrow path in finding the appropriate balance between free speech and respect for community norms, while limiting the potential stakeholder backlash. Finding the appropriate balance is difficult, as the boundaries associated with protecting free speech are ambiguous and evolving. Based on our case review we suggest editorial censorship may be appropriate when it a) is not the first, or only, step in response to growing political correctness and does not reinforce a chilling effect, and b) protects vulnerable stakeholders.

RESOLVING NORMATIVE TENSIONS: HEURISTIC DECISION MAKING PROCESS

Company managers are the "crucial mediators of stakeholder influence" (Fineman & Clarke, 1996, p.715). Given the difficulties of resolving the conflicts arising from incompatible normative demands, managers need a system to prioritize impacts on stakeholders and evaluate trade-offs. Heuristics have been developed in the fields of business ethics, strategic and operations management. For example, Donaldson (1985) developed a heuristic process for resolving the conflicts in multinational decision making; Waldkirch, et. al. (2009) applied heuristics for defining the role of business in a state's social security system; Hamilton et. al. (2009) proposed a heuristics of six questions to help managers to resolve cross-cultural ethical conflicts; and Mitchell et. al.'s (1997) theory of stakeholder identification and salience is based on a set of propositions which are in fact simple rules (i.e., heuristics) for prioritizing the interests of stakeholders in strategic decisions. We suggest a heuristic decision making process based on as structured analysis of previous situations of competing normative expectations should lead to appropriate and justifiable decisions. Although managers over time develop heuristics for making strategic decisions (Bingham & Eisenhardt, 2011), we are suggesting a more formal process.

The first step in developing the heuristic decision making process is to list the consequences to the publisher and the stakeholder identified in vignettes (see Table 2). The consequences were listed according to the degree of harm created or prevented (i.e., violated or respected negative injunctions), and the level of social value produced for individuals and the overall group. Next, a numerical range of values from an absolute negative value of – 5 to the worst harm that could be created (AH), and +5 to the most beneficial consequence (AB) was applied to the ranking to assign values to each consequence. A separate list was created for the company and the stakeholders. The assignment of values to any one consequence could be debated across stakeholder groups (e.g., citizen groups, government officials or public administrators - Lindblom, 1959, p.81); yet, the failure to rank consequences and assign values leaves decision makers in a quagmire of inaction.

The second step is to calculate a sum of the costs and value for each alternative within a decision scenario and then compare the value of each alternative (see Table 3). The next section illustrates the application of the heuristic to some of the publisher responses in the vignettes.

Should Free Speech be Supported?

An example of a publisher visibly reinforcing the principles of free speech (i.e., positive duty) to counter a growing "chilling effect" was Jyllands-Posten's decision to publish the Islamic cartoons. Table 4 illustrates the application of the heuristic to the decision of the Jyllands-Posten's editors. The benefits and costs of this alternative summed to -5. In contrast, the benefits and costs of publishing the cartoons summed to a +3; therefore, the summed values from the heuristics support the conclusion that adopting the positive duty approach created the greatest overall benefit for the publisher and stakeholders.

Should the Publisher Support Government Censorship?

Two important contextual factors influencing the appropriateness of government censorship is whether the censorship is to protect vulnerable groups or individuals in authority positions. A case where the publisher was expected to follow censorship regulations to protect vulnerable groups was Yahoo! operating in France. Applying the heuristic to this situation (see Table 5) led to a sum of benefits and costs for conforming to the French government regulations of a +1, and for ignoring the regulations a -3; thus the application

Table 2. Estimated values of consequences of publishers' decisions

Value	Consequences to Stakeholders	Value	Consequences to Publisher
(-5)	Loss of human life because of publisher's decision	(-5)	End of publishing business
(-4)	Misuse of distribution channels or information leading to community harm	(-3)	Retaliation from offended stakeholders
(-3)	Vulnerable stakeholders experience direct injury (e.g., injustice)	(-3)	Loss of independence in decision making
(-3)	Development of erroneous interpretations	(-3)	Violation of regulatory standards leading to fines and loss of legal license
(-3)	Reinforcement of authority positions	(-2)	Loss of social license to operate in certain communities
(-3)	Reinforced chilling effect	(2)	Protection of social license to operate within certain communities
(-3)	Restricted access to information/materials	(2)	Opportunity to work within a political system
(-3)	Invisible communication barriers	(2)	Support from stakeholders who work to support free speech
(-2)	Stakeholder retaliation to publication	(2)	Protection of staff and assets from stakeholder retaliation
(2)	Access to information/materials	(2)	Minimized stakeholder backlash
(2)	Minimized stakeholder retaliation	(2)	Maintain independent decision making
(3)	Attract attention to government censorship	(3)	Fulfill duty to protect free speech
(3)	Increased awareness of global dynamics or evolving problem	(5)	Flourishing independent publishing industry
(3)	Protection of vulnerable groups from direct injury		
(5)	Open and free speech in every community		

of the heuristic supports Yahoo! conforming to the government regulation.

A contrasting decision on whether to support government censorship regulations is Wikipedia's refusal to conform to the Chinese government's expectations for censorship. In this case the censorship regulations are designed to protect existing government authorities rather than to protecting vulnerable groups. There were only two alternatives available to Wikipedia, at least at the time of the decision, one based on a positive duty of supporting the principles of free speech by not conforming to the government's censorship requirements and not operating in China, and one based on market principles of operating in conformance to the standards of the legitimate government. These alternatives were analyzed through the heuristics in Table 6. The most costly alternative to society and the company was to conform to the government's censorship requirements.

An interesting contrast to Wikipedia's decision is that of Google. Wikipedia's legitimacy and viability would be reduced if there was a loss of trust in the ability to post accurate information. Thus, Wikipedia focused on the positive duty of fully supporting free speech and the importance

Table 3. General heuristic decision making process for comparing stakeholder expectations

Discrete Alternative	Consequences to Stakeholders		Consequences to Publisher		Integral Score Σ
	Benefits	Costs	Benefits	Costs	
Pure Negative Injunction Alternative	$+ bs_1$	$- cs_1$	$+ bc_1$	$- cc_1$	Σ_1
Intermediate Alternative	$+ bs_2$	$- cs_2$	$+ bc_2$	$- cc_2$	Σ_2
…					
Intermediate Alternative	$+ bs_{N-1}$	$- cs_{N-1}$	$+ bc_{N-1}$	$- cc_{N-1}$	Σ_{N-1}
Pure Positive Duty Alternative	$+ bs_N$	$- cs_N$	$+ bc_N$	$- cc_N$	Σ_N

Table 4. Heuristic analysis of original decision to publish controversial cartoons

Discrete Alternative	Consequences to Stakeholders		Consequences to Publisher		Integral Score Σ
	Benefits	Costs	Benefits	Costs	
1. Negative Injunction: Avoid controversy by not publishing commentary	(2) Minimize stakeholder retaliation	(- 3) Reinforce chilling effect (-3) Invisible communication barriers	(2) Protect staff and assets from stakeholder retaliation	(- 3) Loss of independence in decision making	$\Sigma_1 = -5$
2. Positive Duty: Publish cartoons and commentary on a belief that free speech is being limited through political correctness	(3) Increased understanding of evolving problems	(-2) Stakeholder retaliation to publication*	(3) Fulfill duty to protect free speech (3) Maintain independence in decision making	(-2) Loss of social license in certain communities (-3) Retaliation from offended stakeholders	$\Sigma_2 = +2$

Note.

* We calculated the costs as if the editors were making the decision prior to the unanticipated loss of human life, assuming at that point no one would have predicted a loss of human life.

of having access to uncensored information. In contrast, Google's legitimacy is influenced by its ability to provide access to as many information sources as possible, and to have as many individuals as possible using the search engine. Google justified its acceptance of government censorship through a combined positive duty/negative injunction by providing access to information in a more and effective manner and notifying individuals when information had been censored and thus making the censorship visible (see Table 7). This action changed the outcome of the heuristics, leading to more benefit than harm for the publisher and the Chinese Internet users.

Should the Publisher Engage in Editorial Censorship?

An example of editorial censorship is the decision of North American publishers not to republish the actual cartoons in the original Jyllands-Posten article. These editors believed that readers were aware of the cartoons, and re-posting them would only lead to more backlash and harm to vulnerable individuals. Their focus was on furthering commentary on the growing political correctness around Islamic issues (see Table 8). In contrast to the heuristics for the original publication (see Table 4), the creative decision of North American publishers to both raise the issue and mitigate harm to vulnerable groups by publishing editorial comment without the cartoons themselves was better for this given decision context. However, this editorial censorship option became possible only after original publishing of cartoons in Denmark.

These last two applications highlight that not all decisions are "either/or decisions", that publishers need to explore mitigating steps that increase the flow of information without causing additional harm to vulnerable groups. The impact of specific mitigating actions is dependent on the context; however, we suggest that they should lead to increased transparency of decisions by key stakeholders, cooperating with competitors and not-for-profits to reduce externalities, and ensuring vulnerable stakeholders have a voice.

IMPLICATIONS AND FUTURE RESEARCH DIRECTIONS

The next step is to examine alternatives available to the company and organize them in terms of their potential to limit harm created by the

Table 5. Heuristic analysis of restriction nazi-based websites

Discrete Alternative	Consequences to Stakeholders		Consequences to Publisher		Integral Score Σ
	Benefits	Costs	Benefits	Costs	
1. Negative Injunction: Protect vulnerable groups by restricting access to symbols of hatred	**(3)** Protect vulnerable stakeholders from direct injury	**(-3)** Restricted access to needed information/ materials	**(2)** Protection of social licenses to operate in specific communities	**(-1)** Some loss of independence in decision making*	$\Sigma_1 = +1$
2. Positive Duty: Support free speech	**(2)** Access to needed materials or communication channels	**(- 3)** Vulnerable groups experience direct injury	**(3)** Fulfill duty to protect free speech	**(-3)** Violate regulatory standards leading to fines and loss of legal licence ** **(- 2)** Loss of social license to operate in certain communities	$\Sigma_2 = -3$

Note.
* Since the loss of independence in decision making is caused by reasonable regulatory requirements aimed at protecting the vulnerable groups, we cut the costs in more than half
** The intent of the government regulation, in this case, is to protect vulnerable groups, not reinforce position of authority.

company's operations (i.e., negative injunction), limit harm and contribute to social welfare, and pure contributions to social welfare without attention to potential harms created by company's operations (i.e., positive duty). Once the alternatives have been organized a numerical scale for comparing benefits and costs, such as -5 to 5 should be applied to evaluate each alternative in terms of its beneficial impact and costs both to social and corporate value. Each alternative can then be compared to an overall impact, to provide guidance on more effective alternatives in terms of minimizing costs while creating value for both the company and society.

Table 6. Heuristic analysis of Wikipedia's alternatives

Discrete Alternative	Consequences to Stakeholders		Consequences to Publisher		Integral Score Σ
	Benefits	Costs	Benefits	Costs	
1. Pure Negative Injunction: Conform with Chinese government's requirements, employing censorship	**(1)** Stakeholders can access needed materials/ communication channels*	**(- 3)** Restricted access to needed information **(-3)** Reinforce positions of abusive authority **(-3)** Invisible barriers to accessing information are created	**(2)** Have opportunity to work within a political system to push for change	**(-2)** Loss of social license to operate in certain communities **(-3)** Loss of independence in decision making	$\Sigma_1 = -11$
2. Pure Positive Duty: Refuse to conform, no government censorship	**(3)** Attract attention to government censorship	**(- 1)** Stakeholders develop erroneous interpretations **	**(3)** Fulfill duty to protect free speech	**(-3)** Loss of legal license to operate in China	$\Sigma_2 = +2$

Note.
* Even if Wikipedia conformed to censorship requirements, citizens still had access to a better search engine; however, we cut the benefit in more than half because of the censorship
** Given Chinese citizens would not have access to accurate information on some topics, we put this in as a cost, but reduced the impact because it was only on limited set of topics.

Table 7. Heuristic analysis of google's alternatives

Discrete Alternative	Consequences to Stakeholders		Consequences to Publisher		Integral Score Σ
	Benefits	Costs	Benefits	Costs	
1. Negative Injunction: Conform with Chinese government's requirements, employing censorship	**(1)** Access to needed information/ communication channels*	**(-3)** Restricted access to needed information **(-3)** Reinforce positions of abusive authority **(-3)** Invisible communication barriers	**(2)** Opportunity to work within political system	**(-2)** Loss of social license to operate in certain communities **(-3)** Loss of independence in decision making	**Σ_1 = -11**
2. Combined Solution: agree to censor, but inform the users about censored material	**(1)** Access to information/ communication channels* **(3)** Attract attention to government censorship**	**(-3)** Restricted access to needed information	**(2)** Opportunity to work within political system	**(-1)** Loss of independence in decision making***	**Σ_2 =+2**
3. Positive Duty: Refuse to conform, no government censorship, with forecasted blockade in China	**(3)** Attract attention to government censorship	**(- 1)** Develop-ment of erroneous interpretations **** **(-3)** Restricted access to needed information	**(3)** Fulfill duty to protect free speech **(2)** Gaining support from stakeholders who work to support free speech	**(-3)** Loss of legal license to operate in China	**Σ_3 = +1**

Note.

* Even if Google conformed to censorship requirements, citizens still had access to a better search engine; however, we cut the benefit in more than half because of the censorship

** This is an important outcome because it provides more information to the Chinese citizens about the activities of the Chinese government

*** Although Google, in this case, loses some independence in decision making, it is not equivalent to the loss experienced in fully complying with the Chinese government's expectations

**** Given Chinese citizens would not have access to accurate information on some topics we put this in as a cost, but reduced the impact because it was only on limited topics

Our approach, however, has a set of limitations to be addressed in further studies. Firstly, the numerical values assigned to parameters within heuristics tables are merely educated approximation of costs and benefits made by bounded rational decision makers. At this point we are comfortable with the assignment based on comparison to benchmarks of worst and best outcomes, given the difficulties in developing objective, quantifiable measures for benefit and costs. Recognizing these limitations, we recommend that further exploration of decision dilemmas to identify processes for identifying more objective calculations of assigned values.

Secondly, bounded rationality of decision makers implies the possibility of systematic biases in their judgment. The proposed heuristics will only work if individuals are making efforts to truly look for objective information. We anticipate that these biases can be intentionally mitigated if the decisions are made by diverse group of people, employing some form of group decision making rules (e.g., finding consensus, averaging, etc.), a field requiring further theoretical and empirical investigation with regards to decision making when facing ethical dilemmas.

DISCUSSION

Decisions in the publishing industry illustrate difficulties in resolving competing normative expectations. The publishers' alternatives in the three types of controversial publications were

Table 8. Heuristic analysis of following decision of north american publishers to republish the cartoons

Discrete Alternative	Consequences to Stakeholders		Consequences to Publisher		Integral Score Σ
	Benefits	Costs	Benefits	Costs	
1. Pure Negative Injunction: Avoid controversy by not publishing commentary	**(2)** Minimize negative stakeholder backlash	**(- 3)** Reinforce chilling effect **(-3)** Invisible barriers to accessing information	**(2)** Protection of staff and assets from stakeholder retaliation	**(- 3)** Loss of independence in decision making	$\Sigma_1 = -5$
2. Combined Negative Injunction and Positive Duty: Raising the topic without reprinting the actual offensive cartoons	**(3)** Increased understanding of global dynamics		**(3)** Fulfill duty to protect free speech **(2)** Gain support from stakeholders who support free speech	**(-2)** Loss of social license in certain communities	$\Sigma_2 = +6$
3. Pure Positive Duty: Publish cartoons and commentary on a belief that free speech is being limited through political correctness	**(3)** Increased understanding of global dynamics	**(-5)** Loss of human life because of publisher's decision *	**(3)** Fulfill duty to protect free speech **(2)** Gain support from stakeholders who support free speech	**(-2)** Loss of social license in certain communities **(-2)** Injury to staff and assets from stakeholder retaliation	$\Sigma_2 = -1$

Note.
* At this point the threat of loss of human life was real.

to support the principles of free speech, support government censorship, or engage in editorial censorship (see Table 1). Although most publishers had codes of ethics that publicly stated their support of free speech, we suggest publishers needed to go beyond a simplistic view of supporting free speech to consider which of the three alternatives led to an appropriate balance of respecting negative injunctions or engaging in positive duties. Identifying an appropriate action requires an analysis of responsibilities to society/stakeholders (negative injunctions or positive duties), potential impact on stakeholders or social welfare (e.g., risk of harm to vulnerable groups, reinforcement of political correctness, loss of transparency), potential impact on the publisher (e.g., loss of license to operate, stakeholder backlash, loss of access to decision makers) and an ability to initiate mitigating actions (Martin, 2008).

One approach to balancing the harm versus benefits created by support of free speech is to identify ways to reduce the harm while enhancing the benefits that could be achieved. We suggest that whenever a publication decision has the potential to create harm, the publisher has a duty to understand the injustice and identify ways to reduce or balance it (Heish, 2004). A critical example of a publisher creating mitigating actions is Google's operations in China. When Google negotiated the contract to operate in China and conform to censorship requirements, the negotiators did not inform the Chinese government they planned to identify all filtered information for the Chinese users. By taking this action, it could be argued that Chinese users were more informed about the government's censorship than if Google had decided to not enter China. The impact of specific mitigating actions is dependent on the context; however, we suggest that they should lead to increased transparency of decisions by key stakeholders, cooperating with competitors and not-for-profits to reduce externalities, and ensuring vulnerable stakeholders have a voice.

The application of the heuristics will not lead to absolute correct decisions or actions, and companies and management teams will vary in the weightings applied to the different consequences, leading to a choice of different alternatives. However, it provides a method for improving the understanding of the complexities in balancing stakeholder expectations, consistently comparing different alternatives within a company, and a better understanding of the potential mitigating actions to reduce the negative impact of any given action. Decisions in the publishing industry were used to develop the heuristics; however similar heuristics can be applied to analyze competing stakeholder expectations in other industries. Most industries are dominated by three or four issues, (e.g., use of child labor in the apparel industry, operating on indigenous lands in the resource extractive industry or sustainable products in the consumer industry) where competing stakeholder interests create difficult decisions.

The methodology followed to identify relevant vignettes and the ranking of consequences is included in Appendix. The first step in developing a heuristic process that is relevant to a particular issue or industry is to review past decisions or actions taken by companies attempting to manage the competing stakeholder interests. The analysis of these decisions should focus on the competing stakeholder interests and outcomes to identify the expectations/responsibilities of companies, consequences to stakeholders of the different outcomes, consequences to the company and any potential mitigating actions. This information can then be captured in a chart similar to Table 1.

REFERENCES

Agle, B. R., Mitchell, R. K., & Sonnenfeld, J. A. (1999). Who matters to CEOs? An investigation of stakeholder attributes and salience, corporate performance, and CEO values. *Academy of Management Journal*, *42*, 507–525. doi:10.2307/256973

Austen, I. (2007, February 19). Canadian company offers nude photos via cellphone. *The New York Times*.

Becerra, M. (2009). *Theory of the firm for strategic management: Economic value analysis*. Cambridge, UK: Cambridge University Press. doi:10.1017/CBO9780511626524

Bernard, D. (2010). What is Wikileaks? *News.com*. Retrieved from http://www.voanews.com/english/news/what-is-wikileaks--99239414

Bernstein, D. E. (2003). You can't say that: Canadian thought police on the march. *National Review Online*. Retrieved from http://www.nationalreview.com/script/comment/ bernstein 200312020910

Bingham, C. B., & Eisenhardt, K. M. (2011). Rational heuristics: The 'simple rules' that strategists learn from process experience. *Strategic Management Journal*, *32*, 1437–1464. doi:10.1002/smj.965

Brammer, S., & Millington, A. (2004). The development of corporate charitable contributions in the UK: a stakeholder analysis. *Journal of Management Studies*, *41*(8), 1411–1434. doi:10.1111/j.1467-6486.2004.00480.x

Carter, S. M. (2006). The interaction of top management group, stakeholder, and situational factors on certain corporate reputation management activities. *Journal of Management Studies*, *43*(5), 1145–1176. doi:10.1111/j.1467-6486.2006.00632.x

Charter, D., & Richards, J. (2009). The European Union wants to block Internet searches for bomb recipes. In *Freedom of expression*. Farmington Hills, MI: Greenhaven Press.

D'Orlando, F. (2009). The demand for pornography. *Journal of Happiness Studies*.

Darnall, N., Henriques, I., & Sadorsky, P. (2010). Adopting proactive environmental strategy: The influence of stakeholders and firm size. *Journal of Management Studies*, *47*(6), 1072–1094. doi:10.1111/j.1467-6486.2009.00873.x

Doh, J. P., & Guay, T. R. (2006). Corporate social responsibility, public policy, and NGO activism in Europe and the United States: An institutional-stakeholder perspective. *Journal of Management Studies*, *43*(1), 47–73. doi:10.1111/j.1467-6486.2006.00582.x

Donaldson, T. (1985). Multinational decision-making: Reconciling international norms. *Journal of Business Ethics*, *4*(4), 357–366. doi:10.1007/BF00381779

Doyle, T. (2004). Should web sites for bomb-making be legal? *Journal of Information Ethics*, *13*(1), 34–37. doi:10.3172/JIE.13.1.34

Dubbink, W. (2004). The fragile structure of free-market society: The radical implications for corporate social responsibility. *Business Ethics Quarterly*, *14*(1), 23–46. doi:10.5840/beq20041412

Economist. (2007, October 13). International: The tongue twisters, civil liberties: Freedom of speech. *The Economist*.

Economist. (2010, December 2). Wikileaks unpluggable: How WikiLeaks embarrassed and enraged America, gripped the public and rewrote the rules of diplomacy. *The Economist.*

Fathi, N. (2010, February 11). Iran disrupts Internet communications. *The New York Times*.

Fineman, S., & Clarke, K. (1996). Green stakeholders: Industry interpretations and response. *Journal of Management Studies*, *33*(6), 716–730. doi:10.1111/j.1467-6486.1996.tb00169.x

Foulds, L. R. (1983). The heuristic problem-solving approach. *The Journal of the Operational Research Society*, *34*(10), 927–934.

Friedman, A. L., & Miles, S. (2002). Developing stakeholder theory. *Journal of Management Studies*, *39*(1), 1–21. doi:10.1111/1467-6486.00280

Friedman, M. (2002). Capitalism and freedom (40th Anniversary Ed.). Chicago: The University of Chicago Press.

Galt, V. (2007, February 21). Telus hangs up on mobile porn service. *The Globe and Mail*, p. A1.

Gorgan, C., & Brett, J. (2006). *Google and the government of China: A case study of cross-cultural negotiations*. Kellogg School of Management, Northwestern University.

Greenley, G. E., & Foxall, G. R. (1997). Multiple stakeholder orientation in UK companies and the implications for company performance. *Journal of Management Studies*, *34*(2), 259–284. doi:10.1111/1467-6486.00051

Hamilton, J. B., Knousse, S. B., & Hill, V. (2009). Google in China: A manager friendly heuristic-model for resolving cross cultural ethical conflicts. *Journal of Business Ethics*, *86*, 143–157. doi:10.1007/s10551-008-9840-y

Harshaw, T. (2010). The hunt for Julian Assange. *Opinionator, NYTBlogs*. Retrieved from http://opinionator.blogs.nytimes.com/2010/12/03/the-hunt-for-julian-assange/

Heish, N. (2004). The obligations of transnational corporations: Rawlsian justice and duty of assistance. *Business Ethics Quarterly*, *14*, 643–661.

Hill, C. W. L., & Jones, T. M. (1992). Stakeholder-agency theory. *Journal of Management Studies*, *29*(2), 131–154. doi:10.1111/j.1467-6486.1992.tb00657.x

Internet Watch Foundation (IWF). (2008). *IWF statement regarding Wikipedia webpage*. Retrieved from http://www.iwf.org.uk/about-iwf/news/post/251-iwf-statement-regarding-wikipedia-webpage

Julian, S. D., Ofori-Dankwa, J. C., & Justis, R. T. (2008). Understanding strategic responses to interest group pressures. *Strategic Management Journal*, *29*, 963–984. doi:10.1002/smj.698

Kahn, F. R. (2007). Representational approaches matter. *Journal of Business Ethics*, *73*, 77–89. doi:10.1007/s10551-006-9199-x

Lady Chatterley's Lover. (n.d.). Retrieved from http://en.wikipedia.org/wiki/LadyChatterley'Lover's Plot

Le Menestrel, M., Hunter, M., & de Bettignies, H. (2002). Internet e-ethics in confrontation with an activists' agenda. *Journal of Business Ethics*, *39*, 135–144. doi:10.1023/A:1016348421254

Lenat, D. B. (1982). The nature of heuristics. *Artificial Intelligence*, *19*, 189–249. doi:10.1016/0004-3702(82)90036-4

Lessing, D. (2006, July 15). Testament of love. *The Guardian*.

Lindblom, C. (1959). The science of muddling through. *Public Administration Review*, *19*(2), 79–88. doi:10.2307/973677

MacInnis, L. (2009). U.N. body adopts resolution on religious defamation. *Reuters*. Retrieved from http://www.reuters.com/article/2009/03/26/us-religion-defamation-idUSTRE52P60220090326

Maitra, I., & McGowan, M. K. (2007). The limits of free speech: Pornography and the question of coverage. *Legal Theory*, *13*, 41–68. doi:10.1017/S1352325207070024

Malik, K. (2009). A marketplace of outrage. *New Statesman (London, England)*, *138*(4940), 40–42.

Martin, K. E. (2008). Internet technologies in China: Insights on the morally important influence of managers. *Journal of Business Ethics*, *27*(4), 315–324.

Morphy, E. (2008, December 8). British ISPs block Wikipedia page, reigniting 30-year-old child porn controversy. *ECT Newsnet*.

Nunziato, D. C. (2009). *Virtual freedom: Net neutrality and free speech in the internet age*. Stanford, CA: Stanford Law Books, Stanford University Press.

Oliver, C., & Shinal, J. (2006). Google will censor new China service. *MarketWatch*. Retrieved from http://www.marketwatch.com/story/google-builds-censorship-into-china-search-service

Pajunen, K. (2006). Stakeholder influences in organizational survival. *Journal of Management Studies*, *43*(6), 1261–1288. doi:10.1111/j.1467-6486.2006.00624.x

Pan, P. P. (2006, February 19). The click that broke a government's grip. *Washington Post*.

Pasquale, F. (2006). Rankings, reductionism, and responsibility. *Cleveland State Law Review*, *54*, 115–140.

Peloza, J., & Falkenberg, L. (2009). The role of collaboration in achieving corporate social responsibility objectives. *California Management Review*, *51*(3), 95–113. doi:10.2307/41166495

Pfeffer, J., & Salancik, G. R. (1978). *The external control of organizations: A resource dependence perspective*. New York: Harper & Row Publishers.

Ramachander, S. (2008). Internet filtering in Europe. In *Access denied: The practice and policy of global internet filtering*. Cambridge, MA: The MIT Press.

Rheault, M., & Moghaed, D. (2008, Summer). Gallup presents…cartoon and controversy: Free expression or Muslim exceptionalism in Europe? *Harvard International Review*, 68–71.

Rose, F. (2006, February 19). Why I published those cartoons. *The Washington Post*. Retrieved from http://www.washingtonpost.com/wp-dyn/content/article/2006/02/17/AR2006021702499.html

Scheffler, S. (2001). *Boundaries and allegiances: Problems of justice and responsibility in liberal thought*. Oxford, UK: Oxford University Press.

Silver, E. A. (2004). An overview of heuristic solution methods. *The Journal of the Operational Research Society*, *55*, 936–956. doi:10.1057/palgrave.jors.2601758

Simon, J. G., Powers, C. W., & Gunnemann, J. P. (1983). The responsibilities of corporations and their owners. In *Ethical theory and business* (2nd ed.). Englewood Cliffs, NJ: Prentice-Hall, Inc.

Spencer, R. (2009). Muhammad and Aisha, a love story. *Middle East Quarterly*. Retrieved from http://www.meforum.org/2010/the-jewel-of-medina

Spiegelman, A. (2006, June). Drawing blood: Outrageous cartoons and the art of outrage. *Harper's*, 43–51.

Stone, C. D. (1975). *Where the law ends: The social control of corporate behavior*. New York: Harper & Row Publishers.

Time. (1930). *National affairs: Decency squabble*. Retrieved from http://www.time.com/time/magazine/article/0,9171,738937,00.html

UN Global Compact/Office of the United Nations High Commissioner for Human Rights (UNHCHR). (2004). *Embedding human rights in business practice*. New York: UN.

Waldkirch, R. W., Meyer, M., & Homann, K. (2009). Accounting for the benefits of social security and the role of business: Four ideal types and their different heuristics. *Journal of Business Ethics*, *89*, 247–267. doi:10.1007/s10551-010-0392-6

Wall Street Journal. (2008, April 24). Google Brazil turns in user data amid child-pornography inquiry. *The Wall Street Journal*, p. B9.

Wettstein, F. (2010). For better or for worse: Corporate responsibility beyond do not harm. *Business Ethics Quarterly*, *20*(2), 275–283. doi:10.5840/beq201020220

Wikileaks. (n.d.). *What is Wikileaks*. Retrieved from http://wikileaks.org/About.html

Wikipedia. (2011). *Blocking of Wikipedia by the People's Republic of China*. Retrieved from http://en.wikipedia.org/wiki/Censorship_of_Wikipedia#China

Wikipedia. (n.d.). *Wikipedia: About*. Retrieved from http://en.wikipedia.org/wiki/Wikipedia:About

Wu, E. (2006). Liberating Wikipedia in China (almost). *Fortune International*, *154*(9), 14–16.

Zakaria, F. (2008). *The post-American world*. New York: W.W. Norton & Company.

KEY TERMS AND DEFINITIONS

Heuristics: An efficient problem solving approach, implying using simplified methods for finding acceptable solutions (rather than optimal ones). Heuristical solutions are intended to work in majority of cases.

Negative Injunctions: Dominant today business ethics doctrine, prescribing businesses the responsibilities of the "moral minimum": any private business's *raison d'être* is maximizing shareholders' wealth, with laws and legal requirements setting the boundaries for unacceptable actions.

Positive Duties: An alternative to negative injunctions business ethics doctrine, explicitly requiring companies to improve the world, deserving by this means the right to exist, through deliberate accomplishments and honest choices. This view of proactive business's engagement in resolving the social problems is held by many business ethicists, environmental and social activists.

Stakeholder: Any person who can influence an organization, or who is influenced by its actions.

Stakeholder Conflict: Incompatibility of stakeholders' demands, when satisfying the requirements of one stakeholder (or stakeholder group) directly contradicts the interests of the other stakeholder (or stakeholder group).

APPENDIX

The Analytic Method

We explored our research question, "how can a manager most effectively resolve competing normative stakeholder expectations" through five stages of information gathering and qualitative analysis (see Table 9). The final outcome was a heuristic decision making process for evaluating alternatives based on respecting negative injunctions or engaging in positive duties. In the *first stage* we focused on establishing normative responsibilities associated with supporting the principles of free speech. We reviewed philosophical discussions of free speech (e.g., Hare & Weinstein, 2009), universal principles or standards (e.g., UN Declaration of Human Rights), and discussions of corporate social responsibilities (e.g., Friedman & Miles, 2002; Porter and Kramer, 2011). Based on our review we ended negative injunctions and positive duties associated with publishers support of free speech (see Table 9).

In the *second stage* of research we reviewed journal and media articles, as well as books on freedom of speech, to identify situations where publishers had had conflicting stakeholder expectations, based on normative responsibilities, on the appropriateness of a publication or use of information (see Table 9). We specifically sought controversial situations where differing viewpoints were discussed, there was ambiguity as to what was acceptable content and/or there had been stakeholder reactions to the publication (or lack of). We focused on finding examples of controversial cases involving books and magazine articles, and Internet search engines. [These publishing businesses are classified in the NAICS code 51("Information"), particularly the Publishing Industries (511) and Internet Publishing and Broadcasting and Web Search Portals (51913).] Book publishing has a long history of controversy associated with publication decisions, and books still remain a dominant source of new ideas, exploration of values and changes in societal standards. On the other hand, Internet search engines have recently become the dominant conduit of information for research and public discussion (Pasquale, 2006). In this broad review we identified three categories of controversial publications: pornography, hate speech and government censorship.

In the *third stage* we focused our search on identifying specific examples of controversial publications in these three categories. We initially used the terms pornography, obscenity, hate speech, freedom of expression, sedition, political criticism, and censorship to identify specific examples. As specific examples were identified, we used the names of publishing companies, publication titles and labels associated with the controversies to identify more sources (see Table 9). As we explored specific examples we found more search terms to use and new situations to explore. For example, the Jyllands-Posten cartoons were mentioned in a number of articles on hate speech, and as we found articles discussing the cartoons we found additional search terms, such as cultural criticism, and political chill. Applying these additional search terms led to the identification of other cases (e.g., Jewel of Medina, Nazi publications).

After identifying the specific examples we developed illustrative vignettes in each of the three categories of controversial publications. We chose examples for the vignettes where we were able to triangulate existing data through multiple descriptions of the situation (i.e., at least four articles in different media providing similar information), and where we had sufficient information to identify the stakeholder expectations (e.g., rationale or justification for why publishers should align with a given stakeholder), consequences to the identified community or stakeholder for a given decision, potential

Table 9. Stages of analytic method

Stages	Sources of Information	Outcomes
ONE Establishing normative business responsibilities for freedom of speech • Review of corporate social responsibility frameworks • Review of normative principles for business ethics • Review of philosophical and analytical articles on freedom of speech • Review of international standards for human rights and freedom of speech	o Search for academic articles containing, human rights standards/ principles, principles of freedom of speech, normative expectations for business o Primary non nocere /Negative Injunction/ Article 19 – Declaration of Human Rights/ o UNCHR (2004) – Embedding Human Rights in Business Practices/Principle of Contribution o Positive duties based on moral mission (Mulligan, 1993); societal value (Porter & Kramer, 2011)	Two normative approaches: Negative injunctions o Protect vulnerable groups o Respect social and cultural norms Positive duty o Contribute to the positive evolution of social and cultural norms o Provide communications channels for controversial viewpoints o Report abuses and/or criminal users
TWO Identifying types of controversial publications involving freedom of speech • Review journal and media articles and books on freedom of speech to identify types of publications that have competing views on their appropriateness and/or ambiguous boundaries of acceptability	o Academic journals (e.g., Journal of Business Ethics, Journal of Information Ethics) o Editorial articles in media print (e.g., The New York Times) o Books on freedom of speech (e.g., Nunziato, 2009)	Identified three types of controversial publications o Obscenity versus pornography o Hate speech versus cultural criticism o Sedition versus political criticism
THREE Developing illustrative vignettes for each type of controversial publication and the social context associated with each case • Using the search terms pornography, obscenity, hate speech, freedom of expression, sedition, political criticism, publisher names, publication titles, not-for-profit organizations • Identifying social context for each identified case (e.g., Google in China) o Positive and negative social consequences for stakeholders o Positive and negative social consequences for publisher o Mitigating factors	o Electronic sites such as Wikipedia not-for-profit web sites o Published business case, studies, case research in academic journals o Media articles and editorials on specific case situations or decisions o Books on pornography, hate speech and government censorship	Identification and development of illustrative cases for each category o Obscenity versus pornography ▪ Lady Chatterly's Lover ▪ Telus Adult Content ▪ Google in Brazil o Hate speech versus cultural criticism ▪ Satanic Verses/Jewel of Medina ▪ Neo-Nazi sites and Yahoo! ▪ Jyllands-Posten Danish Cartoons o Sedition versus cultural critcism ▪ Google and Yahoo! In China ▪ Wikipedia ▪ Wikileaks
FOUR Identifying publisher alternatives for responding to competing stakeholder expectations Creating summary lists of consequences for stakeholders and company and rank according to costs followed by benefits Ranking consequences or outcomes of alternatives associated with publishers' actions: Assigning numerical values to consequences according to ranking	o Identified illustrative vignettes	• Summary table of decision alternatives with associated consequences for stakeholders and publishers o Support free speech o Support government censorship o Engage in editorial censorship • Comparative listing of costs and benefits that can be included in heuristic for identifying most appropriate alternative
FIVE Developing heuristic process for comparing consequences of respecting negative injunctions and engaging in positive duties	o Management research articles on heuristics, such as Donaldson (1985); Waldkrich et. al., (2009); Hamilton et.al.(2009) o Vignettes and summary list of consequences	• Heuristic process for evaluating benefits and costs to stakeholders of alternatives based on respecting negative injunction and engaging in positive duties • Examples applications of heuristics to illustrative vignettes

stakeholder reactions, consequences to the publisher and any additional actions of the publishers. We were able to develop nine vignettes, three in each category of controversial publications (see Table 9).

In the *fourth stage* of our analysis we reviewed the illustrative vignettes to categorize the responses of publishers. Publishers adopted three responses: fully support the principles of free speech, support government censorship, or engage in editorial censorship (see Table 9). We summarized our review in Table 1, which lists the justifications, consequences to stakeholders, consequences to the publisher and mitigating actions. Next, we created another table by ranking the consequences to stakeholders and publishers from the most harmful to the most beneficial (Table 2). A relative value (i.e., based on the ranking of a consequence) was assigned to each consequence.

The *fifth stage* of the research involved the development of heuristics, or a series of steps or rules, based on experience and judgement that guide decision makers towards plausible or reasonable solutions (Foulds, 1983; Lenat, 1982; Silver, 2004). Based on the findings from Stages One to Four we developed a heuristic process to compare the negative and positive social consequences of publisher's alternatives when there are competing stakeholder expectations. This heuristics is applied to five of the illustrative vignettes in the findings section, Stage Five: Development of a Comparative Heuristics.

Chapter 5
Towards a Subjectively Devised Parametric User Model for Analysing and Influencing Behaviour Online Using Neuroeconomics

Jonathan Bishop
Centre for Research into Online Communities and E-Learning Sytems, European Parliament, Belgium

Mark M. H. Goode
Cardiff Metropolitan University, UK

ABSTRACT

The quantitative-qualitative and subjectivity-objectivity debates plague research methods textbooks, divide academic departments, and confuse post-modernists as to their existence. Those from the objective-quantitative camps will usually demand methods assume parametric principles from the start, such as homogeneity and normal distribution. Many of the subjective-qualitative camps will insist on looking and the individual meanings behind what someone is saying through their narratives and other discourses. The objective-quantitative camps on the other hand think anything that does not involve systematic acquisition and analysis or data cannot be valid. This chapter presents an approach to derive a parametric user model for understanding users that makes use of the premises and ideals of both these camps.

INTRODUCTION

Subjectivity is scorned by many scientists from the materialist traditions of positivism and materialism, but it could be argued that however much we try to fool ourselves scientific inquiry in inherently subjective. Whether it is research participants who will answer a questionnaire differently each time they take it, or the researcher choosing between Verimax or Quartimax rotation based on the result they want, subjectivity is unavoidable. It is important therefore to move away from the perhaps

DOI: 10.4018/978-1-4666-5071-8.ch005

delusionary view that objectivity is desirable and needed. To hold onto such a view will mean that inaccurate models for understanding the way people work will mistakenly assume it is possible for all things to be, when they never are. What might be on a person's mind one minute will be out of it the next, and models of behaviour will have to take account of the chaos in the world and our minds that disrupt the internal and external environment on a moment by moment basis.

BACKGROUND

Two decades after the introduction of the World Wide Web, advocates of user-centred design (UCD) have began to accept that the understanding of Internet users goes beyond making interfaces easy to use. Nielsen (1993) raised the importance of designing interfaces that users would enjoy using as well as being able to use and one increasingly popular way of doing this is gamification. Gamification is the use of game design elements and game mechanics in non-game contexts (Deterding, Sicart, Nacke, O'Hara, & Dixon, 2011; Deterding, 2012; Domínguez et al., 2013).

Recently, gamification is widely used for increasing users' interaction and engagement in variety of domains such as business and marketing, health and wellness, education and training, corporate and vocational training, public policy and government (Hsu, Chang, & Lee, In Press).

Social motivations, especially related to social influence and whether the users find reciprocal benefits from using gamification, are strong predictors for how gamification is perceived and whether the user intends to continue using the service (Hamari & Koivisto, 2013). Compound Identity Theory for understanding self-concept in virtual environments. Compound Identity Theory suggests that a social actor consists of six 'selves' that in turn are made up of three 'component-selves'.

Parametric User Modelling Using Neuroeconomics

Gamification is not simply a one-dimensional system where a reward is offered for performing a certain behaviour; rather, it takes into consideration the variety of complex factors which make a person decide to do something (Birch, 2013). One approach to modelling behaviour is to produce generic models of users through understanding them in detail. One approach to build and describe homogenised users is through parametrics.

Parametric modelling has most frequently been spoken about in human-computer interaction in relation to programming concepts such as objects and classes (Aggarwal, 2003; Szewczyk, 2003). Equally parametric testing of users to derive statistical models are usually developed using so-called objective techniques like questionnaires and other so-called quantitative approaches. The authors however argue that subjectivity is inherent in research studies both from the participant and the researcher. However many people claim they are being objective in looking at a dataset, when it is necessary for models to be constructed – by the researcher – materialist paradigms go out of the window as the researchers need to rely on their mind to construct a model or influence a dataset (e.g. through rotation), meaning research bias is inevitable. The rest of this chapter will look at how qualitative research approaches can be used to devise a testable parametric model for analysing and influencing behaviour in online communities.

The Role of Interviews

Interviews are one of the most established methods of inquiry in various research disciplines. An interview allows for the collection of data based on the spontaneous reactions of others to questions. Whilst a questionnaire uses preformed questions that can result in researcher bias, interviews can be better at correcting this, providing questions are open ended and seek to elicit the interviewees opinions as opposed to confirm the researcher's point of view. The problem with open-ended interviews is that it can sometimes be difficult to code the transcripts with a framework that is universally applicable, again without relying on preconceived ideas. Even those based on the 'framework' approach, which seems to pull the codes from the idealism of the researcher, who is making a judgement, and which may confirm a preconceived mental model, whether the researcher wants to admit it or not.

The author therefore proposes a new approach, using pre-defined modal verbs, which are supported in the interview with pronouns. For instance an interviewee might be asked as a follow up question; "Would you say that you don't like that type of navigational structure?" Or they might be asked: "How much would you say this represents *your* view on a scale of 1 to 7 – 'I don't like navigational structures that get in the way of interaction.'" As can be seen from Table 1 there are a number of pronoun supported modal verbs that can be used to elicit responses and provide a coding framework at the same time. It might be that the pronoun supported modal verbs could be a coding into a different a priori model, or even the associations derived after the coding. It might seem clear however that the words allow for a more objective interview seeking the interviewees opinions and not the interviewers.

Table 2 presents an approach for conducting interviews. As can be seen a number of pronouns and verbs are placed together, which will enable the research more easily entice the participant into expressing statements that can be clearly coded with the M-MARS ethnographical approach (Bishop, 2011a) or a similar approach.

The Role of Subjective Data Capture

According to Štogr (2011), monitoring activities and gathering of data from MUVEs is challenging task. They argue that even in simple VLEs and wikis, these do not properly store all available data about behaviour of users and/or changes of content. Štogr (2011) argues that in most cases analysing these datasets can provide enough information for completion assessment, but cannot be used for modelling simulated experiences. The most promising MUVEs are those where one has full access to both datasets and tools that can be used for "replaying" past activities Štogr (2011) concludes.

Consumer Attitudes, Personality and Behaviour Investigations

Investigating consumer attitudes, personalities and behaviour (CAPB) is a challenging task and completely parametric models seem to throw up the same answers to the wrong questions. Whilst quantitative survey designs can be used to define the operational aspects involved in the decision-making processes of consumers, on its own these studies lack the reality of subjectivity in human interactions. The design of CAPB surveys need to take into account that traditional designs for assessing consumer attitudes do not always take into account technological changes in commerce, which may be approached differently by a consumer because of different opinions towards it. Whilst parametric methods for determining the attitudes, personality and behaviour make claim to generalisability due to the randomly allocated participants that take part in them, all this can offer is a rough and ready insight into the user group they represent. One could argue to the contrary that a clear picture of the factors that form part of the Parametric User Model will develop following subjective analyses and not as a result of parametrically derived ones which rely too heavily on researcher bias.

Website Metrics

Website audiences are broad and come to them freely, often with no registration or prior identification, most data gathered comes from web metrics

Table 1. The selves and component-selves of Compound Identity Theory

Selves	Binary Forces	Component-Selves	Characters
Substantial Self	Order-Chaos	Actual Substantial Self	Big-Man / Troll
		Perceived Substantial Self	Sceptic / Cynic
		Ideal Substantial Self	Patriarchal / Humorous
Synthetic Self	Vengeance-Forgiveness	Actual Synthetic Self	E-Venger / MHBFY Jenny
		Perceived Synthetic Self	Fascist / Pacifist
		Ideal Synthetic Self	Dangerous / Timid
Social Self	Social-Antisocial	Actual Social Self	Flirt / Snert
		Perceived Social Self	Follower / Antagonist
		Ideal Social Self	Orthodox / Pariah
Situational Self	Surveillance-Escape	Actual Situational Self	Lurker / Elder
		Perceived Situational Self	Loner / Catalyst
		Ideal Situational Self	Stranger / Bouncing Coconut
Symbolic Self	Creative-Destructive	Actual Symbolic Self	Wizard / Iconoclast
		Perceived Symbolic Self	Enthusiast / Detractor
		Ideal Symbolic Self	Assiduous / Vanguard
Saturated Self	Existential-Thanatotic	Actual Saturated Self	Chatroom Bob / Ripper
			Striver / Rejector
			Exotic / Pitied

(which reveal behaviour but not attitudes or intent) or from surveys and interviews of a self-selecting group of users (Carson, Kanchanaraksa, Gooding, Mulder, & Schuwer, 2012). In the source selection, methods such as web metrics can be used to evaluate the information source, and inputs from industry experts are essential to providing a complete picture (Xie, Wang, & Chen, 2012).

As can be seen from Table 3 there are a number of factors that can influence involvement. Reactionary influences result in low involvement – because not much effort is needed to achieve the behaviour. Ulterior motives means in the situation

Table 2. Possible interview approaches

Parameter / Orientation	Communication orientated	Experience orientated	Achievement orientated
Recurrent Needs	Relationships (I want/don't want)	Recognition (I can/can't)	Responsibility (I think/I don't think)
Situational Constraints	Group Dynamics (I offer/don't accept)	Antecedent State (I need/don't need)	Prospects (I am/I am not)
Compound Identity	Social Self (I shall / shall not)	Situational Self (I must/must not)	Symbolic Self (I'll always/I'll never)
Situated Actions	Cooperative (I will / I won't)	Impulsive (I should/shouldn't)	Cautious (I like/don't like)

Table 3. Elements to build into website matrics

Involvement	Considerations	Examples
Reactionary	1 aspect: Needs and constraints	Rapid clicking of a mouse, facial expressions, heart rate.
Ulterior	2 aspects: Needs and constraints	Clicking on adverts that financially support a website one uses.
Angular	3 aspects: Needs, constraints and identity	Not willing to change one's attitudes, believe or practices for others regardless of the impact it has on them.
Flow	1 aspect: High level involvement	Speed of transition from one page to the next and one contribution to another.

that a person will do things for their own ends or the websites they use, such as clicking on adverts to increase income for this site (Table 4). This may be seen as a form of trolling, as the person is causing harm to another – the advertiser – for their own or that site's benefit. Angular influences are those where the person will not change, such as where an elder won't change to accommodate new members of the group or point out 'flaws' in others arguments, if they don't match their own. In terms of Flow this has a direct impact on involvement. Flow is high when a person has little involvement in a task though increased effort. Flow is low when involvement is high as the person's actions are not fluid because of the extra effort being put into the task at hand.

Observational Recordings

Observational recordings can provide a useful insight into how users make use of a system and interact with all the elements of it. To devise and test a parametric user model one might first determine the factors that influence the decision-making processes of social actors in a particular user group by testing several dependent variables that may affect the way a social actor uses a virtual environment (e.g. VR-based, iTV, Web).

One approach that might to be used to object the effectiveness of a parametric user model in a live setting would be a study that follows a quasi-experimental research design in which the experimenter controls the independent variables within the virtual environment the participant is using. The participants in such a study may not need to be randomly allocated, as they will either be a member of one user group or another. Using a video prototyping technique, certain elements of the virtual environment are changed (e.g. navigation, graphics, etc) and the effect of this on the participant recorded. This is followed up by qualitative interviews in which the participant is asked to discuss their experiences using the system. Factors that might be observed can be seen in Table 5.

Card-Sorting

It is well known that producers of a technology may have in mind a group of individuals as potential users while designing, producing and marketing it (Wang, Tucker, & Rihll, 2011). It is known that targeting of specific niches as a viable marketing strategy to overcome the disadvantage of smaller network size (Morsillo, 2011). A review of practices of marketing towards diverse ethnics by Cui (1997) revealed several perspectives: traditional marketing, separated marketing, integrated marketing and multicultural marketing. This model

Table 4. Elements to elicit from websites matrices

SARM (Angular Responses)	Karen Horney (Orientation)	Mantovani (Situated Actions)
Cooperative	Compliant	Cooperation
Impulsive	Aggressive	Conflict
Cautious	Detached	Negotiation

Table 5. Parameters to observe to test model

Parameters	Environmental	Capital	Time
Navigation Style (video)	How does a social actor search for information in a specific environment?	How does the amount of capital affect how a social actor navigates an environment?	How does the amount of time affect the navigational style of a social actor?
Involvement (video)	How involved will a social actor be in specific environments?	How involved will a social actor be if they are limited by the amount of money they have?	How involved will a social actor be if they are limited by the amount of time they have?
Satisfaction (video/questionnaire)	How satisfied will a social actor be if certain environmental factors are changed?	How satisfied will a social actor be if they require capital to participate?	How satisfied will a social actor be if the amount of time they have is limited?
Memory (questionnaire)	How much is a social actor able to remember certain elements of the interface in specific environments?	How does the amount of capital a social actor has affect how much they can remember about the shopping experience?	How much is a social actor able to remember about the environment if they are limited by time?

has application to the history of Q-methodology and the challenges it has faced and is facing. Doing a paper prototype exercise to familiarize participants with a new technological application does not prevent participants from forming their own mental models (Slegers & Donoso, 2012). This is confirmed by the fact that paper prototyping exercises that are done before doing card sorting exercises may result in deeper insights into the participants' mental models which is useful for (re)designing application interface structures (Slegers & Donoso, 2012).

The Role of Literature Reviews

The adage, 'If you see one rat then there must be a hundred more nearby,' is no truer than of literature reviews. However original a piece of research's findings it is possible to situate it among other works. Whether this be because they are trying to solve specific problems that are similar, that they are opposing points of view, or simply that the new research goes beyond that in existence previously, literature reviews provide a basis to confirm the findings of research or aid in the refutation of earlier research.

For Confirmation

Models of human behaviour that are subjectively devised that do not rely on parametric premises like homogeneity to be constructed are unlikely to be replicated in the literature. This is because of the generally accepted view that qualitative research, which is usually subjective, cannot be generalised. However the conclusions of qualitative research might be confirmed by another study, whether based on parametric rules or otherwise. For instance in terms of customer satisfaction, whether one uses interviews or questionnaires it is likely that the customer view of whether they are likely to give returned custom will be the same, whether it is measured through words or numerical scales. The literature can therefore be useful in confirming the findings of a new piece of research, even if those findings have not been reported before – because of the underlying premises or resulting conclusions.

For Refutation

It could be argued that as sometimes the data may not fit the theory that one must seek to rebalance that inconsistency. The literature can be effective at showing a particular derived fact not to be generalisable.

BREAKING THE DIVIDE OF OBJECTIVITY V SUBJECTIVITY – THE ROLE OF PARAMETRIC VALIDATION

It is known that whether a test is parametric or non-parametric that having a good theory on which to base one's assumptioms can avoid pitfalls of data collection and analysis. The Consumer Resource Exchange Model (CREM) is based upon the theory that consumers seek to manage resources in order to engage in exchange activities directed at achieving goals (Bristow & Mowen, 1998). The resources in the model have been operationalised as available, accumulative assets that are readily transmittable or exchangeable between individuals (ibid). This might be one model for testing a subjective approach that looks at individual difference. Another framework – The SARM Model – includes the elements of Situational Constraint Theory (SCT) as a basis for explaining the factors that restrict a social actors ability to achieve their goals (Peters, O'Connor, & Eulberg, 1985; Pritchett, 2009). Situational constraints are "a set of circumstances that is likely to influence the behaviour of at least some individuals, and that is likely to reoccur repeatedly in essentially the same form" (Peters et al., 1985). This definition helped researchers build an approach to gather applicable research about situational constraints of a job or task that has or will interfere with past or present work performance (Pritchett, 2009).

Hybrid versions of these models can perhaps be found in (Zikmund & Scott, 1973), which was in existence thirty to forty years before these models. The factors making up all these models are presented in Table 6.

These models can be seen to be an example of what might make up an effective model for analysing and influencing behaviour online through gamification. As can be seen from Figure 1 the SARM model can be seen to reflect external representations – the elements that join the disruption of an equilibrium in a person's mind, body or the world (x) with internalised concepts that are associated with that (y). These are reflected in the neuroeconomic modelling conducted in Bishop (2011b), which are reproduced in Table 7, as well as being integrated with regards to the above in Figure 1, which is below that table.

Figure 1 presents a potential model for tasking in external stimuli and transforming it in such a way that it would be possible to predict the acts of a person in any given situation using the equations from Bishop (2011b). Equation 1 is what is used to transfer one type of cognition (x) which has an impact on another (y) into what is called a 'phantasy,' which is measured on a scale of between -5 and +5. This is to reflect that these phantasies can be constructed subjectively though q-sorting, while also having a clear numerical value for devising a parametrically testable model. These phantasies are then combined to produce a Pression using Equation 2 (Bishop, 2011b). The more phantasies that are integrated into Equation 2, such as from multiple stimuli from the internal and external environment, the more difficult it is for the person to resist the impact of them

Table 6. Possible models for constructing a subjectively devised parametric user mode

SARM (Situational Constraints)	CREM (Motivations)	William Zikmund (Perceived Risk)
Environmental	Physical	Physical
Capital	Financial	Financial
Group Dynamics	Social	Social
Antecedent State		Psychological
Prospects		Opportunity
Time	Informational	Time

Figure 1. A model for measuring the flow of information through the human brain and body

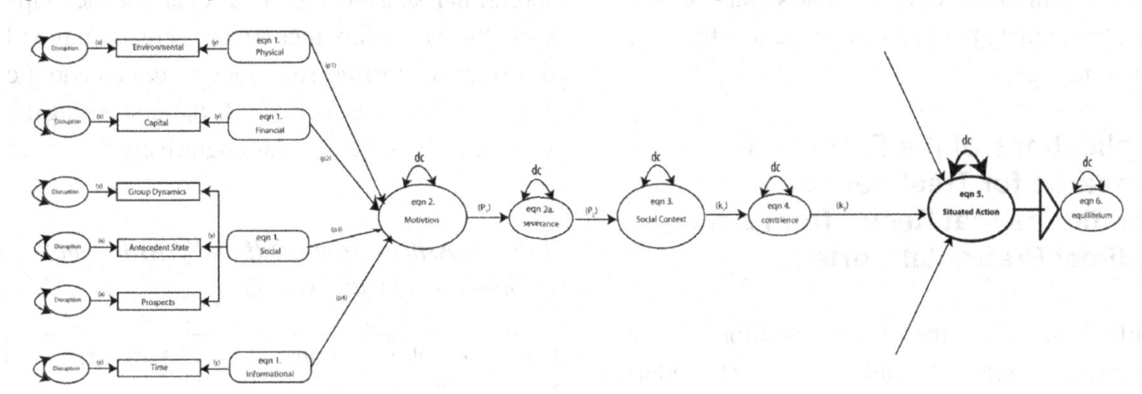

In Figure 1 it can be seen that Equation 3 from Bishop (2011b) has been surrounded by two new equations named as Equations 2a and Equation 4. These equations have not yet been defined, but the role of them are to measure how the Pression is transformed to create a social context as discussed by (Mantovani, 1996a; Mantovani, 1996b). Equation 3 creates a 'knol' which is an instruction generated by the brain and the stronger the knol (+infinity) the easier it is for information to flow through the synapses. The weaker the knol (-infinity) the more difficult it is for a person to perform the task at hand.

Potential Frameworks for Devising Parametric User Models

With the above model in Figure 1 it will be necessary to map existing theories onto its various components so one can better understand the various thought processes that can be influenced by a gamification system to produce desirable behaviours. Table 8 presents an amalgamation

Table 7. Phantasy construction from interaction between pre-frontal cortex detachments/interests and other cognitions

x cognition	x_1	y cognition	y_1	$\bar{z}$	c	Pre-frontal cortex function
Detachment	3	Goal	3	36	-9.6	Problem-solving
Detachment	0	Plan	0	0	-2.4	Self-control
Detachment	0	Value	0	12	-2.4	Conscience
Detachment	0	Belief	0	0	-3.6	Working Memory
Detachment	0	Interest	0	30	-6	Empathy
Detachment	0	Detachment	0	18	-3.6	Deception
Interest	0	Goal	0	45.5	8.9	Problem-solving
Interest	0	Plan	3	1.5	4.3	Self-control
Interest	0	Value	0	20.5	3.9	Conscience
Interest	0	Belief	4	2	6.4	Working Memory
Interest	0	Interest	0	50.5	9.9	Empathy
Interest	0	Detachment	0	30	-6	Deception

of various models from the authors that could be used to populate the fields in the parametric user model in Figure 1.

Implications of the Proposed Research for Treatment of Conditions that Result from a Sub-Optimal Prefrontal Cortex

Table 7 showed some of the cognitions that a parametic user model could use in order to understand and influence human behaviour. The ones featured, which are linked to interests (persons of things important to us) and detachments (persons and things we want to avoid), are based primarily in the prefrontal cortex (Bishop, 2011b; Bishop, 2012). Table 9 presents the links between the neuro-economic factors in Bishop (2012). and the prefrontal cortex functions in (Bishop, 2011b).

Understanding the Concept of a 'Sub-Optimal Prefrontal Cortex' (SOPFC)

If one applied the equations discussed in relation to Figure 1 it would be possible to calculate the degree to which someone is sub-optimised in the prefrontal cortex (SOPFC) when operating at the maximum brain capacity for them. Called a 'knol,' this measurement of the brain's processing speed can using the research described in this chapter allow for the integration of people into an environment using gamification, regardless of their ability. For instance, intellectual development disorders have a physical basis within the brain, which are at present unchangeable. An important thing to note in this regard is the idea that someone with intellectual impairments is slow (i.e. retarded) is misplaced. As one can see from Figure 2 the impact of a sub-optimal prefrontal cortex can have a number of adverse impacts on an individual.

Bipolar disorders can be considered to result from a suboptimal prefrontal cortex where when the increase in dopamine peaks at the same time serotonin is at its lowest. The dopamine then falls with the serotonin then rising, creating mood disorders from this serotonergic-dopamenergic asynchronicity (see below). By understanding the way a person with Bipolar cognitively functions

Table 8. Aspects of the self and possible models for measuring value system

Self Type (Value System)	Source	Variables
Substantial Self (Opportunity)	L. C. Harris & Goode (2010)	Aesthetic Appeal [Originality of Design *[Acceptance, Openess]*, Visual Appeal *[Validity, Certainty]*, Entertainment Value *[Ingratiation, Information]*
Synthetic Self (Understanding)	L. C. Harris & Goode (2010)	Financial Security [Perceived Security *[Claim, Community, Security]*, Ease of Payment *[Impermanence, Imitation, Focus]*]
Social Self (Relevance)	L. C. Harris & Goode (2010)	Layout & Functionality [Usability *[Consistency]*, Relevance of Information *[Backing, Ambivalence]*, Customisation *[Applicability]*, Interactivity *[Inspiration, Encouragement]*]
Situational Self (Aspiration)	L. C. Harris & Goode (2004)	Trust [Service Quality *[Unsatisfactoriness]*, Personal Value *[Warrant, Fairness]*, Satisfaction *[Instigation]*, Loyalty *[Challenge, Recognition]*]
Symbolic Self (Choice)	Goode & Moutinho (1996)	Overall Satisfaction [Overall Expectations *[Conclusion]*, Perceived Risk *[Introspection]*, Recommend to Others *[Confirmation]*, Confidence *[Efficacy]*, Frequency of Use *[Self-Direction]*, Full Use of Services *[Approval]*
Saturated Self (Expression)	Jamal & Goode (2001)	Brand Preference [Education *[Evidence]*, Occupation *[Interaction]*, Gender *[Cooperation]*, Income *[Reciprocity]*, Age *[Tolerance]*, Marital Status *[Complacence]*]

Table 9. The effect of dopamine, serotonin, flow and involvement on prefrontal cortex functioning

Factor	Self-control	Conscience	Working Memory	Empathy	Deception
Dopamine	Increasing dopamine can improve focus	Increasing dopamine can produce thoughts of guilt due to greater awareness	Increased dopamine can overload working memory or enable more creative thinking.	Increasing dopamine can impair empathy because a person may be focussed more on their own thoughts. Reducing dopamine can lead to disinterest in others.	Increased dopamine makes deception difficult because of over focus. Reducing dopamine can make deception easier due to being more relaxed.
Serotonin	Reducing serotonin can reduce task-focussed anxiety	Increasing serotonin can exacerbate feelings of guilt due to anxiety around unwanted thoughts.	Increased serotonin can impair working memory by making it difficult to focus because of resultant anxiety.	Increased serotonin can improve empathy where another person is also experiencing it.	Increased serotonin can make deception more difficult due to task focussed anxiety.
Flow	Increased flow can reduce self-control	Increased low can reduce awareness of guilt or consequences of actions.	Increased flow can improve effectiveness of working memory	Increased flow can make conversations more fluid and empathy easier.	Increased flow can make it easier to deceive others because of an ability to avoid distractions.
Involvement	Increased involvement can improve task-focus	Increased involvement can make it difficult to avoid conscience	Increased involvement depends on working memory	Increased involvement can make empathy difficult due to lack of fluidity of thinking.	Increased involvement can make masking deceptive behaviour more difficult.

it can be possible to intervene so that the factors that induce the asynchonicity, such as external and internal representations (Figure 2) can be identified so they are easier to avoid. Depression can also be seen to exist where greater demand for serotonin results in an increase in involvement by a person trying to avoid something that is difficult to deal with, because of among other things the effect of their sensory perception (Figure 2). Antisocial personality disorder is a condition where specific behaviours result in unsatisfying outcomes for others are based on the sufferer's own selfish and greedy wants. Antisocial personality disorder is know to be linked to abusive behaviours on the Internet (Bishop, 2013). A person with another condition, anti-social personality disorder, is likely to have increased serotonin levels when having thoughts of superiority and at the same time increased serotonin levels. This is likely to result in psychotic and neurotic mental states over a longer period, and the models discussed in this chapter could be used to reduce the serotonin and moderate the dopamine. The same would apply to a related condition called 'operational defiant disorder.' Another disorder, intermittent explosive disorder (IED) could be conceptualised as a sudden elevation in serotonin and dopamine levels when a person is at equilibrium. For instance if others engage in repetitive behaviours they they know a person disapproves of, when they enact those behaviours when the person with IED is otherwise moderate will become enraged, possibly due to past memories of being forced out of their equilibrium state. This research and concepts in this chapter could be used to train a person how to prevent elevations in dopamine and serotonin through understanding the situation better. As discussed in Bishop (2012) if one is not suited to an environment it either has to be adapted or one might want to be removed from it. As has been recently pointed out, not being exposed to a stimuli can reduce the effect

Figure 2. Factors that ccontribute to a sub-optimal prefrontal cortex (SOPFC)

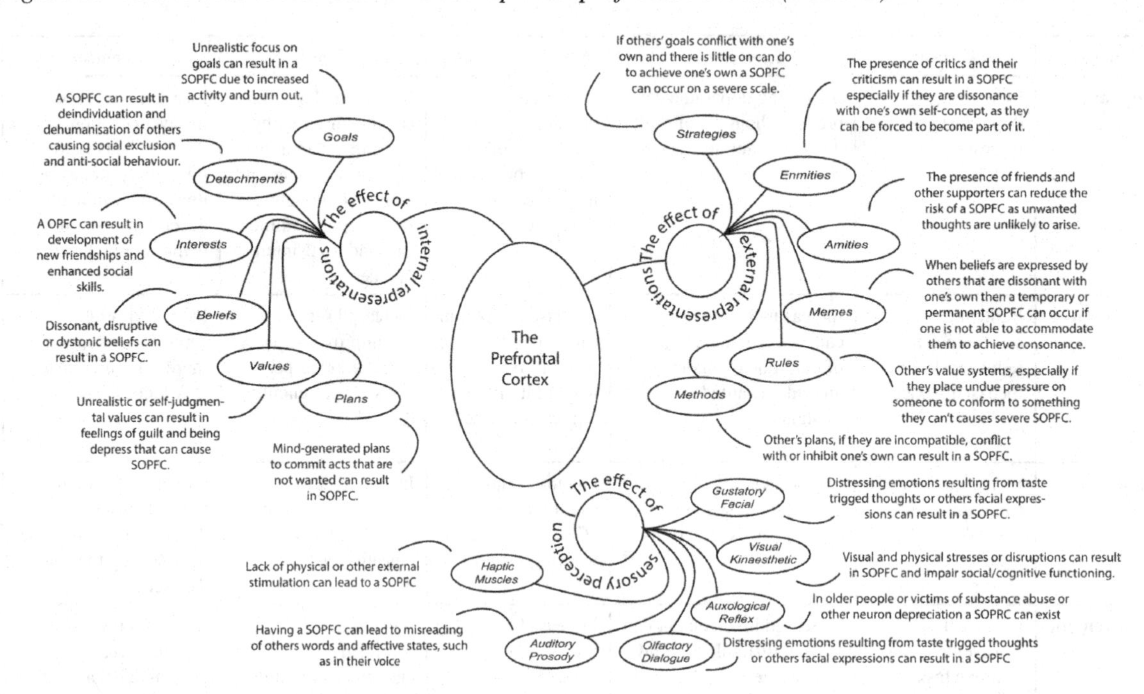

on the phantasies causing problematic behaviours (Crockett et al., 2013). However, the research in this paper and Bishop (2012) suggest that it can be possible to reduce the effect on these phantasies so that they have little or no effect.

Understanding Serotonergic-Dopamenergic Asynchronicity (SDA)

Serotonergic-dopamenergic asynchronicity is the term to describe the fact that a persons serotonin and dopamine levels are not in tune and causing psychiatric problems as opposed to offering an optimal flow to enable a person to be productive. Emerging research based on transcranial direct current stimulation (TDCS) have been touted as a means treating alcohol dependency (Silva et al., 2013). This approach to manipulating the prefrontal cortex is unsound. Like electro-convulsive therapy it might provide for an immediate therapy but a repetition of the same intervention will be needed. The only way an electrical intervention like this could work is by trying to erase the phantasy that causes serotonergic-dopamaneric asychronicity, but this would not remove the others thoughts that resulted from it. The system presented in Bishop (2012) called 'MEDIAT' could be enhanced with the research proposed by this chapter to enable people to independently regulate their thoughts so they can deal with them and use the phantasy to their advantage through the reprogramming of it.

Figure 3 shows how different mental processes can be affected by serotonergic-dopamenergic asynchronicity. Two of these in particular that have not been explored much elsewhere are cognitive structures and structures per se, which were originally considered by Bishop (2007). For instance, gambling disorders can be linked to the needed to avoid activating an unwanted phantasy resurfacing, as with any other addiction where one might be dependant on a substance. When a traumatic phantasy activates – such as after observing a related stimulus – one's serotonin levels rise to suppress it and one's dopamine levels rise to focus attention on it, which causes psychotic and/

or neurotic symptoms. The compulsion to use a substance or other stimulus will only have efficacy for suppressing the phantasy for so long and thus addictions form when more of that stimulus is needed to have the same effect in suppressing the serotonin arising from the phantasy. The research proposed in this paper could result in MEDIAT (Bishop, 2012) being improved to help people identify and deal with the traumatic phantasy that is causing their anxiety and need for a substance to produce a pleasurable experience to restore serotonergic-dopameneric synchronicity. Table 10 shows a list and description of conditions where MEDIAT might help influence serotonergic involvement and dopaminergic flow.

Implications and Future Directions

The literature that has dripped out in 2013 in relation to the prefrontal cortex offers no improvement on that done by Bishop (2011b). and Bishop (2012). This chapter, however, has provided a non-exhausted lists of the various ways in which mental disorders can be caused by serotonergic-dopaminergic asynchronicity (SDA), which result from a sub-optimal prefrontal cortex (SOPFC). The MEDIAT system described in detail in a different paper can assist with helping any of these conditions where the brain's reaction to a phantasy (i.e. an episodic/traumatic memory) is for the person with that phantasy in their mind to perform obsessive, compulsive or narcisstic behaviours. A person may develop rituals to take their mind off their phantasies typically seen in people with autism and obsessive compulsive disorder. It is clear that research in the areas of neuroeconomics, e-therapy and user customisation will be in the forefront of research in years to come and this chapter should go someway to providing direction for those.

Figure 3. Factors that can cause serotonergic-dopamenergic asynchronicity (SDA)

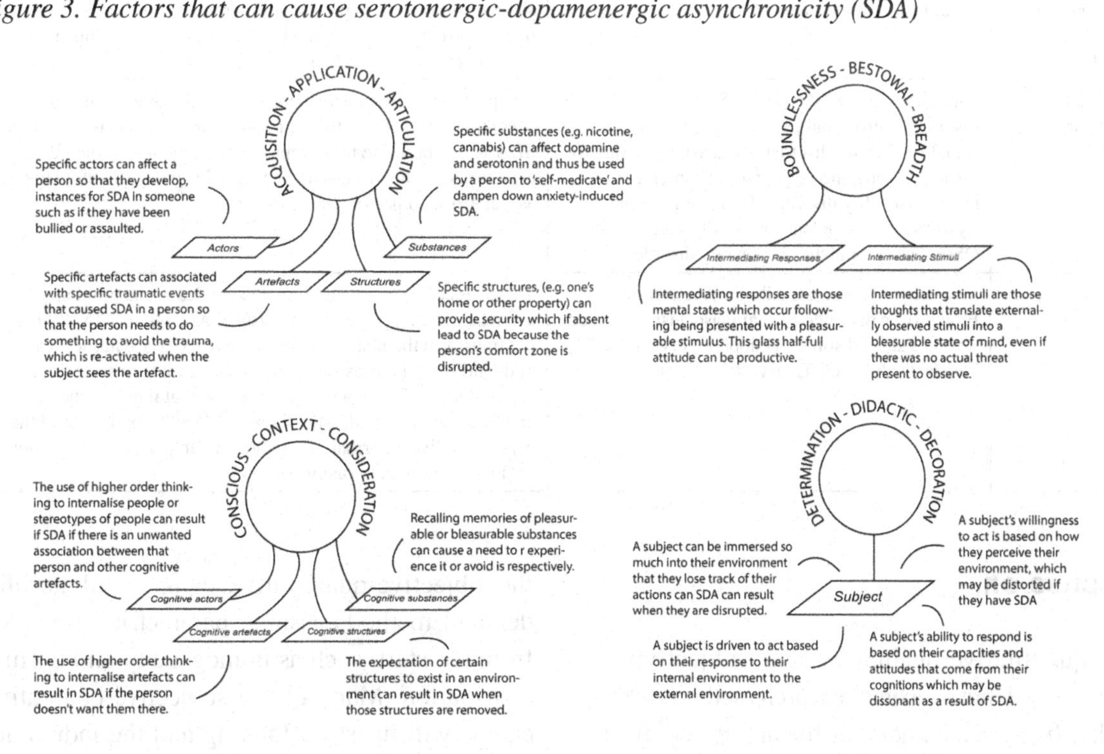

Table 10. Links between recognised medical diagnoses and SDA/SOPFC

Neurodevelopmental disorders	Intellectual developmental disorder, Communication disorders, Attention-deficit/Hyperactivity disorder, Specific learning disorder, Tic disorders.	Many learning impairments can be put down to poor neuro-response plasticity with information lacking the necessary flow between synapses. Many communication impairments are down to phantasies impairing access to essential brain functioning, such as those in the prefrontal cortex (e.g. self-control, empathy, problem solving, working memory).
Schizophrenia spectrum and other psychotic disorders	Delusional disorder, Brief psychotic disorder, Schizophreniform disorder, Schizophrenia, Schizoaffective disorder, Substance/Medication-induced psychotic disorder, Psychotic disorder due to another medical condition, Catatonia	Psychotic conditions can be seen as a reaction to high dopamine and serotonin levels, either through psychiatric shock or more gradual impairment. SDA is at its most severe making treatment a long-term activity. Functions of the prefrontal cortex, such as conscience can be problematic in Schizo-typal disorders due to misplaced guilt.
Trauma- and stressor-related disorders	Reactive attachment disorder Disinhibited social engagement disorder, Posttraumatic stress disorder, Adjustment disorder	Trauma is one of the main causes of SDA as when a flow of trauma induced stimuli enters the brain then the brain inserts phantasies between the synapses to capture the trauma and prevent damage to brain functions.
Dissociative disorders	Dissociative identity disorder, Dissociative amnesia, Depersonalization/Derealization disorder, Other specified dissociative disorder	Traumatic childhood memories can result in altered prefrontal cortex functioning. Self-control is a core part of the prefrontal cortex that can affect DDs.
Elimination disorders	Enuresis, Encopresis	May be linked to prefrontal cortex functions like self-control and empathy. SDA might create an anxiety where person thinks task has to be dealt with or where doing so might be harmful.
Gender dysphoria	Congenital adrenogenital disorder, postransitu-ational delay	A trauma-induced SDA might inhibit the psycho-sexual development of a person in such a way to produce abnormal levels of testosterone or estrogen. Dealing with the SOPFC might allow the flow of information so that a person can develop the neurological connectors that were stalled in development.
Disruptive, impulse-control, and conduct disorders in DSM-5	Oppositional defiant disorder, Intermittent explosive disorder, Antisocial personality disorder, Pyromania, Kleptomania	May be linked to the prefrontal cortex, which is responsible for self-control, deception and empathy. For instance if there is a sudden variation in dopamine and serotonin, such as because of an external force changing the environment, then this might result in extreme behaviour.
Personality disorders in DSM-5	Paranoid personality disorder, Schizoid personality disorder, Schizotypal personality disorder, Antisocial personality disorder, Borderline personality disorder, Histrionic personality disorder, Narcissistic personality disorder, Avoidant personality disorder, Dependent personality disorder, Obsessive-compulsive personality disorder	The prefrontal cortex is responsible for conscience, empathy, deception, and self-control. It might be argued that anyone with SDA will experience the symptoms of a personality disorder, even if not all the symptoms. A SOPFC is likely therefore to be associated with personality disorders.
Paraphilic disorders	Voyeuristic disorder, Exhibitionistic disorder, Frotteuristic disorder, Sexual masochism disorder, Sexual sadism disorder, Pedophilic disorder, Fetishistic disorder, Transvestic disorder	These sets of disorders are prime examples of the problems that can result from a SOPFC that exists as a result of SDA. By helping transform the phantasies between a persons synapses leading to the prefrontal cortex (such as sexual abuse they suffered) it is possible to develop a thought regieme that removes the guilt around a memory so that the focus on it can be reduced and the person can live a normal life without a compulsion to harm others in an unwanted sexual way.

Discussion

The quantitative-qualitative and subjectivity-objectivity debates plague research methods textbooks, divide academic departments, and confuse post-modernists as to their existence. Those from the objective-quantitative camps will usually demand methods assume parametric principles from the start, such as homogeneity and normal distribution. Many of the subjective-qualitative camps will insist on looking and the individual meanings behind what someone is saying through their narratives and other discourses. Subjectivity is scorned by many scientists from the materialist traditions of positivism and materialism, but it could be argued that however much we try to fool ourselves scientific inquiry is in reality inherently subjective. Whether it is research participants who will answer a questionnaire differently each time they take it, or the researcher choosing between Verimax or Quartimax rotation based on the result they want, subjectivity is unavoidable.

This chapter has provided an approach to measuring subjective judgements so as to dervive a model that can be used in a parametric way. Parametrics are, however, not the be-all-and-end-all of science. In parametric studies it is assumed that the sample that data is collected from is representative of the wider society. This can be called the 'society fallacy.' The idea that data from a small group of people can be used to describe a wider strata in which they fit is unsound. The 'ecological principle,' which states that all a study of a small group can do is provided enough information so that by removing the commonalities it is possible to see the individual characteristics of members in that group. That group is an entity in itself, and to say that one group will be reflective of another group, whilst producing interesting statistics is not likely to be reflective of the reality where one group in a county will be nothing like a group in the same country. One can thus argue there is no such thing as 'society' only 'societies.' That is that as each person in a country is different and observes and is influenced by the environment differently, then there is no single society, as each person constructs their own society. So with the added use of the Internet we must be a collection of network societies as opposed to one society. So whilst the model proposed in this chapter can lead to a parametric user model, the question might be – should it?

REFERENCES

Aggarwal, C. C. (2003). Towards systematic design of distance functions for data mining applications. In *Proceedings of the Ninth ACM SIGKDD International Conference on Knowledge Discovery and Data Mining,* (pp. 9-18). ACM.

Birch, H. (2013). *Motivational effects of gamificaiton of piano instruction and practice*. Paper presented at the Dean's Graduate Student Research Conference 2013. New York, NY.

Bishop, J. (2007). Increasing participation in online communities: A framework for human–computer interaction. *Computers in Human Behavior, 23*(4), 1881–1893. doi:10.1016/j.chb.2005.11.004

Bishop, J. (2011a). *The equatrics of intergenerational knowledge transformation in technocultures: Towards a model for enhancing information management in virtual worlds.* (Unpublished MScEcon Thesis). Aberystwyth University, Aberystwyth, UK.

Bishop, J. (2011b). *The role of the prefrontal cortex in social orientation construction: A pilot study.* Paper presented to the BPS Welsh Conference on Wellbeing. Wrexham, UK.

Bishop, J. (2012). Taming the chatroom bob: The role of brain-computer interfaces that manipulate prefrontal cortex optimization for increasing participation of victims of traumatic sex and other abuse online. In *Proceedings of the 13th International Conference on Bioinformatics and Computational Biology (BIOCOMP'12)*. BIOCOMP.

Bishop, J. (2013). The effect of deindividuation of the internet troller on criminal procedure implementation: An interview with a hater. *International Journal of Cyber Criminology, 7*(1), 28–48.

Bristow, D. N., & Mowen, J. C. (1998). The consumer resource exchange model: An empirical investigation of construct and predictive validity. *Marketing Intelligence & Planning, 16*(6), 375–386. doi:10.1108/02634509810237587

Carson, S., Kanchanaraksa, S., Gooding, I., Mulder, F., & Schuwer, R. (2012). Impact of OpenCourseWare publication on higher education participation and student recruitment. *International Review of Research in Open and Distance Learning*, *13*(4), 19–32.

Crockett, M. J., Braams, B. R., Clark, L., Tobler, P. N., Robbins, T. W., & Kalenscher, T. (2013). Restricting temptations: Neural mechanisms of precommitment. *Neuron*, *79*(2), 391–401. doi:10.1016/j.neuron.2013.05.028 PMID:23889938

Cui, G. (1997). Marketing strategies in a multi-ethnic environment. *Journal of Marketing Theory and Practice*, 122-134.

Deterding, S. (2012). Gamification: Designing for motivation. *Interaction*, *19*(4), 14–17. doi:10.1145/2212877.2212883

Deterding, S., Sicart, M., Nacke, L., O'Hara, K., & Dixon, D. (2011). Gamification: Using game-design elements in non-gaming contexts. In *Proceedings of the 2011 Annual Conference on Human Factors in Computing Systems*. New York, NY: IEEE.

Domínguez, A., Saenz-de-Navarrete, J., De-Marcos, L., Fernández-Sanz, L., Pagés, C., & Martínez-Herráiz, J. (2013). Gamifying learning experiences: Practical implications and outcomes. *Computers & Education*, *63*, 380–392. doi:10.1016/j.compedu.2012.12.020

Goode, M. M. H., & Moutinho, L. A. C., C. (1996). Structural equation modelling of overall satisfaction and full use of services for ATMs. *International Journal of Bank Marketing*, *14*(7), 4–11. doi:10.1108/02652329610151331

Hamari, J., & Koivisto, J. (2013). Social motivations to use gamification: An empirical study of gamifying exercise. In *Proceedings of the 21st European Conference on Information Systems*. IEEE.

Harris, L. C., & Goode, M. M. (2004). The four levels of loyalty and the pivotal role of trust: A study of online service dynamics. *Journal of Retailing*, *80*(2), 139–158. doi:10.1016/j.jretai.2004.04.002

Harris, L. C., & Goode, M. M. H. (2010). Online servicescapes, trust, and purchase intentions. *Journal of Services Marketing*, *24*(3), 230–243. doi:10.1108/08876041011040631

Hsu, S. H., Chang, J., & Lee, C. (2013). *Designing attractive gamification features for collaborative storytelling websites*. Cyberpsychology, Behavior, and Social Networking. doi:10.1089/cyber.2012.0492

Jamal, A., & Goode, M. M. H. (2001). Consumers and brands: A study of the impact of self-image congruence on brand preference and satisfaction. *Marketing Intelligence & Planning*, *19*(7), 482–492. doi:10.1108/02634500110408286

Mantovani, G. (1996a). *New communication environments: From everyday to virtual*. London: Taylor & Francis.

Mantovani, G. (1996b). Social context in HCI: A new framework for mental models, cooperation, and communication. *Cognitive Science*, *20*(2), 237–269. doi:10.1207/s15516709cog2002_3

Morsillo, R. (2011). One down, two to go: Public policy in service of an available, affordable and accessible national broadband network for people with disability. *Telecommunications Journal of Australia, 61*(2).

Nielsen, J. (1993). *Usability engineering*. San Francisco: Morgan Kaufman.

Peters, L. H., O'Connor, E. J., & Eulberg, J. R. (1985). Situational constraints: Sources, consequences, and future considerations. *Research in Personnel and Human Resources Management*, *3*, 79–114.

Pritchett, J. E. (2009). *Identification of situational constraints in middle school business information technology programs. (Doctor of Education).* Partial, Georgia: The University of Georgia.

Silva, M. C. d., Conti, C. L., Klauss, J., & Alves, L. G., Cavalcante, Henrique Mineiro do Nascimento, Fregni, F., et al. (2013). Behavioral effects of transcranial direct current stimulation (tDCS) induced dorsolateral prefrontal cortex plasticity in alcohol dependence. *Journal of Physiology, Paris*. PMID:23891741

Slegers, K., & Donoso, V. (2012). The impact of paper prototyping on card sorting: A case study. *Interacting with Computers*, *24*(5), 351–357.

Štogr, J. (2011). Surveillancebased mechanisms in MUVEs (MultiUser virtual environments) used for monitoring, data gathering and evaluation of knowledge transfer in VirtuReality. *Journal of Systemics. Cybernetics & Informatics*, *9*(2), 24–27.

Szewczyk, J. (2003). Difficulties with the novices' comprehension of the computer-aided design (CAD) interface: Understanding visual representations of CAD tools. *Journal of Engineering Design*, *14*(2), 169–185. doi:10.1080/0954482031000091491

Wang, V., Tucker, J. V., & Rihll, T. E. (2011). On phatic technologies for creating and maintaining human relationships. *Technology in Society*. doi:10.1016/j.techsoc.2011.03.017

Xie, X., Wang, Q., & Chen, A. (2012). Analysis of competition in chinese automobile industry based on an opinion and sentiment mining system. *Journal of Intelligence Studies in Business*, *2*(1).

Zikmund, W. G., & Scott, J. E. (1973). A multivariate analysis of perceived risk, self-confidence and information sources. *Advances in Consumer Research. Association for Consumer Research (U. S.)*, *1*(1), 406–416.

KEY TERMS AND DEFINITIONS

Compound Identity Theory: The matching of human attitudes towards the self to stereotypes and emotions.

Neuroeconomics: The fusion of neuroscience and economics to measure and predict human behaviour through the quantification of mental processes.

Q-Methodology: A form of factor analysis for understanding the subjectivity in human opinions and attitudes which unlike traditional factor analysis focuses on classifying people as opposed to their responses to statements.

Serotonergic-Dopamenergic Asynchronicity: The phenomenon where traumatic or other disturbing memories reduce the competence of a person in social, educational and other contexts.

Situational Constraints: The concept that the environment and other actors can have an impact on a person's abilities to achieve their goals.

Section 2
Health and Cognition

Chapter 6
Designing Serious Games for People with Disabilities:
Game, Set, and Match to the Wii™

Lindsay Evett
Nottingham Trent University, UK

Allan Ridley
Nottingham Trent University, UK

Liz Keating
Nottingham Trent University, UK

Patrick Merritt
Nottingham Trent University, UK

Nick Shopland
Nottingham Trent University, UK

David Brown
Nottingham Trent University, UK

ABSTRACT

Serious games are effective and engaging learning resources for people with disabilities, and guidelines exist to make games accessible to people with disabilities. During research into designing accessible interfaces and games, it was noted that people who are blind often report enjoying playing Wii Sports. These games are pick-up-and-play games for casual and non-gamers. They have simplified rules and a natural and intuitive feel. Games designed specifically for players with particular disabilities are often not of interest to other players and take a lot of development time. Because of their niche market, these games are not widely available, developed, or maintained. In contrast, games like Wii Sports are cheap and available, and represent an exciting opportunity as inclusive games. Two blind players were introduced to the games and found Wii Tennis the most accessible. The blind players learned to play the game quickly and easily, found it enjoyable and engaging, and could play competitively against each other, as well as a sighted opponent. Small accessibility enhancements of the existing game could enhance the game for other players. In this paper, implications for the design of accessible, inclusive games are discussed.

DOI: 10.4018/978-1-4666-5071-8.ch006

INTRODUCTION

Brown et al. (2010a) discuss the potential of serious games as effective and engaging learning resources for people with learning and sensory disabilities. A suitable design methodology and its application are detailed, followed by the description of games that have been successfully developed for target groups with a range of disabilities, and an explication of accessibility guidelines. They conclude that there is great potential in the wide range of possible areas of research into, and development of, serious games for supporting people with learning and sensory disabilities, which would contribute greatly to their inclusion in society. An example of such an application is mobile location based learning, developed by Brown et al. (2010b), which aims to support people with disabilities in employment and training.

During research into creating accessible interfaces and accessible serious games, it was noted that people who are blind often enjoy playing games from the Wii Sports bundle. While these games are not necessarily serious games in the sense of Brown et al. (2010a), they can often have a purpose other than just for entertainment, such as for health. To make serious games accessible for people with disabilities, designers usually follow some form of user-centred design process and consider accessibility guidelines, as detailed in Brown et al. (2010a, 2010b). The games in Wii Sports have not been targeted specifically at people with disabilities, but their target audience is non-gamers and casual gamers, which will include people with a wide range of abilities and skills (Casamassina, 2006). The games have simplified rules and are designed to use the Wii™ remote "to provide a natural, intuitive and realistic feel" (Nintendo, 2010). Since the games are designed for a diverse target audience, It would be expected that some sort of Universal Usability approach (e.g., Horton, 2006) was taken and some form of user-centred design involved. Given that people who are blind report playing these games, it is worth evaluating their accessibility.

In the past a few relatively unsophisticated audio games were available for people who are blind; more recently there has been greater interest in audio games for a wider range of reasons (e.g., sound artists, accessibility researchers, mobile games developers). Additionally, people with disabilities are no longer willing to accept being excluded from mainstream technologies (Higginbotham, 2010), and as more and more of them grow up to be digital natives, and legislation against exclusion becomes tougher, greater developments may be expected in this field.

In Human Computer Interaction (HCI), Universal Usability (e.g., Horton, 2006), is emerging as a prominent approach to system design. This approach acknowledges the wide variability in peoples abilities, and, rather than designing for the "average" person, advocates a Design-for-All approach - it has both accessibility and usability in mind with the aim being to "provide for diversity through design rather than accommodation" (Horton, 2006, p. xvi). This and similar approaches to system design will facilitate the development of accessible games.

There are currently some successes in terms of accessible games. For example, Lone Wolf, by GMA games (2008), is a full-featured World War II submarine simulation audio game. It is a complex game based on complex stereo sound which has been designed for blind and visually impaired players. Lack of any visual representation means that the game wouldn't have general appeal for sighted players. Going a step further, Terraformers is playable by players who are blind, who can play the game against sighted opponents (Westin, 2004). Terraformers offers auditory navigation and game information, which blind players can use successfully to navigate its virtual environment and to play the game. It also offers a graphical interface, so that blind and sighted players can play against each other. The team at ICS-Forth has developed a number of accessible games, and uses a Universally Accessible design process (Grammenos et al., 2005; Grammenos & Savidis, 2006).

However, the exciting thing about Wii Sports games is that they are mainstream games which have not been designed specifically or intentionally for players who are blind. Consequently, if playable by people who are blind, they are seriously inclusive – Wii Sports was the best-selling game of all time in 2007 (Nintendo, 2007). The playability of the games by blind players will be explored, and anything that can be learned to contribute to established accessibility guidelines, such as those of Brown et al. (2010a) will be established. Any results will be informative for the Design-for-All approach, and for the future development and design of accessible and inclusive games.

Wii TENNIS

Wii tennis is one of the games in the Wii Sports bundle, which is often packaged in with the Wii™ system. These games have simple controls and primitive graphics, and are "pick up and play" – you just pick up the Wii™ remote (hereafter Wiimote) and swing or gesture as appropriate. The games have simplified rules and are designed to use the Wiimote "to provide a natural, intuitive and realistic feel" (Nintendo, 2010). The games are aimed at casual and non-gamers, and have been tremendously successful as such – Wii Sports has been well received by critics, has won awards and was the best-selling game of all time in 2007 (Nintendo, 2007); 60.67 million copies had been sold worldwide by the end of 2009 (Nintendo, 2009). The whole Wii™ approach has really invigorated this branch of gaming, and captured a whole new audience.

Interestingly, during other work with blind users, it was noted that they often have experience of Wii™ games, and enjoy playing them; games mentioned were boxing and tennis. One member of a target group (player B*) reported playing Wii tennis often with his sighted teenage cousins, and said he was good at it. Consequently, it was decided to explore the potential of Wii Tennis and other Wii™ "pick up and play" games as accessible, inclusive games.

PLAY

Players

- **Player B*:** Player B* is 22, and is profoundly blind, from birth. He can distinguish between light and dark. He has played real blind cricket once. He plays Wii™ tennis on a regular basis and can beat his sighted teenage cousins.
- **Player B1:** Player B1 is 38, recently registered blind. She has some very limited vision and some difficulties with colour and contrast. Flashing lights will usually render her unconscious.
- **Player B2:** Player B2 is 54, and is profoundly blind. He can distinguish between light and dark. He did have sight until his 30s, although he has always had vision problems. He has some experience of using Wii™ technology from his involvement in the Virtual Cane project (Battersby, 2007; Evett et al., 2009).
- **Player S:** Player S is 25, with uncorrected vision. He is an experienced gamer, and has experience of Wii™ Sports.

Initially, player B* and two other blind players (players B1 and B2) were invited to the lab for a knock about. Player B* played a few games with a sighted player (player S) who is familiar with the Wii™ and Wii™ games, and with games in general, and is fairly competitive. Player B* beat player S convincingly. Players B1 and B2 had not played Wii™ games before. Player B2 had some experience of Wii™ controls, as he is part of the team concerned with the design and evaluation of the Virtual Cane (cf. Evett et al., 2009). These two players had a go at all the games in the Wii Sports bundle. Tennis appeared to be the most accessible game. Bowling wasn't bad, but needed some sort of feedback to inform blind players that they were in the zone for bowling down the lane, and to inform them of which pins they had knocked down and which pins remained. Golf and baseball were not easy for them to play (they take a bit of practise for sighted players), and boxing was found to be a lot of fun. Consequently, it was decided to focus on Wii Tennis.

During the initial knock about, player B1 hit most of the balls out; player B2 performed better than B1 but missed a lot of balls. Five weeks later they came back to play more formally. At this second session, player B1 performed a lot better than before, player B2 about the same. They both played against a sighted opponent (S) and against each other. The sighted player and an observer explained what was happening during play when necessary, and explained why balls were hit out or missed. This feedback was continued during all the sessions. The two blind players enjoyed playing the game.

In the third session, 1 week later, and the following two weekly sessions, each player played 3 "sets" of best of 5 matches against the sighted player and against each other. After session 3, the blind players got their own Wii™ (because they enjoyed it so much), and played most days. Over these 3 sessions, the play of the two blind players clearly progressed, and interestingly, appeared to progress in the way that the play of sighted players playing real tennis would. That is, their serves improved, they became more likely to win their service points, the length, number and quality of rallies increased, the number of games to deuce increased, and they hit more winners. B2 in particular developed a mean power serve (B2 served by tossing up; B1 served by pressing the button). During the 5th session, player B1 broke S's serve and won a "set". Player B2 took 2 "sets" from player B1 in both sessions 4 and 5. Tables 1 through 3 show the scores over sessions 3, 4 and 5.

Both the blind players showed a clear and rational progression in their play. Both players enjoyed playing, so much so that they invested in a Wii™. The sighted player played generously, but not so much that he would give complete games away easily; so B1's taking of a set from S is a real triumph. B1 has some sight and, although movement becomes a blur to her, perhaps had some additional cues about when to hit the ball that B2, with no sight, didn't have. B2's winning of sets from B1 is also quite a triumph. Both blind players, and particularly player B2, started producing aces in sessions 4 and 5, and both players played a number of winners against S in session 5.

DISCUSSION

Two blind players have demonstrated that Wii Tennis is a game which is accessible to them; they can play it, and they have reached a good standard of play; their play is still improving. Player B* can regularly beat sighted opponents; players B1 and B2 started having some successes

Table 1. Set scores for session 3

B1 v S	B2 v S	B1 v B2
S 3-0, 3-0, 3-0	S 3-0, 3-1, 3-0	B1 3-1, 3-0, 3-0
Good rallies in some games	Good rallies in some games	Good rallies in some games
	B2 won 1 game	B2 won 1 game
3 deuce games		

Table 2. Set scores for session 4

B1 v S	B2 v S	B1 v B2
S 3-0, 3-0, 3-1	S 3-0, 3-1, 3-1	B1 3-1, 2-3, 2-3
Good rallies in some games	Good rallies in some games	Good rallies in some games
B1 won 1 game	B2 won 2 games	B1 won 7; B2 won 7; B2's first sets
4 deuce games		

against player S in sessions 4 and 5. Play in this game is natural, intuitive and direct. The sounds and auditory feedback of the game are very effective. As it stands, it is an example of an accessible, inclusive game, and the fact that it is mainstream, readily available and cheap is a big bonus, and a significant addition to the range of accessible, inclusive games available to people who are blind. The accessibility of the game could be improved. In the sessions reported here, some aspects of what was going on in the game and some feedback was provided by the sighted player and the observer; however, these aspects could have been learnt by the blind players over time, although it would have taken longer. Players will interact and advise each other on play as a normal part of these games, so this additional feedback isn't unusual. The one notable feature which would help play for people who are blind would be for the ball to be called out verbally. This is indicated by the reaction of the crowd and by the scoring but this is quite indirect. Additionally, the setting up of the game could be made more accessible, most simply by adding self-voicing.

The game follows the fundamental HCI design principle of direct manipulation, and conforms to the relevant Brown et al. (2010a) accessibility guidelines for serious games (such as good colour contrast; simple, consistent organisation; avoid flashing lights etc.). There are instances where the game should employ alternatives (e.g., verbal "out" as well as visual "out") and should consider assistive technology (e.g., text-to-speech, screen-readers for setting up process, but self- voicing would work well here). The rumble feature could be used to provide alternative forms of feedback. The game is playable as it is, but could be enhanced.

A set of games, VI Fit, has been developed by researchers at SUNY and at the University of Nevada. The VI Fit research project developed some exergames that could be played without visual feedback, with the aim of using them to increase the participation of users with visual impairments in physical activity to improve their health. All VI Fit games can be downloaded for free and played using the Wiimote, through which vibrotactile and audio cues are provided. Two of the games, VI bowling and VI tennis are versions of the Wii™ games with some aspects

Table 3. Set scores for session 5

B1 v S	B2 v S	B1 v B2
S 3-0, 3-0, 2-3	S 3-0, 3-0, 3-0	B1 3-1, 2-3, 1-3
Good rallies in some games	Good rallies in some games	Good rallies in some games
B1 won 3 games B1 takes a set from S Some winners against S	Some winners against S	B1 won 6 games, B2 won 7 games and 2 sets
8 deuce games		

removed and some additional cues (VI Fit, 2010). VI tennis was evaluated in terms of amount of exercise achieved with audio vs. audio + tactile cues; exercise achieved was the same for both, but better scores and greater rated enjoyment were recorded when tactile cues were included. The players were 13 children who are blind at a special summer camp (Morelli et al., 2010). The children who played the game liked the vibrotactile cues, as they meant they didn't have to concentrate so much on the audio, and felt this helped them to learn. The vibrotactile (rumble) cues helped them plan when to hit the ball. In order to do so, the game was slowed significantly. Player B1 considered that it was so slow that it detracted seriously from the playability and enjoyability of the game. Additionally, the visual representation of the game was removed, as were the auditory responses of the audience, which both B1 and B2 found to provide informative, and entertaining, feedback.

The results presented in this paper demonstrate that additional cues are not necessary for successful learning and play of Wii Tennis. The Morelli et al. (2010) testing was done at a camp for people with disabilities, and within that context the children were happy to play VI tennis. Wii Tennis might be a bit harder to learn at first, but it is reasonable to assume children would prefer to play the same games as everyone else, and with anyone else. The faster Wii Tennis game is more exciting than VI tennis. Further, the lack of visual representation in VI tennis means that it is not inclusive. Not only would sighted players probably not be very interested in playing a much slower game with no visual form, but also, sighted carers, companions and players would find it hard to follow what is going on and get involved in the play. Importantly, some people who are blind have some sight, as do people with visual impairment, so there are no visual cues for them. Given these considerations and the evidence presented in this paper, Wii Tennis would appear to be the better game for blind players. VI bowling does improve on Wii Bowling, in that vibrotactile cues are used to enable the player to line up in the alley. It also includes some feedback on the number of (but not which) pins are knocked down. In the Wii™ game, there are no auditory cues for either of these factors. This information is easily supplied by any sighted opponents, making the game playable and inclusive. Lack of visual representation again means VI bowling is not inclusive. If these four games are considered, from the experiences of the current players, in terms of accessibility, inclusion, playability and enjoyability, Wii Tennis clearly wins (Table 4).

Rather than design games especially for blind players, the alternative proposition is to take a Design-for-All approach, following HCI principles and guidelines for accessible games. Additionally, game makers, such as Nintendo in this case, could enhance their games to facilitate learning and play by people with disabilities. In the case of Wii Tennis (and Wii Bowling, and some of the Sports Resorts games) this wouldn't be very difficult to achieve, and wouldn't make the games

Table 4. Comparison of Wii Sports games and VI Fit games for players who are blind

	Designed for casual and non-gamers		Designed for players who are blind	
	Wii™ Tennis	Wii™ Bowling	VI tennis	VI bowling
Accessibility	√		√	√
Inclusion	√	√		
Playability	√√√	√√	√	√√
Enjoyability	√√√	√√	√	√√
Rank	1	2.5	4	2.5

appear particularly different; for example, calling balls out verbally is what happens in the real game of tennis, so it's unusual that it doesn't happen in Wii Tennis. The additional features might even enhance the games for people without disabilities. For example, a colleague stated that having balls called out would help his young children to understand the game.

Forrester Research (2004) has shown that the majority of people who use computer systems have some form of disability (57%). They found that a significant number of users of accessibility features (32%) and users of AT products (35%), such as trackballs and screen magnifiers, have no difficulties or impairments. People prefer accessibility options and products because they find them easier to use, more comfortable, or they want to avoid developing a health problem (such as RSI). This research was carried out for the Microsoft Corporation, which proposes designing accessibility into systems (Microsoft, 2009). According to PopCap games, 1 in 5 casual gamers has some form of physical, mental or developmental disability (PopCap, 2008). This research strongly supports the designing for diversity approach, since a significant number of people like accessibility features and AT products, and because a lot of customers have disabilities. In addition, there are laws across the world enforcing accessibility. There has been a recent case reported whereby a visually impaired player is suing Sony Online because he argues that their refusal to implement or facilitate changes to make online games more accessible violates the Americans with Disabilities Act (Sinclair, 2009). Enhancing accessibility would avoid such challenges.

CONCLUSION

In conclusion, this demonstration of the accessibility of Wii Tennis to people who are blind is informative for the future development of inclusive, accessible games. This pick-up-and-play, casual game, and others in the Wii Sports and Wii Sports Resorts packs, conforms to fundamental HCI principles and is designed for an audience with a wide range of skills and abilities. This Design-for-All approach has been shown to result in good accessibility in this case, and in an inclusive game. Games designed solely for the special needs of a particular group can result in exclusive games with limited availability, and often focus on certain attributes while ignoring others, sometimes to the detriment of the enjoyability and playability of the game. The Design-for-All approach offers clear benefits, and could contribute significantly to the design of serious games which are effective and engaging learning resources for supporting people with learning and sensory disabilities.

REFERENCES

Battersby, S. J. (2007, December). Serious games and the Wii™ – a technical report. In *Proceedings of the 27th BCS SGAI Workshop on Serious Games*, Cambridge, UK.

Brown, D. J., McHugh, D., Standen, P., Evett, L., Shopland, N., & Battersby, S. (2010b). Designing location based learning experiences for people with intellectual disabilities and additional sensory impairments. *Computers & Education*, *56*(1), 11–20. doi:10.1016/j.compedu.2010.04.014

Brown, D. J., Standen, P., Evett, L., & Battersby, S. (2010a). Designing serious games for people with dual diagnosis: learning disabilities and sensory impairments. In P. Zemliansky, & D. M. Wilcox (Eds.), *Design and implementation of educational games: Theoretical and practical perspectives*. Hershey, PA: IGI Global. doi:10.4018/978-1-61520-781-7.ch027

Casamassina, M. (2006). *Wii Sports review*. Retrieved from http://Wii™.ign.com/articles/745/745708p1.html

Evett, L., Battersby, S., Ridley, A., & Brown, D. (2009). An interface to virtual environments for people who are blind using Wii technology – mental models and navigation. *Journal of Assistive Technologies*, *3*, 26–34. doi:10.1108/17549450200900013

Fit, V. I. (2010). *VI Fit, exergames for users who are visually impaired or blind.* Retrieved from http://vifit.org

Forrester Research, Inc. (2004). *Accessible technology in computing: Examining awareness, use, and future potential* (pp. 22–41). Cambridge, MA: Forrester Research, Inc.

Games, G. M. A. (2008). *Lone wolf for Windows – A submarine simulation, GMA games.* Retrieved from http://www.gmagames.com/lonewolf.shtml

Grammenos, D., & Savidis, A. (2006). *Unified design of universally accessible games (say what?).* Retrieved from http://www.gamasutra.com/features/20061207/grammenos_01.shtml

Grammenos, D., Savidis, A., & Stephanidis, C. (2005, July). UA-Chess: A universally accessible board game. In *Proceedings of the 3rd International Conference on Universal Access in Human-Computer Interaction*, Las Vegas, NV.

Higginbotham, A. (2010). *Dragging accessible games into the 21st century, BBC – Ouch! (disability).* Retrieved from http://www.bbc.co.uk/ouch/features/dragging_accessible_computer_games_into_.shtml

Horton, S. (2006). *Access by design: A guide to universal usability for web designers.* Berkeley, CA: New Riders.

Microsoft Corporation. (2009). *Engineering software for accessibility*. Redmond, WA: Microsoft Press.

Morelli, T., Foley, J., Columna, L., Lieberman, L., & Folmer, E. (2010, June). VI-Tennis: A vibrotactile/audio exergame for players who are visually impaired. In *Proceedings of the Foundations of Digital Interactive Games Conference*, Monterey, CA (pp. 147-154).

Nintendo. (2007). *Supplementary information about earnings release.* Retrieved from http://www.nintendo.co.jp/ir/pdf/2010/100507e.pdf#page=6

Nintendo. (2009). *Wii Sports Resort.* Retrieved http://Wiisportsresort.com/en/#/home

Nintendo. (2010). *Wii Sports.* Retrieved from http://www.nintendo.co.uk/NOE/en_GB/games/Wii/Wii_sports_2781.html

PopCap Games. (2008). Survey: *'Disabled gamers' comprise 20% of casual video games audience.* Retrieved from http://popcap.mediaroom.com/index.php?s=43&item=30

Sinclair, B. (2009). *Visually impaired gamer sues Sony Online.* Retrieved from http://uk.gamespot.com/news/6239339.html

Westin, T. (2004). Game accessibility case study: Terraformers – a real-time 3D graphic game. In *Proceedings of the 5th International Conference on Disability, Virtual Reality & Associates Technology*, Oxford, UK (pp. 95-100).

KEY TERMS AND DEFINITIONS

Design-For-All: Design-For-All is an approach that acknowledges the wide variability in peoples abilities, and, rather than designing for the "average" person, advocates designing for accessibility and usability with the aim being to "provide for diversity through design, which facilitates the development of accessible games.

Disability: A physical or mental impairment that has a substantial and long-term adverse effect on one's ability to carry out normal day-to-day activities.

Human Computer Interaction: In Human Computer Interaction is emerging as a prominent approach to system design, which acknowledges the wide variability in peoples abilities which has accessibility and usability in mind with the aim being to provide for diversity through design rather than accommodation.

Serious Games: E-Learning or computer based training environments that have been transformed through gamification to make them more engaging and effective for learning.

Terraformers: Terraformers is a game playable by players who are blind, who can play the game against sighted opponents Terraformers offers auditory navigation and game information, which blind players can use successfully to navigate its virtual environment and to play the game.

University Usability: University Usability is an approach to user-centered deisgn that assits in developing games that are designed for a diverse target audience.

This work was previously published in Developments in Current Game-Based Learning Design and Deployment, edited by Patrick Felicia, pp. 136-143, copyright 2013 by Information Science Reference (an imprint of IGI Global) and in the International Journal of Game-Based Learning, Volume 1, Issue 4, edited by Patrick Felicia, pp. 11-19, copyright 2011 by IGI Publishing (an imprint of IGI Global).

Chapter 7
ExerLearning®: Movement, Fitness, Technology, and Learning

Judy Shasek
ExerLearning

ABSTRACT

ExerLearning® provides parents, educators and others with a solid background of the direct connection between regular, rhythmic aerobic activity, balance, eye-foot coordination and academic success. We can increase students' fitness while simultaneously increasing their academic success. Activity breaks have been shown to improve cognitive performance and promote on-task classroom behavior. Today's exergame and related computer technology can seamlessly deliver activity without over-burdening busy teachers in grades K-12. Activity isn't optional for humans, and our brain, along with its ability to learn and function at its best, isn't a separate "thing" perched in our heads. The wiring, the circulation, the connection between mind and body is very real. The brain is made up of one hundred billion neurons that chat with one another by way of hundreds of different chemicals. Physical activity can enhance the availability and delivery of those chemicals. Harnessing technology to that activity is the ExerLearning solution.

READING, WRITING AND EXERGAMES?

Our lifestyles have become ever more sedentary with screens – television, computer and video game – being used for leisure, entertainment, communication, information and a pervasive social-networking culture. With only so much time in the school day it's tough to fit in physical activity, balance training and fitness to counter "screen" time. Research has been done by many independent and university researchers over the past decade. There is adequate evidence on the positive benefits of physical fitness on academic

DOI: 10.4018/978-1-4666-5071-8.ch007

success and cognitive skills. We developed specific ExerLearning strategies that can tap into computer and game technology to easily become an integral part of the school environment.

What is ExerLearning? It is a technology-delivered intervention that interrupts the sedentary practice of learning and classroom procedure. ExerLearning challenges sitting, a desk and conventional computer input devices as the default – or the best – route to cognitive skill development and academic success. According to Dr. John Ratey, author of "Spark – The Revolutionary New Science of exercise and the Brain" says, "Darwin taught us that learning is the survival mechanism we use to adapt to constantly changing environments." ExerLearning adds physical activity known to enhance the workings of the brain to learning.

- ExerLearning can be delivered by technology
- ExerLearning can be led and managed by students
- Technology delivered solutions like exergames and computer peripherals that require standing, balance and rhythmic movement improve students' fitness while simultaneously increasing their academic success.
- ExerLearning meets the needs of the most challenged, challenging or diverse learners without requiring teachers to write additional lesson plans
- ExerLearning sessions should occur for approximately 10 minutes every few hours.

Over-scheduled teachers and over-scheduled school days beg the question, "When will we have time to add 15-45 minutes of daily – or even weekly – physical activity? Preparing healthy, active children for life has taken a backseat to preparing the K-12 student for standardized tests. Educators have been mandated to address both the fitness **and** the test-score issues, but they need help.

Key Concept: In order to add more activity to the learning environment we can tap into technology that can add exactly the sort of physical activity students need and deliver it simultaneously with core content. ExerLearning tackles fitness and academic goals simultaneously-while saving teachers' time – and Districts' money. ExerLearning engages the very students we target while harnessing computer technology to consistently deliver any time any where rhythmic, aerobic activity.

There is an endless list of factors that impact a student's academic achievement. Among those, maybe the one least understood is the impact of regular physical activity. Throughout the development of ExerLearning concepts and practices, research on the benefits of regular rhythmic, aerobic and balance activity has been explored. Today's brain-scanning tools and a sophisticated understanding of biochemistry have led researchers to realize that the mental effects of exercise are far more profound and complex than they once thought.

"Exercise optimizes the brain and the person for learning. It creates the right environment for all of our 100 billion nerve cells up there. Exercise promotes the growth of new brain cells more than anything else we know," says Dr. Ratey. Ratey cites studies showing that exercise promotes the growth of new cells in the hippocampus, an area in the brain associated with memory and learning.

That's just the beginning. This chapter will weave facts and findings from brain research, innovative PE programs, unique peripherals that tie technology to physical activity and solid academic research on learning and test success. Our goal is to provide a solid overview of ExerLearning's potential in K-12 learning environments. Why did we coin the term, ExerLearning? As we work with educators, students, parents and wellness/learning advocates in dozens of states and hundreds of school Districts, we have found that having a vocabulary to explain this ground-breaking work is very important.

ExerLearning allows educators to reach academic goals while integrating aerobic exercise, balance, eye-foot coordination and agility to the learning environment. Physical activity does not have to be restricted to the gym. We love PE and welcome all a school can deliver. Budgets and philosophies differ across the country and there is no denying that we are not meeting PE mandates even in the 44 states where they have been set. Until adequate budgets, schedules, time and resources are allocated so educators can offer students daily, regular physical activity sessions ExerLearning can be a technology-delivered solution. Educators can choose ExerLearning interventions that deliver students prepared to learn via cost and time-effective choices. There is a wide array of products and programs that can provide ExerLearning.

No matter what a school District's stance on standardized testing, it is time for most test-prep practices to get a boost. An extraordinary amount of time and budget dollars are invested in test-prep materials and practices that, sadly, leave many students in a constant of failure. Our current efforts at raising performance for the 30% most challenged and challenging of our students remains frustrating for all. We hope to provide the case for meeting them where their needs and learning style intersect:

- Provide movement during academic work right at the computer using educational software and learning games already integrated in a school's curriculum. Many exergames and FootGaming™ via the FootPOWR™ peripheral can do that.
- Decrease negative behaviors and increase attendance (28% on average – teacher reporting)
- Prepare the brain to learn via the very process and physical development the brain was meant to experience during cognitive tasks

How Can a Teacher Add Exerlearning to the Regular Classroom Environment?

Exergames are video games that use exertion-based interfaces to promote physical activity, fitness, and gross motor skill development. Wii FIT and Wii Sports, Dance Dance Revolution stations, balance balls instead of chairs, FootPOWR computer peripherals instead of the conventional mouse, game cycles and similar exergames can all deliver physical activity that our brains need. The unique aspect of these unusual learning tools is that they consistently deliver endless customizable, compelling, fun and multi-level activity via technology.

In a classroom, students want to play the exergames. The built-in "fun-factor" is useful for teachers. In a very short time students can be taught a summary of brain research connecting physical activity to many benefits deemed valuable by students. Once the students realize that the exergame is not simply a reward for work completed or a diversionary toy, they respect it in new ways.

The next step is for teachers to take the time to develop a set of expect behaviors around the use of the exergames. An effective strategy has been used through a student-led ExerLearning program, Generation FIT, in more than 200 schools over the past five years (http://www.generation-fit.com). The 30% most challenged or challenging students are trained to lead and manage peers in groups of 2-3 using the exergame available in class. Leadership, providing a valuable service, the actual impact of the exergame practice and the understanding of related brain research all serve to positively impact these targeted students.

The student mentors rotate to the exergame stations in a classroom as it suits the teacher. Ideally, the students will enjoy one or more 10-15 minute ExerLearning breaks throughout the school day. Before a particularly difficult assignment, at the completion of a contracted amount of work, or when frustration, stress or other negative emotion threatens to undermine a student's best learning an ExerLearning session can be assigned.

When exergames or unique activity-driven peripherals are added to the classroom the novelty may attract attention from non-participating students. Consistent adherence to expected behaviors is important. In most cases students who need a consequence for unacceptable behavior around the exergaming activity will change quickly when their ExerLearning break is cancelled and they are positioned at the end of the student rotation.

Why Add ExerLearning to the Regular Classroom Environment

Physical activity enhances a student's ability to pay attention. According to Charles Hillman and Darla Castelli (2009), professors of kinesiology and community health at the Neurocognitive Kinesiology Laboratory at Illinois, "physical activity may increase students' cognitive control -- or ability to pay attention -- and also result in better performance on academic achievement tests." The goal of the study was to see if a single acute bout of moderate exercise was beneficial for cognitive function in a period of time afterward.

For each of three testing criteria, researchers noted a positive outcome linking physical activity, attention and academic achievement. Study participants were 9-year-olds (eight girls, 12 boys) who performed a series of stimulus-discrimination tests known as flanker tasks, to assess their inhibitory control. Following the acute bout of walking, children performed better on the flanker task. They had a higher rate of accuracy, especially when the task was more difficult. Along with that behavioral effect there were changes in their event-related brain potentials (ERPs) – in these neuroelectric signals that are a covert measure of attentional resource allocation.

In an effort to see how performance on such tests relates to actual classroom learning, researchers next administered an academic achievement test. The test measured performance in three areas: reading, spelling and math. The researchers noted better test results following exercise. The following should bring smiles to the faces of hard-working teachers concerned about reading achievement and scores.

"When we assessed it, the effect was largest in reading comprehension," Hillman said. In fact, he said, "If you go by the guidelines set forth by the Wide Range Achievement Test, *the increase in reading comprehension following exercise equated to approximately a full grade level.*"

It's not easy for busy teachers to add physical activity required for this type of benefit. Harnessing technology to deliver easily quantified and measured activity data, we suggest ExerLearning. Computer and console delivered exergames can be a valuable learning tool, particularly the easy-to-implement FootPOWR computer peripheral. It makes good sense to harness software and computers already in the classroom for integrating physical activity into the curriculum.

Fit to Learn?

According to a CDC survey (2003), only 3.8 percent of elementary schools, 7.9 percent of middle schools and 2.1 percent of high schools provide daily physical education. A study published in the 2007 issue of Health Economics stated that daily P.E. for high school students declined from 41.6 percent in 1991 to 28.4 percent in 2003. (The survey did not have statistics for middle and elementary schools.)

- 22 percent of schools don't require kids to take any P.E.
- Nearly half -- 46 percent -- of high school students were not attending any P.E. classes when surveyed by the CDC.

In more than 44 states, mandates to provide more physical education are being flouted due to lack of time, space, or competition from academic requirements. Research indicates again and again that schools that don't offer enough P.E. are "cheating" children. *Active Living Research* (ALR) a national program of the Robert Wood Johnson Foundation in an article called, "Active Education" (2007) found that, although more than one-third of U.S. children and teens are considered overweight or obese, schools are increasingly replacing physical education with academic coursework in their push to improve standardized test scores.

However, the report points out that decreased P.E. time is not associated with improved academic performance. In fact, the report indicates:

- Children who are physically active tend to perform at higher levels in the classroom and on standardized tests.
- In addition, active students exhibit fewer behavior problems and better concentration skills.

According to Jerry Gabriel (2001), a "growing body of research [suggests] that physical activity is integral to keeping cognitive processes working on all valves" (p.3). Current educational practices were developed in the early 19th century well before any of the brain research existed. It's time for a change.

On average, the PE mandate requires 150 minutes per week for all elementary students and 225 for middle and high school students. The budgetary support for this mandate spans "not enough" to "next to nothing." Some schools have stretched the "physical activity" point so far that they count the time children use to move from class to class – calling it 30 minutes per day and their fulfillment of the mandate. The impetus for doing that is as understandable as it is pathetic. Sometimes there seems to be no other viable solution.

What are over-scheduled, busy teachers to do? Our reply is, empower tomorrow's fitness and tech-leaders today. Make use of the time, energy and expertise of your students now. Budgets are tight and such increases are not being funded or enforced as needed. Teachers are busy and the school day is already over-scheduled. Technology tools and unique peripherals like the FootPOWR pads can enhance computer technology while saving teachers time and delivering valuable physical activity to students. The concepts behind ExerLearning are ones that we can wrap our minds around. We can begin to connect the dots on strategies that make good sense for lots of students, especially the most challenged and challenging ones.

What if You Knew That Physical Activity Actually Enhanced Academic Success?

While the sedentary habits of our youth are conclusively adding to the overweight crisis, we argue that not including regular and daily physical activity breaks *also* short-changes the *productivity, focus and chance for academic success we're working so hard to deliver.* Consider this: *exercise and balance- activity practice has been shown to directly improve students' cognitive development and academic achievement.* This adds a new dimension to the rationale we hold when we define "what is learning?"

The greatest limitation to the addition of physical activity to the learning environment is over-scheduled and over busy teachers. School schedules are full and test-prep takes preference in terms of time and scheduling. The recent development of the FootPOWR pad, a computer peripheral that can do anything a mouse can do, add activity simultaneously with learning outcomes and seamlessly integrates into the classroom.

What's a FootPOWR Peripheral?

FootPOWR pads look like the dance mats conventionally used in dance video games. The similarity ends there. Added FootPOWR microcontrollers turn the dance mat into a computer peripheral that can do anything a mouse can do. You simply plug the pad into the computer USB port, stand on the pad and move your feet to move the cursor. Suddenly many existing educational software games and hundreds of other software can become physical activity and balance generating interactive tools for your students. By re-inventing the computer "controller" from a mouse or keyboard used while sitting, to a dance-mat type tool that requires rhythmic activity and balance we can positively and easily impact learning and the needs of diverse learners.

Evidence that physical activity affects the brain in ways that improve learning has been mounting from the fields of molecular, cognitive, behavioral, and systems neuroscience, psychology, and directly from field studies performed in schools. That information inspired the development of Generation FIT over five years and led to the development of the FootPOWR pad. Teachers need easy to use tools that deliver results. Schools need fitness tools that can be obtained via reading, math, Title I and technology budgets when PE budgets are not enough. Bridging budgets to deliver physical activity right on the classroom makes sense for both learning and health outcomes.

Imagine how well an educator can do what he or she does best – teaching and facilitating in a learning environment - when all students are *prepared to learn both mentally and physically.* ExerLearning breaks throughout the day can deliver exactly that. When students in the ExerLearning program called, Generation FIT, (http://www.generation-fit.com) lead and managed the use of exergames during the learning day we found that participants:

- Reduced absenteeism by 28%
- Increased leisure reading
- Reduced negative behaviors, frustration and improved mood
- Gained confidence and social/teamwork skills

Programs like Generation FIT and FootGaming which deliver physical activity for a healthy body *and mind* are crucial, especially for the most challenged learners. The mind-body connection is well-known and widely noted. Time and time again research has proven that children who are physically fit score higher on standardized math and English tests than do their less fit peers.

A study, led by Virginia R. Chomitz, PhD (2009), found a significant relationship between physical fitness and academic performance. She and her team examined the test scores of over 1,000 children enrolled in grades 4 to 8 for the 2004 to 2005 academic year. Looking at two sets of figures—the MCAS test (Massachusetts Comprehensive Assessment System test) and physical fitness tests— Chomitz and her fellow researchers found that the likelihood of passing the academic tests improved as the number of fitness tests increased, even when controlling for gender, race/ethnicity, and socio-economic status.

How Can Educators Inject More Sensorimotor Experience Into the Regular Classroom, in After School Programs and at Home?

In times of diminishing financial resources, educators must make hard choices.

- Do dance, recess, exercise breaks and physical education belong in the budget? Are they frills or fundamentals?
- When classroom teachers are over-scheduled and busier than ever is there time for movement, exercise and dance in the regular classroom learning environment?
- We need to invest in test prep strategies, reading interventions, behavior modification and meeting the needs of students with diverse learning styles?
- Students must acquire content and technology skills via increased time interacting with 'screens" – how can we add the physical activity and fitness factors they need to best succeed?

Physical Education, Daily Activity and Learning

As mentioned earlier in this chapter, an astonishingly high percent of K-12 American students do not participate in a daily physical education program. That is not the case at Naperville Central High School (NCHS) where PE4Life founder, Phil Lawler, and Paul Zientarsky have re-invented the process of preparing students to learn. According to Zientarsky (2007), "In our department, we create the brain cells. It's up to the other teachers to fill them up." Go ahead, read that again. It's ***powerful*** stuff. Research meets reality under the heroic leadership of Phil Lawler and Paul Zientarsky in the Naperville, IL school district.

Results and Outcomes

- Students are more focused and energized in class
- Increased confidence and engagement in the learning process
- Test scores improve
- Students and teachers have an understanding of the physiology, brain research and kinesiology behind the effects – and that made all the difference!

NCHS LRPE = MIND and BODY CONNECTION Learning Readiness Physical Education (LRPE) was designed based on research indicating that students who are physically active and fit are more academically alert and experience growth in brain cells or enhancement in brain development; NCHS pairs a PE class that incorporates cardiovascular exercise, core strength training, cross lateral movements and literacy and math strategies with literacy and math classes that utilizes movement to enhance learning and improve achievement.

LRPE students have experienced notable gains in their reading ability and comprehension as well as improvement in math and other courses. Starting their day with physical workouts seems to be "waking up" their brains. The study incorporated in this project is providing the justification for expanding the program so many more students can experience the improvements and achievements of the original group.

Naperville Central High has embraced the premise of ExerLearning with full support of Reading, Math and PE teachers, District budgets and parents. Such a comprehensive program did not happen overnight and every school may not be ready for such an investment just yet. FootGaming is one way to test the waters, so to speak. It could move your students toward the sort of gains made by the LRPE students in Naperville.

Research and recent "Learning Readiness PE" (LRPE) classes at Naperville Central High (IL) indicate that physical activity can impact student performance enough to elevate test scores. "We're putting kids in P.E. class prior to a classes that they struggle in and what we're doing is we're finding great, great results," said Paul Zientarski, who helps run LRPE program at Naperville Central High School. The program was started in response to research showing a link between exercise and increased brain function. He says that he has seen the results.

"Kids who took P.E. before they took the math class had *double the improvement* of kids who had P.E. afterward," Zientarski, explained. (P. Zientarski, personal communication, May, 2009). Naperville Central High School has embraced the idea that working out helps a child learn. There you can find exercise equipment, including FootPOWR peripherals, in some classrooms.

"Their bodies are moving and their brains are thinking and they're engaged - not sitting still trying to memorize something," said Maxyne Kozil, a reading teacher, who believes that kids learn best when they're moving. An example of this is having a student work on her vocabulary while standing on balance boards. "They say having to balance actually helps them to concentrate even better," Kozil said.

It is time for our test-prep practices to get a boost. Our current efforts at raising performance for the 30% most challenged and challenging of our students remains frustrating for all. We hope to provide the case for meeting these students where their needs and learning style intersect:

- Provide movement during academic work right at the computer using educational software, interactive peripherals that require physical activity and exergames
- Provide leadership and collaborative practice for students in need of confidence and social skills
- Decrease negative behaviors and increase attendance (28% on average-teacher reporting)
- Prepare the brain to learn via the very process the brain was meant to experience during cognitive tasks

Do Physical Activity and Physical Education Interfere With Academics?

In many states physical education is done away with because "academic classes" are considered more important. Importance being "you get what you measure" and we measure reading and math scores far more than fitness and wellness choices our students make. A two year study in San Diego that examined standardized test scores revealed that students having a physical education class outperformed those that did not (1992). Researchers have demonstrated that physical activity improves brain function, elevates mood, and promotes learning. Exercise improves blood flow to the brain and spurs cell growth, leading some to compare the brain to a muscle which performs best when the body exercises.

Teachers and administrators are working harder than ever to provide what each student needs to be a successful learner while making the most of skills, aptitudes and learning styles. A variety of funding sources for academics, including No Child Left Behind (NCLB) are being used to their maximum. But are we impacting the students who need it most by implementing strategies that shift the paradigm of what "test prep" and learning look like? Unless regular physical activity breaks are included in the learning environment we are shortchanging cognitive success, focus and productivity.

Activity Breaks Can Improve Cognitive Performance and Classroom Behavior

As referenced time and again in this chapter, many studies involving elementary students, regular physical activity breaks during the school day may enhance academic performance. Introducing physical activity has been shown to improve cognitive performance and promote on-task classroom behavior. During our field testing of Generation in Central Oregon elementary schools we found that students exhibited significantly more on-task behavior and significantly less fidgeting on days with a scheduled activity break than on non activity days. This isn't just a case of a distracting insertion of random "playtime." Short activity breaks during the school day can improve students' concentration skills and classroom behavior.

Judy Shasek, author of this article and developer of Generation FIT and FootGaming, discovered that students as young as age 8 are eager to learn exactly ***why*** regular aerobic activity and balance practice helps a brain to be more productive, to increase BDNF and neurotransmitters and to help them focus. During field studies, we used a simple PowerPoint slideshow called, "Brainy Stuff" (2009). With it and others created by students, hundreds of students have shared complicated brain research with peers and teachers. For many struggling students it is a major relief to understand why learning is often such a struggle and how they can do something that actually ***enjoy*** to increase their academic success (and often their behavior). Many challenged learners are relieved to discover that their lack of "sit still and focus" is linked to a real, physical and brain-driven need to move on a regular basis throughout the learning day. Once we harness technology to deliver that movement practice in an orderly and seamless manner everyone is happier

What Does the Research Say?

A cross-sectional study conducted in 2002 by the California Department of Education demonstrated a strong association between physical fitness and academic performance. Using the Fitnessgram®, a six-faceted measure of overall physical fitness, and students' grades on the SAT-9 state standardized test, nearly one million students in grades five, seven, and nine were evaluated. The consistent finding: not only did those with higher levels of physical fitness score higher on the SAT-9, but there was a positive linear relationship between the number of fitness standards achieved and academic achievement. This result held for boys and girls in both math and reading, but was most pronounced in math. The same study found:

- Ten minutes of rhythmic aerobic exercise before a cognitive task (like reading or math) resulted in better success at that task
- Students who did 10 minutes of rhythmic aerobics before a standardized test, did up to 25% better at that test than students who received 20 minutes of test-specific tutoring.
- School-age children who have a higher level of aerobic fitness processed information more efficiently

Even when time allocated to other subjects is reduced, shifting more curricular time toward physical activity does not negatively affect academic achievement. In one such study (2003), a reduction of two-hundred forty minutes per week of class time, replaced with increased PE, led to higher scores on standardized math examinations. As we review studies that assessed the association between physical activity and academic outcomes among school-aged children, we conclude that there is evidence to suggest that short term cognitive benefits of physical activity during the school day adequately compensate for time spent away from other academic areas" According to Dr. Debbye Turner Bell (2009), researchers are finding that exercise can do more than keep you fit; it can also make you smarter.

"Exercise in many ways optimizes your brain to learn," said Dr. John Ratey (2008), a clinical associate professor of psychiatry at Harvard Medical School in Boston, author of "Spark." Exercise improves circulation throughout the body, including the brain, Ratey explains. Exercise also boosts metabolism, decreases stress and improves mood and attention, all of which help the brain perform better, he said.

"The brain cells actually become more resilient and more pliable and are more ready to link up," he says. It's this linking up that allows us to retain new information. Much of the research on the specific effects of exercise on neurons has been done in the lab. But studies in people also are backing physical activity as a way to keep the brain healthy and our minds sharp.

The research is profoundly relevant to today's health and fitness crisis. At the same time the fitness-overweight dilemma is growing, funding for programs like physical education in schools are being reduced or eliminated. Kids are spending an average of 5.5 hours a day in front of a screen of some sort.

With ExerLearning programs integrated right in classrooms and computer labs, such activity is easy to integrate into the regular school day, not only in structured Physical Education classes. When neurons fire together they "wire together." If we already don't include words like glutamate, serotonin, norepinephrine and dopamine when we talk about learning, hopefully we soon will. ExerLearning students as young as age eight easily banter about, "increasing neurotransmitters," "sending oxygen to your brain," and BDNF as "Miracle Gro" for the brain as reasons they lead their peers in regular sessions of FootGaming and other "exergames." The very students who struggle to read at the second grade level in sixth grade, most readily absorb the ExerLearning philosophy and practice. ExerLearning programs aim to deliver many of those important outcomes for teachers and schools with limited time or budget – right now.

Mandated PE Minutes

As we mentioned earlier in this chapter, more than 44 states have mandated 150 minutes of weekly physical activity for grades K-5 and 225 minutes weekly for grades 6-12. In almost every case, no additional budget, teachers or time is added to the school program. What's a busy teacher to do when presented with this challenge? Schedules are already full and teachers are already over-extended with responsibilities and measurable outcomes in the "academic" areas. We propose that ExerLearning can help support the mandates while providing exactly the sort of physical activities kinesthetic learners and the 30% most challenged students need.

The very students who wiggle, act out, collect repeated absences and are consistently disengaged, perform best with more technology. These kinesthetic learners seem married to their "screens" and eagerly spend *upwards of three hours a day playing video games.* These same very challenging students could provide the solution to the over-busy teacher and not enough physical activity in the learning environment. Technology that can deliver physical activity at the same time students spend working on reading and math software in a computer lab is readily available. Not only can students gain valuable "learning time" by moving as they learn – they will gain a good measure of the mandated activity minutes.

Much research has been done to quantify the physical activity afforded by such an intervention – at little more cost that a high quality mouse. For example, consider this study, "Energy Expenditure of Sedentary Screen Time Compared With Active Screen Time for Children"(2006).

SUMMARY: The team examined the effect of activity-enhancing screen devices on children's energy expenditure compared with performing the same activities while seated. Their hypothesis was that energy expenditure would be significantly greater when children played activity-promoting video games, compared with sedentary video games. Energy expenditure was measured for 25 children aged 8 to 12 years, 15 of whom were lean, while they were watching television seated, playing a traditional video game seated, watching television while walking on a treadmill at 1.5 miles per hour, and playing activity-promoting video games.

They found that energy expenditure more than doubles when *sedentary screen time is converted to active screen time.* Sitting in front of a television, video game, or computer screen has been associated consistently with low levels of physical activity. Weekly screen time for children is as high as 55 hours/week, and the average home in the United States has a television on for 6 hours per day. Although many programs have attempted to separate children from the screen, these activities are highly valued and children are resistant to relinquishing them. An alternative approach is to examine whether sedentary screen time can be converted into active screen time. This is exactly what ExerLearning via FootGaming was created to do – and to do right in the home or classroom.

CONCLUSIONS. Energy expenditure more than doubles when sedentary screen time is converted to active screen time. This is the strategy upon which ExerLearning is built.

ExerLearning helps teachers by harnessing technology to the *physical activity and balance practice need for optimal brain function*. Walking on two feet is very difficult. It takes a lot of balance, coordination, synchronization and timing of muscles. It takes a tremendous amount of motor control to be able to do that. It takes constant output from the brain and constant feedback to the brain. When we are in an upright position these receptors constantly fire back to the brain. They stimulate the brain. *It appears that as humans stood in an upright position or as we became more and more upright, our brain grew larger and larger in response to this constant stimulus of gravity.* Our goal in ExerLearning and its various intervention games and tools is to increase the amount of time students spend standing, moving and balancing under the influence of gravity *while they are learning.*

Decreased stimulation from postural muscles to cerebellum and brain, anything that takes us away from standing and being upright, will affect our brain in an adverse way. It will slow down the temporal processing speed of the brain, or parts of the brain, with resulting "clumsiness" and cognitive developmental delays. ExerLearning's foundation is solidly built upon the need to get students out of the desk or chair and working in an upright, standing position. Simply standing is not very inviting to most of us. Technology allows teachers add the playful fun of "exergames" accessed by controllers that require standing, movement and balance.

Balance – The ExerLearning Bonus Benefit

Everyone knows the five basic senses; seeing, hearing, taste, smell and touch. But there are other senses that are not as familiar including the sense of **movement (vestibular), and sense of muscle awareness (proprioception**). Unorganized sensory input creates a traffic jam in our brain making it difficult to pay attention and learn. To be successful learners, our senses must work together in an organized manner. This is known as **sensory integration**.

The foundation for sensory integration is the organization of tactile, **proprioceptive and vestibular input.** A person diagnosed with ADD or ADHD, due to their difficulty paying attention, may in fact have an immature nervous system causing sensory integration dysfunction. This makes it difficult for him/her to filter out nonessential information, background noises or visual distraction and focus on what is essential. There is a direct relationship between sensory integration, learning and attention.

ExerLearning technology tools, consisting of exergames and activity driven computer peripherals like the FootPOWR pad, provide the development of the vestibular sense. By providing technology delivered activities that provide balance practice students become prepared to learn. **The vestibular sense is important for development of balance, coordination, eye control, attention, being secure with movement, emotional security and some aspects of language development**. Disorganized processing of vestibular input may be seen when someone has difficulty with attention, coordination, following directions, reading (keeping eyes focused on the page or board) or eye-hand coordination.

Ironically, the cerebellum, an area of the brain most commonly linked to movement turns out to be a virtual switchboard of cognitive activity. The first evidence of a linkage between mind and body originated decades ago with Henrietta Leiner and Alan Leiner (1997), two Stanford University neuroscientists. Their research began what would eventually redraw "the cognitive map"

The Leiners' work centered on the cerebellum, and they made some critical discoveries that spurred years of subsequent research. First, the cerebellum takes up just one-tenth of the brain by volume. But it contains over half of all its neurons. It has some 40 million nerve fibers, 40 times more than even the highly complex optical tract. Those fibers not only feed information from the cortex to the cerebellum, but they feed them back to the cortex. If this was only for motor function, why

Table 1.

What does ExerLearning™ Do?	How does that happen?
ExerLearning opens up a direct channel to the brain/mind-	The mind becomes a sponge: absorption, processing, integration, retention, cognition (i.e. LEARNING) all improve.
ExerLearning gets the brain pumping	By getting the heart pumping.
The brain is muscle that can be developed through physical training	Just like the heart the brain can be strengthened via physical training.
ExerLearning adds physical movement	Utilizing the mind-body connection
ExerLearning optimizes the learning environment for diverse learners	Harnesses technology to deliver rhythmic physical activity and valuable fitness factors to the learning process
ExerLearning allows neurogenesis: growth of new brain cells	Exercise enables more blood and proteins to enter the brain

are the connections so powerfully distributed in both directions to all areas of the brain? In other words, this subsection of the brain -- long known for its role in posture, coordination, balance, and movement -- ***may be the ExerLearning hub.***

Students who tip back on two legs of their chairs in class often are stimulating their brain with a rocking, vestibular-activating motion. While it's an unsafe activity, it happens to be good for the brain. We ought to give students more activities that let them move safely while practicing balance skills. Busy teachers may have difficulty planning interventions such as those but when technology and exergames are selected for classroom use students get the balance practice they need. Such interventions, ExerLearning at its best, can change the world of learning for struggling students.

In one field study using dance mat video games with a fourth grade class, an autistic student whose entire left side was affected by cerebral palsy, participated as a program mentor for 10-20 minutes per day using the dance mat changed his balance, coordination, social engagement and enthusiasm for PE. Studies done by neuroscientist Eric Courchesne (1995) of the University of California have shown that autistic children have smaller cerebellums and fewer cerebellar neurons. Courchesne says the cerebellum filters and integrates floods of incoming data in sophisticated ways that allow for complex decision making. Once again, the part of the brain known to control movement is involved in learning. Movement and learning have constant interplay.

Some of the decline in physical activity is due to schools' implementation of strategies designed to improve achievement outcomes. But the theory that spending more time learning academics in the classroom will lead to higher test scores and grades has not been proven. The more brain research is explored the more crucial physical activity proves to be for cognitive tasks. In other words, allotting too little time to physical activity may undermine the goal of better performance, while adding time for physical activity may support improved academic performance.

Children who participated in a Generation FIT-ExerLearning *peer mentoring program* were absent 25% fewer days than the control group.

This is a key measure and critical for decision makers in schools at local and District levels. Absenteeism costs Districts $9-$20 per student per day. In a field study done during the 2004-2005 school year at Vern Patrick Elementary (Redmond, OR), fourth graders who used Generation FIT ExerLearning peer mentoring program were absent fewer days, even during flu season, than they had been in the Fall quarter before the program was begun. They were absent 25% fewer days than other fourth graders not participating in increased daily physical activity. This caught the attention of teachers (more time in class meant more time to make an impact on the student) and District budget staff.

Health is not the only reason children miss school. The most challenged and most disengaged students find numerous and creative ways to be absent. We discovered that many students that fit such a profile made the most improvement in both attendance and engagement in the learning process after being trained as Generation FIT ExerLearning peer mentors and leaders. **Ask any teacher the ramifications of these two changes on the lowest performing students.**

With Districts budgets already tight it would pay to create a preliminary estimate of the potential impact of physical inactivity and related health factors on school funding. The *Executive Summary: Healthy Children, Healthy Schools* predicts the loss in large cities could be $28 million in New York, Chicago could forfeit $9 million and Los Angeles an estimated $15 million. So, obviously, we want our children to get and stay active!

In nine states (California, Idaho, Illinois, Kentucky, Mississippi, Missouri, new York, Tennessee and Texas) collectively serving more than one-third of all students in the US, state funding for schools is determined using the Average Daily Attendance (ADA) methodology. In other words, public education dollars in these states are determined not by how many students are enrolled, but by how many actually show up at school. Student absenteeism can therefore have a negative impact on the school's bottom line. Data from The Finance Project, a nonprofit policy research and technical assistance group, demonstrate how absenteeism can be a significant problem for school budgets. These data suggest that a single-day absence by one student costs a school district in these states anywhere between $9 and $20.

While these figures seem small, they add up quickly. An estimated 16 percent of youth are overweight to a degree that affects their health. One study found that severely overweight students miss (using the median number) one day per month or nine days per year.[1] This type of absentee rate among overweight students in a student population with average prevalence of overweight could lead to a potential loss of state aid of $95,000 per year in an average size school district in Texas, and $160,000 per year in an average California school district.

In the Vern Patrick Elementary study, additional reasons for decreased absenteeism emerged from anecdotal reporting from both students and their parents. Student mentors who managed the day to day operation of the Generation FIT ExerLearning program gained ownership of the technology-delivered game activity. Their foray into ExerLearning included a leadership/peer mentoring piece that changed their attitudes about school attendance and their engagement in the learning process when they were at school. Increased daily attendance by students who were among the most challenged learners were part of the dramatic 25 percent improvement in attendance over the quarter prior to their program participation. Parents reported that students refused to miss school for any reason on the days they were scheduled to use the dance mats and mentor their peer-team.

A Summary- What Happens When We Exercise?

When humans exercise, the body-brain goes into a homeostatic state, balancing brain chemicals, hormones, electricity, and system functions. When the body-brain is out of balance because of poor nutrition and lack of physical activity, the student is not in a good learning state. Movement, physical activity, and exercise change the learning state into one appropriate for retention and retrieval of memory, the effects lasting as much as 30-60 minutes depending on the student. Studies show that just 10 minutes of rhythmic aerobic activity prior to a cognitive task improves academic success.

Physical Activity Provides Enriched Environments

Physical activity in a positive social setting creates an environment conducive for learning.

Being Active Grows New Brain Cells

Aerobic activity releases endorphins, the class of neurotransmitters that relax us into a state of cortical alertness. Exercise also tends to raise levels of glucose, serotonin, epinephrine, and dopamine, chemicals that are known to balance behavior.

Aerobic Fitness Aids Cognition

Researchers found that subjects who were the most aerobically fit had the fastest cognitive responses, measured by reaction time, the speed that subjects processed information, memory span, and problem solving.

Exercise Triggers BDNF

Exercise triggers the release of BDNF a brain-derived neurotropic factor that enables one neuron to communicate with another. (Kinoshita 1997) Students who sit for longer than twenty minutes experience a decrease in the flow of BDNF. Recess and physical education is one way students can trigger sharper learning skills.

Cross Lateral Movement Organizes Brain Functions

Crossing the midline integrates brain hemispheres to enable the brain to organize itself. When students perform cross lateral activities, like dance, sport and most play, blood flow is increased in all parts of the brain making it more alert and energized for stronger, more cohesive learning.

Eye Tracking Exercises and Peripheral Vision Development Helps Reading

One of the reasons students have trouble with reading is because of the lack of eye fitness. When students watch screens their eyes lock in constant distant vision and the muscles that control eye movement atrophy. In video games that provide screens with ever changing patterns and whole-body response to those screens, as in Red Octane's "In the Groove" dance games, eye tracking and expectation skills, peripheral vision are all improved

Balance Improves Reading Capacity

The vestibular and cerebellum systems (inner ear and motor activity) are the first systems to mature. These two systems work closely with the RAS system (reticular activation system) that is located at the top of the brain stem and is critical to our attentional system. These systems interact to keep our balance, turn thinking into action, and coordinate moves. Games and activities that stimulate inner ear motion like Red Octane's "*In The Groove*," are useful in laying the foundation for learning.

Exercise Reduces Stress

Movement can foster self-discipline, improve self-esteem, increase creativity, and enhance emotional expression through social games like FootGaming.

Movement Can Help Reinforce Academic Skills For All Students.

Eighty five percent of school age children are natural kinesthetic learners (Hannaford). Sensory motor learning is innate in humans. Teachers who incorporate kinesthetic teaching strategies reach a greater percentage of the learners. Kinesthetic learners do best while touching and moving. Kinesthetic learners tend to lose concentration if there is little or no external stimulation or movement. To integrate this style into the learning environment educators integrate creative strategies like:

- Using activities that get students up and moving
- Use activities that include music or rhythm
- Give frequent brain breaks that include activity and moving

REFERENCES

Action for Health Kids. (2003). Retrieved from http://actionforhealthykids.org

Courchesne, E. (1995, February). *An MRI study of autism: The cerebellum revisited. Journal of Autism and Developmental Disorders,* 25(1), 19–22. PubMed doi:10.1007/BF02178164

Dietz W.H., Bandini L.G., Morelli J.A., Peers K.F., Ching P.L. Effect of sedentary activities on resting metabolic rate. *American Journal of Clinical Nutition, 59,* 556–559.

Generation Fit. (2007). [Video File]. Video posted to http://www.generation-fit.com

Lanningham-Foster, L., Jensen, T., Foster, R. C., & Redmond, A. B. (2006, December). Energy expenditure of sedentary screen time compared with active screen time for children. *Pediatrics, 118*(6). PubMed doi:10.1542/peds.2006-1087

Leiner, A. C., Leiner, H., & Noback, C. R. (1997). *Cerebellar Communications with the Prefrontal Cortex: Their Effect on Human Cognitive Skills.* Palo Alto, CA: Channing House.

Maloney, A. (2007). *Generation-Fit, a Pilot Study of Youth in Maine Middle Schools Using an "Exerlearning" Dance Video Game to Promote Physical Activity During School.* Retrieved from http://clinicaltrials.gov/ct2/show/NCT00424918

Ratey, J. (2008). *Spark: The revolutionary new science of exercise and the brain.* New York: Little, Brown and Company.

Sallis, J. F., Hovell, M. F., Hofstetter, C. R., & Barrington, E. (1992). Explanation of vigorous physical activity during two years using social learning variables. *Social Science & Medicine, 34,* 25–32. doi:10.1016/0277-9536(92)90063-V PMID:1738853

Schwimmer, J. B., Burwinkle, T. M., & Varni, J. W. (2003, April 9). Health-related Quality of Life of Severely Obese Children and Adolescents. *Journal of the American Medical Association,* 289. PMID:12684360

Shasek, J. (2009). *Brainy Stuff* [PowerPoint slides]. Retrieved from http://www.slideshare.net/inven-TEAM/brainy-stuff-1078097

Shepard, R. J. (1997). Curricular Physical Activity and Academic Performance. *Pediatric Exercise Science.*

Turner Bell, D. (2009, March). *Exercise Gives The Brain A Workout, Too.* Retrieved from http://www.cbsnews.com/stories/2009/01/30/earlyshow/health/main4764523.shtml

KEY TERMS AND DEFINITIONS

Cerebellum: The cerebellum is an area of the brain most commonly linked to movement is like a virtual switchboard of cognitive activity. The cerebellum takes up just one-tenth of the brain by volume. But it contains over half of all its neurons. It has some 40 million nerve fibers, 40 times more than even the highly complex optical tract.

Exercise: Exercise optimizes the brain and the person for learning, creating the right environment for all of our 100 billion nerve cells up there. Exercise promotes the growth of new brain cells more than anything else we know.

Exergames: Exergames are video games that use exertion-based interfaces to promote physical activity, fitness, and gross motor skill development.

ExerLearning: ExerLearning is a technology delivered intervention that interrupts the sedentary practice of learning and classroom procedure. It challenges sitting, a desk and conventional computer input devices as the default – or the best – route to cognitive skill development and academic success.

Sensory Integration: Sensory integration is the principle that to be successful learners need to use all of their senses together in order to avoid unorganised sensory input, which would otherwise create a traffic jam in the brain.

This work was previously published in Design and Implementation of Educational Games: Theoretical and Practical Perspectives, edited by Pavel Zemliansky and Diane Wilcox, pp. 409-423, copyright 2010 by Information Science Reference (an imprint of IGI Global).

Chapter 8
Rehabilitation Gaming

Henk Herman Nap
Stichting Smart Homes, The Netherlands

Unai Diaz-Orueta
INGEMA, Spain

ABSTRACT

A recent innovation in rehabilitation is the use of serious gaming to train motor, cognitive, and social abilities. The main advantages of rehabilitation gaming are related to the motivation to engage in rehabilitation, the objectivity of rehabilitation measurements, and the personalization of the treatment. This chapter focuses on the use and effectiveness of serious gaming in rehabilitation and illustrates the possibilities and strengths in this new and exciting work field. Furthermore, a review of the literature and examples of rehabilitation games are presented. The state-of-the-art technologies and directions for future research are also discussed. Rehabilitation gaming has great potential for today's and future health care, and despite the research gaps, there is increasing evidence that gaming can positively contribute to the rehabilitation and recovery process.

INTRODUCTION

Until recently, both the media and scientists focused mainly on the negative consequences of digital gaming like aggressive behaviour (Ferguson, 2007; Anderson & Bushman, 2001). Fortunately, the tide has turned, and the focus has shifted to the positive effects of digital game play and the powerful, persuasive, and motivating elements of digital games are-aside for entertainment purposes-used for the better: training, learning, and skill acquisition. Digital games that are specifically developed for these purposes are called serious games (Boyle, Connolly, & Hainey, 2011). Serious games have an explicit and carefully thought-out educational purpose and are not intended to be played primarily for amusement (Abt, 1970). Serious games have been recognized and are employed in various fields like the military and education, but have found increasing interest from the health domain,

DOI: 10.4018/978-1-4666-5071-8.ch008

particularly in rehabilitation, partially due to the rise of low-cost embodied gaming.

The benefits of rehabilitation can be translated into a higher quality of life for both patients and their families. In addition, rehabilitation can result in lower costs for additional health care and higher productivity as patients may return to the workforce much faster. Furthermore, health care innovations that enhance rehabilitation could increase the benefits even more. One of the latest innovations in rehabilitation is the use of serious games for cognitive, psychological, motoric, and social rehabilitation. Rehabilitation gaming is a form of mediated rehabilitation, similar to telerehabilitation which is mediated by videophone (see, Popescu, Burdea, Bouzit, & Hentz, 2000) and rehabilitation mediated by Virtual Reality (VR) (see, Difede & Hoffman, 2002; Ready, Gerardi, Backscheider, Mascaro, & Rothbaum, 2010). The main advantages of mediated rehabilitation compared to traditional rehabilitation, in particular game-based rehabilitation, are related to the *motivation* to engage in rehabilitation, the *objectivity* of rehabilitation measurements, and the *personalization* of the treatment.

There is little doubt that digital games are highly *motivating* to play, because of the interactivity and feedback mechanisms that can increase the player's self-efficacy and mastery. Self-efficacy is the belief in one's ability to succeed in specific situations and it determines whether coping behaviour will be initiated, how much effort will be expended, and how long it will be sustained in the face of obstacles and aversive experiences (Bandura, 1977). In addition to enhanced self-efficacy, players can reach a state of optimal experience, which is called 'flow' (Csikszentmihalyi, 1975; Csikszentmihalyi & Csikszentmihalyi, 1988). When players are fully engaged with the task at hand, actions are performed automatically and an optimal balance between skills and challenge is reached. Digital games in rehabilitation can motivate players to continue their training activity and advance in their skills while being 'in the zone.' In contrast, conventional rehabilitation can be a tedious exercise by performing the same movements over and over again and even highly motivated patients and therapists can become unmotivated and tired after numerous repetitions of the same movement. Rehabilitation that is mediated by means of technology can track and translate small unnoticeable advancements in the clients' recovery process into observable (in-game) progresses like scores, bonuses, and level advancements. A rehabilitation game can provide a client with positive feedback when most appropriate, which should preferably be provided on learning goals rather than performance goals (Dweck, 1986). In addition to game related advancements, rehabilitation games, controlled by e.g., a haptic feedback glove like the Rutgers II (see, Popescu et al., 2000), can monitor *objectively* over time the number of repetitions, strength levels, and extension distance. Therapists are highly trained in these measurements; however, the measurements can deviate between therapists and/or may sometimes be obstructed from view. Objective precise measurements and in-game scores can be used as input for the game dynamics and feedback mechanisms. Most game worlds can already be easily adapted by changing the scenery, complexity, avatar, controls, etc. A patient or therapist can use these variables to easily *personalize* the game and make it compatible with the clients' abilities and needs. Not only the patient or therapist can adapt the rehabilitation game, the game itself can also automatically adapt to the clients' progress. For example, decreasing in-game scores which reflect a decline in the rehabilitation phase can be used to adapt the difficulty level and/or in-game speed to lower values to continue progress and possibly increase the clients' motivation to carry on rehabilitating. The whole virtual environment, including the characters, scenery, and even the storyline can be adapted to the player's specific needs and abilities. The flexibility of virtual environments provides a considerable advantage compared to conventional rehabilitation.

Rehabilitation gaming has great potential for today's and future health care. This chapter will focus on the use and effectiveness of serious gaming in rehabilitation. The chapter is not to be exhaustive, but illustrates the possibilities and strengths in this new and exciting work field. Furthermore, the chapter provides a review of the literature, examples of rehabilitation games, technologies in rehabilitation gaming, and directions for future research.

REHABILITATION: FOCUS AREAS

In the following sections, we present the research on rehabilitation gaming in the areas of cognitive, psychological, physical, and social rehabilitation.

Cognitive Rehabilitation: The Role of Digital Games in Cognitive Function

In the latest years, the effectiveness of brain training games, like Nintendo's Brain Training by Dr. Kawashima or Big Brain Academy, and the presence of the 'use it or lose it' hypothesis in our daily life has been assumed as a truth or 'folk psychology.' Beyond the evidence of cognitive training interventions such as the one implemented in the ACTIVE study (Willis et al., 2006), computerized cognitive training and more specifically, the research on the area of digital games targeting cognitive functions is still in its early stages (see also, Salthouse, 2006). Nevertheless, we would like to stress that stimulation and training goes beyond the notion of 'use it or lose it,' which according to Goldberg (2005) should rather be rephrased to 'use it and get more of it.' Digital games that try to provide stimulation and training for the aging or damaged brain should broaden and challenge the players' knowledge and skills.

Digital Games for the Improvement of Attention

In a work of 2003, Green and Bavelier demonstrated that digital game play enhances the overall capacity of the attentional system (the number of items that can be attended), the ability to effectively deploy attention over space, and the temporal resolution of attention (the efficiency with which attention acts over time). Expert action digital game players were found to outperform non-gamers on tasks measuring the spatial distribution and resolution of visual attention, the efficiency of visual attention over time, and the number of objects that can be attended simultaneously. On a similar work, Castel, Pratt, and Drummond (2005) examined the similarities and differences between digital game players and non players in terms of the ability to inhibit attention from returning to previously attended locations, and the efficiency of visual search in easy and more demanding search environments. People who played digital games showed overall faster response time to detect targets, and overall faster response time for easy and difficult visual search tasks compared to non players, which can be attributed to faster stimulus-response mapping. It is likely that these findings may have implications to real life situations. People may benefit from digital game play in their experience in daily life, e.g., in complex and attention demanding tasks like driving in a car or way finding in an airport. As Green and Bavelier (2006) state, players seem to show a reduced cost of divided attention and thereby could outperform non players at detecting items in the periphery, like a child chasing a ball toward the street while driving. Digital game play has the potential to enhance attentional processes, which are highly relevant in successfully performing the Activities of Daily Life (ADL).

Spatial Ability

In visual tasks, spatial ability is the ability to estimate, judge, or predict the relations among figures or objects in different contexts (Elliot & Smith, 1983). Digital game play-in particular first-person shooters (FPS)-can have a positive effect on this ability. Feng, Spence, and Pratt (2007) studied the relation between digital game play and spatial ability and varied the genre of the game in an experiment. The experimental group in their study played a three-dimensional (3D) FPS game called 'Medal of Honor-Pacific Assault' (Electronic Arts) during a training session of 10 hours. The control group played a non-action game called 'Ballance' for the same amount of time. Ballance is a 3D puzzle game in which the player has to steer a ball through a hovering maze of paths and rails with various obstacles. It was found that the training with the FPS action game resulted in substantial enhancements in spatial attention and mental rotation. Females benefited more than men, as such that prior gender differences were reduced or eliminated. On a later work, Barlett, Anderson, and Swing (2009) stated that the ability to mentally rotate or arrange objects in space is related to a number of general learning tests and paradigms. According to the authors, research has shown that digital game play is related to this spatial ability (Barlett et al, 2009). Interestingly, research from Ferguson (2010) showed that playing action games has a more positive effect on visuospatial cognition than playing Tetris, developed by Alexey Pajitnov, which was primarily designed for mental rotation tasks. It is suggested that the findings are probably due to the fast action commonly found in a FPS game (Ferguson, 2010).

Digital Games for Reasoning

Basak, Boot, Voss and Kramer (2008) recently studied the effect of training on a real-time strategy game (Rise of Nations, by BigHuge Games, & Microsoft Game Studios) on older adults' performance in a wide-range of executive control tasks, such as short-term memory, working memory, task switching, and inhibition. Significant benefits were found after 23.5 hours of strategy-based gaming on executive control functions following training. The well performed study was one of few that showed enhancements of cognitive functioning of older adults by playing digital games. The results from a recent study of Owen et al. (2010), published in Nature, suggest that improved cognitive functioning by brain training games only hold for the cognitive tasks that are trained. They found no transfer to more general tests of cognition, even when those tasks were closely related to the trained task. Owen et al. (2010) used a large sample of 11,430 participants ranging from 18 to 60 years of age, yet, no seniors were included who could potentially benefit the most from playing brain training games.

Improving Visual Memory with Digital Games

Ferguson, Cruz, and Rueda (2008) examined whether visuospatial recall of abstract and common objects was related to gender, and whether there was influence of experience with digital game play on their visual memory recall performance. In the study, both the previous exposure to digital games and the exposure to violent digital games were included. The total time spent playing, as well as exposure to violent digital games, predicted increased visual memory recall performance, leaving the door open for the study of positive effects of games with violent-related content (e.g., Re-Mission game for cancer education, Kato et al., 2008, see this chapter). In relation to this, a meta-analytic review was developed by Ferguson (2007), who found that publication bias was a problem for studies of both aggressive behaviour and visuospatial cognition. Once corrected for a publication bias, studies of digital game violence provided support for a higher visuospatial cognition. Additional research is necessary, since the body of evidence is small and effect studies are lacking.

One Step Beyond: Last-Generation Console Gaming for Intervention with Dementia

In the field of dementia rehabilitation, increasing evidence is emerging focusing on technology based solutions versus traditional paper-and-pencil rehabilitation techniques. Boulay et al. (2009) tested the MINWii, a music therapy game to be used in the treatment of people with Alzheimer's disease. The game allowed the patients to improvise or play songs of their choice by pointing with a Wiimote Pistol at a virtual keyboard that was displayed on a TV set. A Nintendo Wiimote Pistol is a hand light gun grip in which a standard Wiimote can be placed as such that it resembles and operates as a pistol. Seven patients with Alzheimer's disease participated in the study of Boulay et al. (2009) and it was shown that the MINWii game was usable and an instant mastery was shown and even a learning effect. The game fostered positive interactions with the caregivers and elicited powerful reminiscence.

Recently, Cherniack (2011) described VR based applications in the identification and treatment of older people with cognitive disorders. According to the author, VR can in potential offer an assessment of function and could enhance the ability to perform activities of daily living in patients with dementia, stroke, and Parkinson's disease. However, scientific evidence is still limited and the performed studies have been small and unblinded. In order to throw some more light over this topic, a recent study from Fernández-Calvo et al. (2011) assessed the efficacy of cognitive training using Nintendo's Big Brain Academy (BBA)-a brain training game-compared to the Integrated Psychostimulation Program (IPP), which is a classical paper-and-pencil cognitive therapy, for patients who are diagnosed with mild Alzheimer's disease. In total, forty-five patients were randomly assigned to three experimental conditions: intervention with BBA; intervention with IPP; and no treatment (NT). Interestingly, from the results it was shown that the group trained with the BBA suffered a lower cognitive decline than the IPP and NT groups. Moreover, the BBA group showed a significantly higher reduction of depressive symptomatology when compared to the IPP and NT groups. The study of Fernández-Calvo et al. (2011) showed the strength and potential of digital games in health treatment, in particular because an off-the-shelf game like BBA outperformed a classical therapy in the training of people with a cognitive impairment.

DIGITAL GAMES' CONTRIBUTION TO PSYCHOLOGICAL HEALTH AND TREATMENT OF PSYCHOLOGICAL DISORDERS

Impact on Self-Efficacy

Griffiths (2002) reported on the educational benefits of digital games, and discussed the study of Thomas, Cahill, and Santilli (1997). In that study, an interactive computer program called 'Life Challenge' was used as a tool to enhance adolescents' perceived self-efficacy on HIV/AIDS prevention programs. To support the enhancement of self-efficacy, Thomas, Cahill, and Santilli (1997) used a prevention program that was based on a time travel adventure game format. In the game, a player can choose a co-player on a journey to various places (e.g., a medieval castle or a Jazz club in space). Players will have to negotiate in tasks related to HIV and Aids related high-risk behaviours. Significant gains were achieved in factual information about safe sex practices and in self-efficacy scores. Despite these positive effects, it is unclear which mechanisms contribute to the success of the prevention program. To gather an insight in the mechanisms that work, fundamental research is needed with small differences in the manipulations, e.g., solely vary the co-player, the game environment, etc.

Another self-efficacy study showed improved treatment adherence and cancer-related self-efficacy and knowledge after an intervention with the 'Re-Mission' game, which can be seen in Figure 1. Remission in cancer treatment means that the treatment is effective for a period of time. The digital game addressed the issues of cancer treatment and care for teenagers and young adults (Kato, Cole, Bradlyn, & Pollock, 2008). The Re-Mission's game play includes destroying cancer cells and managing common treatment-related adverse effects such as bacterial infections, nausea, and constipation by using chemotherapy (Kato et al., 2008). To win, players control a nanobot named 'Roxxi' to ensure strategically that virtual patients engage in positive self-care behaviours, like taking oral chemotherapy to fight cancer cells, and taking antibiotics to fight infection.

Treatment of Anxiety and Anxiety Related Disorders

Exergames are games that combine play and exercise with the goal to motivate players to engage in physical activity (Bogost, 2005) and have been used to improve depressive symptoms and mental health related quality of life (Rosenberg et al., 2010). In the study of Rosenberg et al. (2010), positive effects were found after a twelve week intervention, with three 35-minute sessions a week, using Nintendo's Wii Sports. Positive effects have also been found of gaming on anxiety treatment in a study of Patel et al. (2006). In this study, 112 children participated who were undergoing general anaesthesia for elective surgery. The patients were assigned to one of three groups; parent present; parent present and sedative medication, and parent present and digital game distraction by means of game play on the Gameboy (a handheld digital game device developed by Nintendo). The results showed that digital game distraction decreased anxiety from the baseline measurement, while the other groups showed a significant increase in anxiety. The authors conclude that digital game play on a handheld can be provided as a low cost, easy to implement, and effective means to decrease the anxiety before surgery and during anaesthesia. Another noteworthy project to treat anxiety by means of gaming is 'Relax to Win' of the MindGames team at Media Lab in Europe (McDarby, Condron, & Sharry, 2003). 'Relax to Win' is a two-player competitive game which uses biofeedback to control a dragon in a race. The more a player relaxes, as measured by the galvanic skin response (GSR), the faster the dragon moves, i.e., the player wins by learning to relax. According to the authors, the game has proved to be engaging to children and motivated them to become curious on how they relax and to learn new relaxation skills. Future research and development in relaxation gaming should rather add additional measurements besides GSR. A combination of measurements is preferred because some people have little variation in their GSR between relaxation and stress, while others always have a high conductive skin response, irrespective of their emotional status. Furthermore, an increase in the measured level of conductance cannot explain the type of emotion that is triggered, and therefore future research could combine ECG measurements, heart rate variability, facial expressions, etc. as a more valid indicator of relaxation.

Figure 1. Image from the digital game Re-Mission (© 2006, HopeLab Foundation, Inc. Used with permission)

In a review of Wilkinson et al. (2008), there is some evidence that digital games can also contribute to reduced Attention Deficit Hyperactivity Disorder (ADHD) symptoms, and a report on a study by Pope and Palsson (2001) about the development of a NASA patent on ADHD intervention by biofeedback modulated 'off-the-shelf' digital games. The participating children played a number of games on the Sony Playstation, varying from skateboarding to adventure games. Overall, they had 40 sessions, lasting approximately one hour per session with about an average of two sessions per week. No difference was found in the efficacy of the intervention to reduce inattention and hyperactivity compared to conventional interventions. Yet, the participants stated that game intervention was far more enjoyable than existing methods to reduce ADHD symptoms. In 2011, Nesplora (see, www.nesplora.com) have developed AULA, an evaluation test which employs VR to facilitate diagnosis of Attention Deficit Disorder with and without Hyperactivity. The AULA system analyzes the behaviour of a child in the context of a virtual classroom. The tool is perceived initially as a game, in which the child performs a task while typical distracters of a classroom are presented to him or her, which is demonstrated in Figure 2.

The AULA test evaluates factors determining the existence of ADHD, such as sustained attention, impulsivity, divided visual and auditory attention, excessive motor activity, and a tendency to distraction (by means of a movement sensor). After the test, the system returns an evaluation report that helps the clinician perform a more accurate and reliable diagnosis. Nesplora is also working on 'ISLA CALMA,' which will be commercialized in 2012. ISLA CALMA is an introspection of an island, seen in Figure 3, in which the player must explore and give life to some scenarios that need the interaction with the user; in this way, participants get immersed in a situation demanding their attention, thus minimizing situations of pain and anxiety. The Island will be marketed to be used in dental clinics or in surgical procedures in hospitals, as well as for relaxation techniques' training and for post-traumatic stress disorder treatment.

People who suffer from Post-Traumatic Stress Disorder (PTSD) have usually experienced a highly traumatic 'extraordinary' event with human death(s) involved, like a military battle or terrorist attack. PTSD is generally treated by means of imaginal exposure therapy; however, many patients are unable to regenerate the traumatic event and avoid the trauma, as demonstrated by Difede and Hoffman (2002). These authors showed the effectiveness of VR exposure therapy for PTSD in a case study with a patient who suffered from 'World Trade Center' PTSD, which was triggered by an intense experience or repeated exposure to the WTC attack in New York (September 11, 2001). The computer-generated environment consisted of lower Manhattan (New York), and

Figure 2. Image from AULA software (© 2011, Nesplora S.L.; Used with permission)

Figure 3. Image from ISLA CALMA anxiety and pain distracting software (© 2011, Nesplora S.L.; Used with permission)

the event she re-experienced was the WTC attack. The patient in the study failed to improve after traditional imaginal exposure therapy, yet, after the VR exposure therapy, depression was reduced by 83 percent and PTSD symptoms by 90 percent.

In the field of early interventions on PTSD, Holmes et al. (2009) performed a study in which participants played TETRIS half an hour after viewing a traumatic video. They expected that game play interference with trauma memory consolidation processes would reduce flashback frequency. In order to reach this objective, 40 participants were presented with a 12 minute film of traumatic scenes of injury and death. After this, the participants were randomly assigned to an experimental or control condition. The experimental participants played TETRIS for 10 minutes while control participants sat quietly during the same amount of time. Flashbacks were monitored with a diary, and results indicated that the participants in the TETRIS condition produced a significant reduction in flashback frequency over one week, and these results were in agreement with a clinical measure of PTSD symptomatology. These results imply that non-invasive cognitive interventions like playing the TETRIS game may be used at a crisis intervention after traumatic events.

PHYSICAL REHABILITATION: THE ROLE OF DIGITAL GAMES IN PHYSICAL RECOVERY

The use of games to enhance physical therapy and to motivate people's engagement towards rehabilitation procedures is increasingly gaining attention, partially due to the introduction of the low-cost Nintendo Wii platform. Hurkmans, Ribbers, Streur-Kranenburg, Stam, and Van den Berg-Emons (2011), examined whether embodied digital game play, such as Nintendo Wii Sports, could provide the required energy expenditure necessary for health improvement purposes. Ten chronic stroke patients participated in the study, who were instructed to play Wii Sports tennis and boxing for 15 minutes each. There was a 10 minute break between the games, which were played in a randomized order. Measures of oxygen uptake during exercise and rest were taken and physical activity was classified according to the American College of Sports Medicine guidelines and the American Heart Association guidelines. Results showed that, except for one patient in the tennis group, chronic stroke patients played Wii Sports tennis and boxing at moderate intensity, which is a sufficient energy expenditure for health maintenance and improvement purposes. In the following paragraphs we focus on evidence related to health related conditions of high prevalence in our society: stroke, cerebral palsy balance, and pain treatment.

Stroke and Post-Stroke Limb Rehabilitation

In a recent meta-analysis, Saposnik and Levin (2011) tried to determine the added benefit of VR technology on arm motor recovery after stroke. Thirty-five studies were identified, of which 12 met the inclusion/exclusion criteria totalling 195 participants. Among them, there were five randomized clinical trials and seven observational studies with a pre-/post-intervention design. An improvement of Fugl-Meyer, which is a measurement of motor impairment, was used as the primary outcome and the secondary outcomes included improvement in motor function measured by the Wolf Motor Function Test (WMFT), Box and Block Test, and the Jebson-Taylor Hand Function Test. Interventions were delivered within four to six weeks in nine of the studies and within two to three weeks in the remaining three. From the meta-analysis it appeared that there was a significant higher chance of improvement in motor strength for patients randomized to VR systems. Furthermore, there was a significant improvement in motor impairment and improvement in motor function outcomes from the observational studies. Eleven of the 12 studies showed a significant benefit toward VR for the selected outcomes and the authors conclude that VR and digital game applications are potentially useful technologies that can be combined with traditional rehabilitation for upper arm improvement post-stroke.

Colombo et al. (2007) designed two robot devices and used simple game elements-difficulty level of the task and feedback on performance-for the rehabilitation of upper limb movements of chronic stroke patients. In the study, a one degree of freedom (DoF) wrist manipulator and a 2 DoF elbow-shoulder manipulator were designed to be used for the treatment of the study participants, who suffered from chronic stroke. Visual feedback was provided on a TFT flat screen by three coloured circles. The participants had to follow a circular path for the wrist device and a square or a more complex path for the elbow-shoulder device. During the treatment, the device provided visual and auditory feedback to the patient and their task scores were displayed on the screen. By providing feedback on performance, the patients' interest remained high during the training session. In addition, precise measures were obtained of the patients, which provided the therapist with the possibility to present positive feedback for their efforts to enhance the patients' motivation and devotion to the training.

Recently, the Rehabilitation Gaming System (RGS) project started on the use VR for post-stroke limb rehabilitation. An image of the RGS Spheroids game can be seen in Figure 4. The RGS can capture the movements of the arms by means of a camera that is positioned on the top of a display. The camera detects colour patches that are located on the elbows and wrists of the patient. Furthermore, a pair of data gloves (optic fibre) measure finger flexion. Da Silva, Bermudez, Badia, Duarte, and Verschure (2011) studied the clinical impact of the RGS VR system on time to recovery after acute stroke. Eight acute stroke patients used the system during 12 weeks in addition to conventional therapy and eight served as a control group who performed a time matched alternative treatment. After the therapy, between-group comparisons showed that the RGS group displayed significantly improved performance in the speed of the impaired arm, matched by better performance in the arm subpart of the Fugl-Meyer Assessment Test and the Chedoke Arm and Hand Activity Inventory. Furthermore, the RGS treatment group presented a significantly faster improvement over time for all clinical scales during the therapy. From the study, it appeared that rehabilitation with the RGS facilitates the functional recovery of the upper extremities. The authors suggest that rehabilitation gaming by means of the RGS is a promising tool for stroke neurorehabilitation.

Accordingly, Yong Joo et al. (2010) studied the feasibility of Nintendo Wii exercises in addition to traditional rehabilitation of patients with post-stroke upper limb weakness. In total, 16 participants - inpatients within three months after a stroke with upper limb weakness-received six training sessions over two weeks. The participants played Nintendo's Wii Sports, including boxing, bowling, tennis, golf, and baseball. All participants found Nintendo Wii gaming enjoyable and comparable to traditional training. There were small but statistically significant improvements in the Fugl-Meyer Assessment and Motricity Index scores; hence, Nintendo Wii appears to be a feasible device as an *addition* to conventional rehabilitation therapies for patients with post-stroke upper limb weakness. Mouawad, Doust, Max, and McNulty (2011) have also investigated the effectiveness of the Nintendo Wii in therapy for post-stroke rehabilitation. In their study, seven patients (post-stroke) and five healthy people who served as controls undertook one hour of therapy on 10 successive weekdays. From the study it was shown that functional motor ability improved for the post-stroke patients and a transfer of functional recovery to everyday activities of daily life was found.

Cerebral Palsy

A relevant improvement made in the combination of low-cost digital gaming and rehabilitation of cerebral palsy (i.e., motor impairments resulting from lesions of the brain) comes from the work of Deutsch, Borbely, Filler, Huhn, and Guarrera-Bowlby (2008). This team studied the feasibility and outcomes of using Wii Sports (boxing, tennis, bowling, and golf) to augment the rehabilitation of a patient with cerebral palsy. The patient performed 11 training sessions, of which two included other players. The sessions lasted between 60 to 90 minutes. Improvements in visual-perceptual processing ranged from a 4th percentile change in form constancy to a 70th percentile change in visual discrimination. Postural control improved in terms of greater loading on the lower extremities; other improvements were reported in centre-of-pressure sway decrease, more symmetry of medial-lateral weight distribution, and an increase in the ambulation with forearm crutches.

Qui et al. (2009) used the Haptic Master and rehabilitation simulations, to allow two children with hemiparetic cerebral palsy to interact with virtual environments. The participants performed a battery of clinical testing and kinematic measurements of reaching. A number of virtual simulations were used during the training, like a bubble explosion simulation and a car race. The car race presented the patient with a race track and three other competing cars. To accelerate or decelerate the car, patients had to either use a force forwards or backwards and the car could be turned by pronating or supinating their forearm. The participants completed nine hours of training in three weeks in which no negative responses to treatment were reported. One participant showed an overall performance improvement on the functional aspects of the testing battery and the

Figure 4. Image of the Spheroids game from the Rehabilitation Gaming System Project (© 2011, RGS Project Consortium; Used with permission)

other participant improved in the upper extremity active range of motion and in kinematic measures of reaching movements.

Another study on the rehabilitation of cerebral palsy was performed by Huber et al. (2010), who developed a pilot study which was designed to examine the feasibility of home rehabilitation by means of digital gaming to address hand impairments in patients with cerebral palsy. Three patients participated in the study and trained in their home environment for about 30 minutes a day, a couple of times a week, over a 6 to 10 month period. The participants wore a sensing glove and played custom-developed games on a modified Sony PlayStation 3, which were specifically designed for the purpose of accommodating the participants' limited range of motion and to enhance the finger range and motion speed. Three virtual reality finger exercises were applied of which one training the range of motion and two finger velocity exercises. Significant improvements were found in finger range of motion, which were related to self- and family-reported improvements in ADL. The authors encourage the development of new rehabilitation games, including games to train independent finger movement, endurance, power output, and force exertion. Again, although the results are promising, only a limited number of people participated in the aforementioned studies, and therefore future experimental research is needed with larger samples and controls to validate the potential of robotics and gaming in rehabilitation.

Balance

A widely used method for training balance by means of digital games comes from the use of Nintendo's Wii Fit with a balance board. The balance board has four pressure sensors under all four corners of the board, similar to a regular weight scale. The Wii Fit is a relatively new product marketed to improve-among others-balance and general fitness. Nitz, Kuys, Isles, and Fu (2010) examined the marketing claims of the Wii Fit in improving balance, strength, flexibility, and fitness for healthy people. Ten females participated in the study who were aged between 30 to 58 years. The training intervention involved a 30 minute session, twice a week for a total of 10 weeks. The participants were assessed before and after the intervention. Clinical measures for balance and mobility were collected, as somatosensory measures and cardiovascular measures. Results showed that balance and lower limb muscle strength improved significantly. Yet, changes in touch, vibration, proprioception, cardiovascular endurance, mobility, weight change, activity level, and well-being were not found. Hence, Wii Fit training was found to have an immediate effect on the strength and balance of the participants. Future studies, with more statistical power, should try to find additional support for these claims (Nitz et al., 2010).

In a related study, Graves et al. (2010) compared the physiological cost and enjoyment of exergaming on Nintendo's Wii Fit with aerobic exercise in three different age groups (adolescents, young adults, and older adults). In the study, cardiorespiratory and enjoyment measures were compared between the three age groups. The participants performed handheld (inactive) digital gaming, Wii Fit activities (e.g., yoga, balance, aerobics), and brisk treadmill walking (approx. 5 km/h) and jogging. For all groups, the energy expenditure and heart rate of Wii Fit activities were greater than handheld gaming but lower than the treadmill exercise. However, the heart rate during Wii aerobics fell below the recommended intensity for maintaining cardiorespiratory fitness. It is interesting to note that the group enjoyment was higher for the Wii Fit balance and aerobics activities than for the treadmill walking and jogging. As a conclusion, this study showed that Nintendo's Wii Fit is experienced as an enjoyable exergame for different age groups and that Wii Fit stimulates light to moderate intensity activity. It is apparent that the Wii Fit is an enjoyable platform for physical training, yet, its role as a valid rehabilitation tool requires further research. Lange, Flynn, Proffitt, Chang, and Rizzo (2010) also explored the use of digital game consoles such as the Nintendo Wii Fit as rehabilitation tools. According to the authors, case studies have demonstrated that the use of digital games may be beneficial for balance rehabilitation, yet, today's commercial off-the-shelf games lack compatibility with the specific training needs that are necessary to meet therapy goals. Therefore, Lange et al. (2010) developed a game to be played with the Nintendo Wii Fit balance board. The game design was based on focus group data and observations with patients that specifically targeted weight shift training. In the game, a player had to move a balloon to collect falling stars and to avoid falling rocks by changing the weight on the balance board in the direction they wanted the balloon to move. The usability of the prototype was evaluated by a number of clinicians and people with neurological injuries. The feedback was overall positive and the preliminary research provided support for the development of a rehabilitation game that targets the key requirements for training weight shift.

PAIN TREATED WITH VIRTUAL REALITY

The use of Virtual Reality (VR) games in pain treatment is based on a distraction of the thought processes in the brain, including pain experiences. Das, Grimmer, Sparnon, McRae, and Thomas (2005) examined the additional (positive) effect of playing a VR game combined with routine pharmacological pain treatment on procedural pain in children with acute burn injuries. The software that was used in the study was based on the game 'Quake' (ID software) and a head mounted display (HMD) with tracking system and mouse was used to interact with the virtual world. The simulation provided the children with the sense of being on a track on which they could use a pointer to aim and shoot monsters. The children had burns to more than 3 percent of their body surface area and required dressing changes. In the randomized trial, seven children acted as their own controls, in a series of 11 trials. The participants had to score their average pain experience using the Faces self-report pain scale at the end of each phase of a dressing change. The parents or carers and nurses were interviewed at the same time to gather their perceptions on the anxiety, pain perception, and the use of VR during the treatment. From the results it appeared that the average pain scores were lower when VR was combined with pharmacological treatment, than when medication was used alone. A subsequent randomized study by Rutter, Dahlquist, and Weiss (2009) examined whether the distraction of VR reduced cold pressor pain in adults. Twenty-eight adults participated and underwent one baseline cold pressor trial and one VR distraction trial in randomized order each week over 8 weeks in total. In the VR distraction trial, participants played the game 'Finding Nemo,' level 'Catch Dory' (Traveller's Tales), on the Sony Playstation 2 (PS2). The game world was viewed through a HMD with integrated headphones and controlled by a PS2 controller which was mounted to a table to allow participants to manipulate the controller with one hand while the other hand was in the cold pressor (i.e., water). The authors also found that VR distraction decreased pain, since the pain threshold and tolerance increased and pain intensity decreased. Both studies (Das et al., 2005; Rutter et al., 2009) provide support for the positive additional effect of VR distraction in pain treatment.

SOCIAL REHABILITATION GAMING AND GAMES FOR SOCIAL SKILLS AND PRO-SOCIAL BEHAVIOUR LEARNING

One of the main motivations to play digital games together is the possibility to interact with and meet others (Jansz & Martens, 2005; Staiano & Calvert, 2011). In respect to players' preferences, multiplayer and group game play is preferred over solitary play by preadolescent children (Chin A Paw, Jacobs, Vaessen, Titze, & van Mechelen, 2008) and older people have also been found to prefer co-located multiplayer gaming, however, not mediated over the Internet without the possibility for social interaction (Gajadhar, Nap, de Kort, & IJsselsteijn, 2010). From a number of experimental studies, it appeared that more enjoyment is experienced when playing against human co-players compared to a computer generated player (Gajadhar, de Kort, and IJsselsteijn, 2008; Gajadhar, de Kort, and IJsselsteijn, 2009). According to the authors, the possibility to communicate and interact with a co-player while playing adds to the fun and involvement experienced in digital game play (Gajadhar, de Kort, and IJsselsteijn, 2008; Gajadhar, de Kort, and IJsselsteijn, 2009). Social play is also receiving increasing attention in physical training (e.g., seniors with the Wii Fit; Aarhus, Grönvall, Larsen, & Wollsen, 2011) and in rehabilitation to reconnect patients in their social environments and to provide an additional incentive to engage in rehabilitation (Van den Hoogen, IJsselsteijn, & de Kort, 2009; Vanacken et al. 2010).

Vanacken et al. (2010) studied game-based arm rehabilitation and social play of multiple sclerosis (MS) patients. The value of force-feedback assisted rehabilitation of the upper extremities in MS patients was examined. Furthermore, it was studied how such technologies (i.e., a VR Environment) can be applied in a self-motivating way, providing the patients with training tasks to be carried out and monitoring their progress and success rate. The added value is the inclusion of a collaborative rehabilitation environment, opening up the possibility of social play. One of the co-play games is the BalancePump game; a two player game in which the goal is to collect all stars by hitting them with a ball, which can be seen in Figure 5. The ball can be moved by lifting the ends of the beam and each end is controlled by one of the players involved in the game. A player performs a pumping gesture to move his/her end upwards as such that the ball can roll towards a star. The relative or friend that participates as a co-player in the game uses a Wiimote, while the patient uses the Haptic Master (a 6 degrees of freedom force controlled robot) with gimbal. The authors demonstrated how collaboration between a patient and relative can be used for social play in rehabilitation gaming. According to the authors, social rehabilitation, in particular with relatives, friends, or other patients could potentially enhance the motivation for training (Vanacken et al., 2010). We foresee future experimental research in which the relation between the presence of a co-player and the effectiveness, motivation, and enjoyment in rehabilitation is studied.

Aside from studies on the enjoyment and motivational factors of social play, Durkin and Barber (2002) studied the relation between digital game play and social behaviour. Adolescents who played digital games scored more favourably than non-players on several measures such as family closeness, activity involvement, friendship network, and (less) disobedience to parents. Furthermore, Greitemeyer and Osswald (2009) examined if playing pro-social digital games reduces aggressive cognitions, and found that playing a prosocial digital game reduced the hostile expectation bias and decreased the accessibility of antisocial thoughts compared to playing a neutral game. However, these benefits cannot be attributable only to non-violent digital games. Ferguson (2010) states that the so called Massively Multiplayer Online Role Playing Games (MMORPGs) such as World of Warcraft (Blizzard Entertainment) allow for complex social interactions to occur within the game world, and that those social connections can be very meaningful to those involved. In relation to this, Ferguson and Rueda (2010) examined the causal effects of digital game playing on aggressive behaviour, hostile feelings, and depression, with 103 young adults. The participants were provided with a frustration task and then they were randomized

Figure 5. Collaborative BalanceBump training task (© 2011, Hasselt University - Expertise centre for Digital Media; Used with permission)

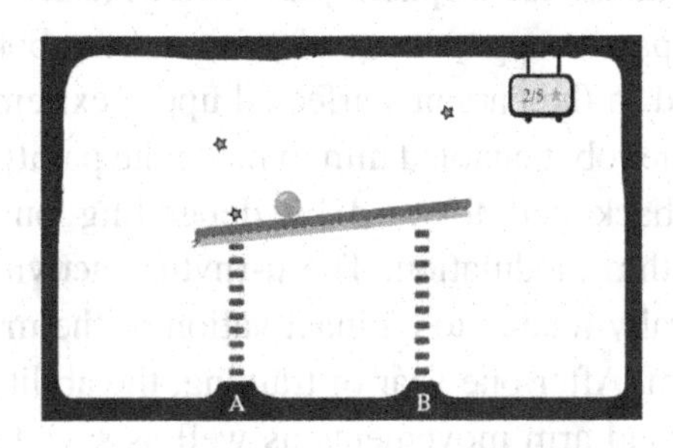

to play a nonviolent game, a violent game with a good versus evil theme (i.e., playing the bad or the good guy), or they played no game. The authors found no evidence that short-term exposure to violent video games either increased or decreased aggressive behaviour. Yet, a history of real-life violent video game-playing was predictive of decreased hostile feelings and decreased depression. It is suggested that violent games could, at least for some individuals, reduce depression and hostile feelings in players through mood management.

REHABILITATION GAMING– FUTURE TECHNOLOGIES

For decades, doctors and therapists have explored and used a variety of technologies to enhance the rehabilitation process, and in the previous literature review, a number of systems have already been introduced, like Nintendo's Wii with balance board, the Rehabilitation Gaming System, and robot devices like the Haptic Master. With the introduction of low-cost 3D motion sensing, like Microsoft's' Kinect and Sony's Playstation Move, new possibilities arise for rehabilitation training. In the following sections, we will shortly discuss some of the systems and future technologies that have received little attention in the previous sections and discuss some studies that support the use of these systems for rehabilitation purposes.

3D Motion Sensing

One of the latest advancements in rehabilitation gaming is the commercially available SilverFit platform that operates via 3D motion sensors. SilverFit provides a hardware and software platform to facilitate physical rehabilitation after e.g., a stroke, a fall, or a cardiac incident. The solution uses a time-of-flight (TOF) camera that can monitor the body movement of a player in three dimensions. TOF monitors the distance to a 3D object by measuring the absolute time a light pulse needs to travel from a source into the 3D scene and back after reflection. TOF is different from the infrared 'LightCoding' technique used by Microsoft Kinect, which projects a light pattern into a 3D scene, which is viewed after reflection by camera(s). A distortion of the light pattern on the object allows for computing the 3D structure (Castaneda & Navab, 2011).

The SilverFit system can trace body posture adjustments, arm movements, standing up, sitting down, walking, etc. within a 5x5 meter area. The input is converted into game elements shown on a High Definition (HD) flat screen or via beamer (Rademaker, Van der Linden, & Wiersinga, 2009). Figure 6 shows a rehabilitation session with the SilverFit system. The patient was recovering from a hip surgery, who was afraid to walk with a stick without support. During the session, she picked flowers in the rehabilitation game, with increasing speed in all directions. Afterwards, the patient acknowledged that she was not consciously aware about her walking, because "she had to pick flowers" (Pieter van Foreest, 2011). The state of the patient during the session is similar to a 'flow' state described earlier in this chapter, and the example shows the strength of rehabilitation gaming, because the patient performed movements she was afraid of and would otherwise not perform.

Another, yet more expensive, 3D motion system that is used in rehabilitation is CAREN (Computer Assisted Rehabilitation Environment) for the medical treatment of military and civilian patients. The CAREN system allows a patient to stand on a computer driven motion platform surrounded by multiple projections of a VR world. A number of cameras track the patient's position via markers that are placed on his/her body. CAREN offers a test environment with means of almost unlimited exploratory behaviours for patients and a strong tool for motor control research (van der Eerden, Otten, May, & Even-Zohar, 1999).

Brain-Computer Interfaces

Research on Brain-Computer Interfaces (BCIs) started in the 1970s and has recently seen an increasing interest in rehabilitation and digital gaming. BCIs can use a variety of information types from the user's brain activity to execute actions and tasks. For example, sensory perception, motor control, imaginations, stress levels, and workload can be translated into computer actions (Krepki, Curio, Blankertz, & Müller, 2007). Most BCIs use an Electroencephalography (EEG) in real-time for the interaction with a computer, which can be seen in Figure 7. To control a game by means of brain activity, for example to walk forward, a player may be instructed to think about moving his legs. The EEG data that is associated with this thought will be correlated with walking forward. According to Nijholt, Plass-Oude Bos, and Reuderink (2009) any computer action could in potential be controlled by certain brain activity, although an appropriate mapping is feasible, e.g., it is more convincing to mentally move your legs for a virtual walk than performing a mental rotation task for walking.

A BCI application that has currently great potential involves rehabilitation to regain motor control lost from diseases such as a stroke (Moore Jackson & Mappus, 2010). A case study of Broetz et al. (2010) showed that a combination of BCI training with goal-directed, active physical therapy may improve the motor abilities of chronic stroke patients. The patient in his/her study could not extend his fingers, hand, or arm for any relevant activity of daily life. During the training, brain activity was identified by means of an EEG and a magnetoencephalogram (MEG) to drive an orthopaedic apparatus (orthosis) and a robot attached to the patient's affected upper extremity, and the robot enabled him to move the paralyzed arm back and forward by depending on the μ-rhythm modulation. The μ-rhythm activity is generally related to an inactivation of the motor system. After one year of training, the ability of hand and arm movements as well as speed and safety of gait improved significantly. A positive relation was found between increased ipsilesional μ-power and the motor improvement. The authors suggest that a rehabilitation intervention by BCI could support chronic stroke patients who lack residual finger extension (Broetz et al., 2010), yet, there is still limited evidence for the effects.

Rehabilitation Gaming Integrated in Smart Homes: Future Prospects

Aside from the potential of new technologies in rehabilitation gaming, we foresee a future in which these types of systems are integrated in smart living; as such that rehabilitation can take place at home. This is particularly valuable because the population is aging: in 2050, more than a third of the European population will be aged over 60 years (UN, 2008), and it is likely that the workforce will not be able to support the people who are retired, in pensions and healthcare. These future prospects demand for solutions that require fewer costs with less people. In potential, telemedicine and eHealth solutions, integrated in a smart home, could reduce the workforce costs. Furthermore, these solutions can provide health care professionals the possibility to interact (e.g., via Vo-IP) and provide medical care to people who live in remote rural areas and who may have difficulty travelling to care providers, in particular specialized care.

Game-based telerehabilitation (Lange, Flynn, & Rizzo, 2009) and in-home rehabilitation gaming

Figure 6. Rehabilitation by means of the SilverFit at Zorginstellingen Pieter van Foreest, Physiotherapy 'De Naaldhorst', Netherlands (© 2011, Pieter van Foreest; Used with permission)

(Huber et al., 2010) have received wide spread attention and are increasingly applied in addition to regular telemedicine and rehabilitation solutions. However, a largely unexplored area for rehabilitation gaming is the *integration* in smart living. The infrastructure of a smart home can use and integrate any information derived from sensors, actuators, and interpreters to support independent living. The information that can be derived from a rehabilitation game, e.g., motor improvements and in-game scores, can be sent in real-time to care professionals who can monitor and react to deviations. Within the smart home, persuasive agents or avatars displayed on a television or mobile phone could prompt and motivate users to train whenever they should. Furthermore, movement and presence sensors that detect low activity could also be used to remind the user to exercise or just to perform a desired level of activity. For optimal rehabilitation, it is crucial that a patient is well nourished and a smart home can support patients to prevent malnourishment. Half of all

Figure 7. A player controlling the World of Warcraft game, using brain activity and the mouse and keyboard (© 2011, Anton Nijholt; Used with permission)

stroke patients are malnourished on admission to the inpatient rehabilitation service (Finestone & Greene-Finestone, 1999) and the factors that are associated with malnutrition include poorer functional outcome and prolonged rehabilitation stay. Different types of sensors in the kitchen can track the intake of food and drinks and could prevent malnutrition by reports to the care providers who can act accordingly, but also to (in-game) persuasive agents who remind the patient to drink and eat when necessary. Telemedicine devices like an online blood pressure monitor and weight scale can also provide an indication of malnutrition. In a study of Parker et al. (2009) it was found that rehabilitation at home showed no disadvantage compared to day hospital rehabilitation. Although it was found that home-based rehabilitation was not cheaper than day hospital rehabilitation, it is expected that the prices will change with time. The market of eHealth and telemedicine products is relatively new and is still serving a niche market. Insurance companies are more and more willing to contribute to the costs of these products and services, and therefore it is expected that these products and services will become cheaper in the future ahead. Future studies could focus on reducing the costs in rehabilitation gaming while maintaining high quality care. Furthermore, future research is necessary to explore the possibilities of integrating game elements in the home rehabilitation process, as an incentive to train and to enhance the enjoyment during rehabilitation. In addition, social rehabilitation with family and friends could enhance the effects even further.

CONCLUSION

Rehabilitation games make use of the motivating features of entertainment games that trigger the playful mind and induce challenge, fantasy, control, cooperation, and competition. As shown by the numerous studies reported in this chapter, the entertainment value and feedback mechanisms of games enhance the motivation to engage in rehabilitation, and some studies even show an advancement compared to conventional rehabilitation methods. Furthermore, with the rise of low-cost rehabilitation devices we expect that rehabilitation gaming could in potential reduce the increasing costs in health care, especially when the treatment can be performed in the home environment. A small number of rehabilitation games and systems have found their way to market, yet, the challenge of commercialization remains. For the future, it will be important that people in health gaming research, business, and government closely work together to find ways for funding and commercialization possibilities. Furthermore, sufficient effort should be spent to ensure that rehabilitation games and technologies meet the health and social needs of all people. Although there is an increasing body of evidence for the benefits of rehabilitation gaming, there is a constant need for further exploration and validation. According to Lieberman et al. (2011) there are a number of gaps in research on the efficacy of exergames for rehabilitation, for example there is a need for studies with randomized controlled methods to identify how digital game play influences behaviour. Furthermore, there is still limited evidence for an actual transfer of the activities trained during digital game play to ADL. In respect to the game design and the apparatus for rehabilitation gaming, we should take age related preferences and abilities into account since a number of studies showed that older people have specific gaming needs and differ in their abilities from young adults (De Schutter, 2011; IJsselsteijn, Nap, de Kort, & Poels, 2007; Nap, de Kort, & IJsselsteijn, 2009; Pearce, 2008). Qui et al. (2009) also reported about age specific preferences. The rehabilitation games that were used in their study were specifically designed for the rehabilitation of adults and a number of these games did not resonate well with children (Qui et al., 2009). Future studies in rehabilitation gaming should rather take age specific needs and abilities into account, to increase the positive return from rehabilitation gaming even more. Adaptive and personalized games could potentially overcome the obstacles to successful and rewarding game play. A challenge that needs further exploration in rehabilitation gaming, particularly at home, is to ensure the safety and health during training.

Despite the research gaps, we became inspired by the increasing work on rehabilitation gaming and the potential these games have to train motoric, cognitive, and social abilities. For the future, we see room for social rehabilitation gaming, the integration of rehabilitation games in smart living, new embodied gaming technologies, and gamification of real-life activities that could enhance the rehabilitation process even further.

ACKNOWLEDGMENT

Support from the Lifelong Learning Programme of the European Commission, the LEAGE project, and Stichting Smart Homes is gratefully acknowledged.

REFERENCES

Aarhus, R., Grönvall, E., Larsen, S. B., & Wollsen, S. (2011). Turning training into play: Embodied gaming, seniors, physical training and motivation. *Gerontechnology (Valkenswaard), 10*(2), 110–120. doi:10.4017/gt.2011.10.2.005.00

Abt, C. C. (1970). *Serious games*. New York, NY: The Viking Press.

Anderson, C. A., & Bushman, B. J. (2001). Effects of violent video games on aggressive behavior, aggressive cognition, aggressive affect, physiological arousal, and prosocial behavior: A meta-analytic review of the scientific literature. *Psychological Science, 12*(5), 353–359. PubMed doi:10.1111/1467-9280.00366

Bandura, A. (1977). Self-efficacy: Toward a unifying theory of behavioural change. *Psychological Review, 84*, 191–215. PubMed doi:10.1037/0033-295X.84.2.191

Barlett, C. P., Anderson, C. A., & Swing, E. L. (2009). Video games effects–Confirmed, suspected, and speculative. A review of the evidence. *Simulation & Gaming*, *40*(3), 377–403. doi:10.1177/1046878108327539

Basak, C., Boot, W. R., Voss, M. W., & Kramer, A. F. (2008). Can training in a real-time strategy video game attenuate cognitive decline in older adults? *Psychology and Aging, 23*(4), 765–777. PubMed doi:10.1037/a0013494

Bogost, I. (2005). *The rhetoric of exergaming*. The Georgia Institute of Technology. Retrieved August 11, 2011, from http://www.exergamefitness.com/pdf/The%20Rhetoric%20of%20Exergaming.pdf

Boulay, M., Benveniste, S., Boesplug, S., Jouvelot, P., & Rigaud, A. S. (2009). A pilot usability study of MINWii, a music therapy game for demented patients. [PubMed]. *Technology and Health Care, 19*(4), 233–246. PMID:21849735

Boyle, E., Connolly, T. M., & Hainey, T. (2011). The role of psychology in understanding the impact of computer games. *Entertainment Computing*, *2*(2), 69–74. doi:10.1016/j.entcom.2010.12.002

Broetz, D., Braun, C., Weber, C., Soekadar, S. R., Caria, A., & Birbaumer, N. (2010). Combination of brain-computer interface training and goal-directed physical therapy in chronic stroke: A case report. *Neurorehabilitation and Neural Repair, 24*(7), 674–679. PubMed doi:10.1177/1545968310368683

Castaneda, V., & Navab, N. (2011). Time-of-flight and kinect imaging: Labcourse. Retrieved August 10, 2011, from http://campar.in.tum.de/twiki/pub/Chair/TeachingSs11Kinect/2011-DSensors_LabCourse_Kinect.pdf

Castel, A. D., Pratt, J., & Drummond, E. (2005). The effects of action video game experience on the time course of inhibition of return and the efficiency of visual search. *Acta Psychologica, 119*, 217–230. PubMed doi:10.1016/j.actpsy.2005.02.004

Cherniack, E. P. (2011). Not just fun and games: Applications of virtual reality in the identification and rehabilitation of cognitive disorders of the elderly. *Disability and Rehabilitation. Assistive Technology, 6*(4), 283–289. PubMed doi:10.3109/17483107.2010.542570

Chin A Paw, M., Jacobs, W., Vaessen, E., Titze, S., & van Mechelen, W. (2008). The motivation of children to play an active video game. *Journal of Science and Medicine in Sport, 11*, 163–166. PubMed doi:10.1016/j.jsams.2007.06.001

Colombo, R., Pisano, F., Mazzone, A., Delconte, C., Micera, S., Chiara Carrozza, M., et al. (2007). Design strategies to improve patient motivation during robot-aided rehabilitation. *Journal of Neuroengineering and Rehabilitation, 4*(3). PubMed

Csikszentmihalyi, I., & Csikszentmihalyi, M. (1988). *Optimal experience: Psychological studies of flow in consciousness*. New York, NY: Cambridge University Press. doi:10.1017/CBO9780511621956

Csikszentmihalyi, M. (1975). *Beyond boredom and anxiety*. San Francisco, CA: Jossey-Bass.

Da Silva, C. M., Bermudez, I., Badia, S., Duarte, E., & Verschure, P. F. (2011). Virtual reality based rehabilitation speeds up functional recovery of the upper extremities after stroke: A randomized controlled pilot study in the acute phase of stroke using the rehabilitation gaming system. [Epub ahead of print]. *Restorative Neurology and Neuroscience*.

Das, D. A., Grimmer, K. A., Sparnon, A. L., McRae, S. E., & Thomas, B. H. (2005). The efficacy of playing a virtual reality game in modulating pain for children with acute burn injuries: A randomized controlled trial [ISRCTN87413556]. *BMC Pediatrics, 5*(1). PubMed doi:10.1186/1471-2431-5-1

De Schutter, B. (2011). *De betekenis van digitale spellen voor een ouder publiek.* (Unpublished doctoral dissertation). Catholic University of Leuven, Leuven, Belgium.

Deutsch, J. E., Borbely, M., Filler, J., Huhn, K., & Guarrera-Bowlby. (2008). Use of a low-cost, commercially available gaming console (Wii) for rehabilitation of an adolescent with cerebral palsy. *Physical Therapy, 88*(10), 1196–1207. PubMed doi:10.2522/ptj.20080062

Difede, J., & Hoffman, H. G. (2002). Virtual reality exposure therapy for World Trade Center post-traumatic stress disorder: A case report. *Cyberpsychology & Behavior, 5*(6), 529–535. PubMed doi:10.1089/109493102321018169

Durkin, K., & Barber, B. (2002). Not so doomed: Computer game play and positive adolescent development. *Applied Developmental Psychology, 23*(4), 373–392. doi:10.1016/S0193-3973(02)00124-7

Dweck, C. S. (1986). Motivational processes affecting learning. *The American Psychologist, 41*, 1040–1048. doi:10.1037/0003-066X.41.10.1040

Eliot, J., & Smith, I. M. (1983). *An international directory of spatial tests*. Windsor, UK: NFER-Nelson.

Feng, J., Spence, I., & Pratt, J. (2007). Playing an action video game reduces gender differences in spatial cognition. *Psychological Science, 18*(10), 850–855. PubMed doi:10.1111/j.1467-9280.2007.01990.x

Ferguson, C. J. (2007). The good, the bad and the ugly: A meta-analytic review of positive and negative effects of violent video games. *The Psychiatric Quarterly, 78*, 309–316. PubMed doi:10.1007/s11126-007-9056-9

Ferguson, C. J. (2010). Blazing Angels or Resident Evil? Can violent video games be a force for good? *Review of General Psychology, 14*(2), 68–81. doi:10.1037/a0018941

Ferguson, C. J., Cruz, A., & Rueda, S. (2008). Gender, video game playing habits and visual memory tasks. *Sex Roles, 58*, 279–286. doi:10.1007/s11199-007-9332-z

Ferguson, C. J., & Rueda, S. M. (2010). Violent video game exposure effects on aggressive behavior, hostile feelings, and depression. *European Psychologist, 15*(2), 99–108. doi:10.1027/1016-9040/a000010

Fernández-Calvo, B., Rodríguez-Pérez, R., Contador, I., Rubio-Santorum, A., & Ramos, F. (2011). Eficacia del entrenamiento cognitivo basado en nuevas tecnologías en pacientes con demencia tipo Alzheimer. [PubMed]. *Psicothema, 23*(1), 44–50. PMID:21266141

Finestone, H. M., & Greene-Finestone, L. S. (2002). The role of nutrition and diet in stroke rehabilitation. *Topics in Stroke Rehabilitation, 6*, 46–66.

Gajadhar, B. J., de Kort, Y. A. W., & IJsselsteijn, W. A. (2008). Shared fun is doubled fun: Player enjoyment as a function of social setting. In P. Markopoulos, B. de Ruyter, W. IJsselsteijn, & D. Rowland (Eds.), *Fun and games* (pp. 106–117). New York, NY: Springer. doi:10.1007/978-3-540-88322-7_11

Gajadhar, B. J., de Kort, Y. A. W., & IJsselsteijn, W. A. (2009). Rules of engagement: Influence of social setting on player involvement in digital games. *International Journal of Gaming and Computer-Mediated Simulations, 1*(3), 14–27. doi:10.4018/jgcms.2009070102

Gajadhar, B. J., Nap, H. H., de Kort, Y. A. W., & IJsselsteijn, W. A. (2010). Out of sight, out of mind: Co-player effects on seniors' player experience. In *Proceedings of Fun and Games '10 the 3rd International Conference on Fun and Games*, (pp. 74-83).

Goldberg, E. (2005). *The wisdom paradox: How your mind can grow stronger as your brain grows older*. New York, NY: Penguin Group.

Graves, L. E., Ridgers, N. D., Williams, K., Stratton, G., Atkinson, G., & Cable, N. T. (2010). The physiological cost and enjoyment of Wii Fit in adolescents, young adults, and older adults. [PubMed]. *Journal of Physical Activity & Health, 7*(3), 393–401. PMID:20551497

Green, C. S., & Bavelier, D. (2003). Action video game modifies visual selective attention. *Nature, 423*(6939), 534–538. PubMed doi:10.1038/nature01647

Green, C. S., & Bavelier, D. (2006). The cognitive neuroscience of video games. In P. Messaris, & L. Humphreys (Eds.), *Digital media: Transformations in human communication* (pp. 211–223). New York, NY: Peter Lang.

Greitemeyer, T., & Osswald, S. (2009). Prosocial video games reduce aggressive cognitions. *Journal of Experimental Social Psychology, 45*(4), 896–900. doi:10.1016/j.jesp.2009.04.005

Griffiths, M. (2002). The educational benefits of videogames. *Education for Health, 20*(3), 47–51.

Holmes, E. A., James, E. L., Coode-Bate, T., & Deeprose, C. (2011). Can playing the computer game "Tetris" reduce the build-up of flashbacks for trauma? A proposal from cognitive science. [PubMed]. *PLoS ONE, 4*(1), e1453–e1458. PMID:19127289

Huber, M., Rabin, B., Docan, C., Burdea, G. C., Abdelbaky, M., & Golomb, M. R. (2010). *IEEE Transactions on Information Technology in Biomedicine, 14*(2), 526–534. PubMed doi:10.1109/TITB.2009.2038995

Hurkmans, H. L., Ribbers, G. M., Streur-Kranenburg, M. F., Stam, H. J., & Van den Berg-Emons, R. J. (2011). Energy expenditure in chronic stroke patients playing Wii Sports: A pilot study. *Journal of Neuroengineering and Rehabilitation, 8*(1), 38. PubMed doi:10.1186/1743-0003-8-38

IJsselsteijn, W. A., Nap, H. H., de Kort, Y. A. W., & Poels, K. (2007). Digital game design for elderly users. In B. Kapralos, & M. Katchabaw (Eds.), *Proceedings of Future Play '07 the 2007 conference on Future Play* (pp. 17–22). doi:10.1145/1328202.1328206

Jansz, J., & Martens, L. (2005). Gaming at a LAN event: The social context of playing video games. *New Media & Society, 7*(3), 333–355. doi:10.1177/1461444805052280

Kato, P. M., Cole, S. W., Bradlyn, A. S., & Pollock, B. H. (2008). A video game improves behavioral outcomes in adolescents and young adults with cancer: A randomized trial. *Pediatrics, 122*, e305–e317. PubMed doi:10.1542/peds.2007-3134

Krepki, R., Curio, G., Blankertz, B., & Müller, K. R. (2007). Berlin brain-computer interface-the HCI communication channel for discovery. *International Journal of Human-Computer Studies, 65*(5), 460–477. doi:10.1016/j.ijhcs.2006.11.010

Lange, B., Flynn, S., Proffitt, R., Chang, C. Y., & Rizzo, A. S. (2010). Development of an interactive game-based rehabilitation tool for dynamic balance training. *Topics in Stroke Rehabilitation, 17*(5), 345–352. PubMed doi:10.1310/tsr1705-345

Lange, B., Flynn, S. M., & Rizzo, A. (2009). Game-based telerehabilitation. [PubMed]. *European Journal of Physical and Rehabiliation Medicine, 45*(1), 143–151. PMID:19282807

Lieberman, D. A., Chamberlin, B., Medina, E., Franklin, B. A., Sanner, B. M., & Vafiadis, D. K. (2011). The power to play: Innovations in getting active summit 2011. A science panel proceedings report from the American Heart Association. *Circulation, 123*(21), 2507–2516. PubMed doi:10.1161/CIR.0b013e318219661d

McDarby, G., Condron, J., & Sharry, J. (2003). Affective Feedback–Learning skills in the virtual world for use in the real world. Retrieved August 15, 2011, from http://medialabeurope.org/mindgames/publications/publicationsAAATEAffectiveFeedback2003.pdf

Moore Jackson, M., & Mappus, R. (2010). Applications for brain-computer interfaces. In D. S. Tan, & A. Nijholt (Eds.), *Brain-computer interfaces*. London, UK: Springer-Verlag. doi:10.1007/978-1-84996-272-8_6

Mouawad, M. R., Doust, C. G., Max, M. D., & McNulty, P. A. (2011). Wii-based movement therapy to promote improved upper extremity function post-stroke. A pilot study. *Journal of Rehabilitation Medicine, 43*(6), 527–533. PubMed doi:10.2340/16501977-0816

Nap, H. H., de Kort, Y. A. W., & IJsselsteijn, W. A. (2009). Senior gamers: Preferences, motivations and needs. *Gerontechnology (Valkenswaard), 8*(4), 247–262. doi:10.4017/gt.2009.08.04.003.00

Nijholt, A., Plass-Oude Bos, D., & Reuderink, B. (2009). Turning shortcomings into challenges: Brain-computer interfaces for games. *Entertainment Computing, 1*(2), 85–94. doi:10.1016/j.entcom.2009.09.007

Nitz, J. C., Kuys, S., Isles, R., & Fu, S. (2010). Is the Wii Fit a new-generation tool for improving balance, health and well-being? A pilot study. *Climacteric, 13*(5), 487–491. PubMed doi:10.3109/13697130903395193

Owen, A. M., Hampshire, A., Grahn, J. A., Stenton, R., Dajani, S., & Burns, A. S. et al. (2010, April 20). Putting brain training to the test. *Nature*. doi:10.1038/nature09042 PMID:20407435

Parker, S. G., Oliver, P., Pennington, M., Bond, J., Jagger, C., & Enderby, P. et al. (2009). Rehabilitation of older patients: Day hospital compared with rehabilitation at home. A randomised controlled trial. [PubMed]. *Health Technology Assessment, 13*(39), 1–168. PMID:19712593

Patel, A., Schieble, T., Davidson, M., Tran, M. C. J., Schoenberg, C., Delphin, E., & Bennett, H. (2006). Distraction with a hand-held video game reduces pediatric preoperative anxiety. *Paediatric Anaesthesia, 16*, 1019–1027. PubMed doi:10.1111/j.1460-9592.2006.01914.x

Pearce, C. (2008). The truth about baby boomer gamers: A study of over-forty. *Games and Culture, 3*(2), 142–174. doi:10.1177/1555412008314132

Pieter van Foreest. (2011). *Digitaal revalideren bij Fysiotherapie De Naaldhorst*. Retrieved August 10, 2011, from http://www.pietervanforeest.nl/PieterVanForeest/Over+Pieter+Van+Foreest/Actueel/Persberichten/Digitaal+revalideren+bij+Fysiotherapie+De+Naaldhorst.htm

Pope, A. T., & Palsson, O. S. (2001). *Helping video games 'rewire our minds'*. Paper presented at Playing by the Rules Conference, Chicago, IL.

Popescu, V. G., Burdea, G. C., Bouzit, M., & Hentz, V. R. (2000). A virtual-reality-based telerehabilitation system with force feedback. *IEEE Transactions on Information Technology in Biomedicine, 4*(1), 45–51. PubMed doi:10.1109/4233.826858

Qiu, Q., Ramirez, D. A., Saleh, S., Fluet, G. G., Parikh, H. D., Kelly, D., & Adamovich, S. V. (2009). The New Jersey Institute of Technology robot-assisted virtual rehabilitation (NJIT-RAVR) system for children with cerebral palsy: A feasibility study. *Journal of Neuroengineering and Rehabilitation, 6*(40). PubMed

Rademaker, A., van der Linden, S., & Wiersinga, J. (2010). SilverFit, a virtual rehabilitation system. *Gerontechnology (Valkenswaard), 8*(2), 119.

Ready, D. J., Gerardi, R. J., Backscheider, A. G., Mascaro, N., & Rothbaum, B. O. (2010). Comparing virtual reality exposure therapy to present-centered therapy with 11 U.S. Vietnam veterans with PTSD. *Cyberpsychology, Behavior, and Social Networking, 13*(1), 49–54. PubMed doi:10.1089/cyber.2009.0239

Rosenberg, D., Depp, C. A., Vahia, I. V., Reichstadt, J., Palmer, B. W., Kerr, J., et al. (2010). Exergames for subsyndromal depression in older adults: A pilot study of a novel intervention. *The American Journal of Geriatric Psychiatry, 18*(3), 221–226. PubMed doi:10.1097/JGP.0b013e3181c534b5

Rutter, C. E., Dahlquist, L. M., & Weiss, K. E. (2009). Sustained efficacy of virtual reality distraction. *The Journal of Pain, 10*(4), 391–397. PubMed doi:10.1016/j.jpain.2008.09.016

Salthouse, T. A. (2006). Mental exercise and mental aging: Evaluating the validity of the "use it or lose it" hypothesis. *Perspectives on Psychological Science, 1*, 68–87. doi:10.1111/j.1745-6916.2006.00005.x

Saposnik, G., & Levin, M. (2011). Virtual reality in stroke rehabilitation: A meta-analysis and implications for clinicians. *Stroke, 42*(5), 1380–1386. PubMed doi:10.1161/STROKEAHA.110.605451

Sherry, J. L. (2001). The effects of violent video games on aggression. A Meta-analysis. *Human Communication Research, 27*(3), 409–431.

Staiano, A. E., & Calvert, S. L. (2011). Exergames for physical education courses: Physical, social, and cognitive benefits. *Child Development Perspectives, 5*(2), 93–98. PubMed doi:10.1111/j.1750-8606.2011.00162.x

Thomas, R., Cahill, J., & Santilli, L. (1997). Using an interactive computer game to increase skill and self-efficacy regarding safer sex negotiation: Field test results. *Health Education & Behavior, 24*(1), 71–86. PubMed doi:10.1177/109019819702400108

UN. (2008). *World population prospects, the 2008 revision*. United Nations Department of Economic and Social Affairs, Population Division. Retrieved November 17, 2011, from http://www.un.org/esa/population/publications/wpp2008/wpp2008_highlights.pdf

Van den Hoogen, W. M., IJsselsteijn, W. A., & de Kort, Y. A. W. (2009). Yes Wii can! Using digital games as a rehabilitation platform after stroke-The role of social support. In *Proceedings of 2009 Virtual Rehabilitation International Conference* (pp. 195).

Van der Eerden, W. J., Otten, E., May, G., & Even-Zohar, O. (1999). CAREN-computer assisted rehabiliation environment. [PubMed]. *Studies in Health Technology and Informatics, 62*, 373–378. PMID:10538390

Vanacken, L., Notelaers, S., Raymaekers, C., Coninx, K., Van den Hoogen, W., IJsselsteijn, W., & Feys, F. (2010). Game-based collaborative training for arm rehabilitation of MS patients: A proof-of-concept game. *Proceedings of GameDays, 2010*, 65–75.

Wilkinson, N., Ang, R. P., & Goh, D. H. (2008). Online video game therapy for mental health concerns: A review. *The International Journal of Social Psychiatry, 54*(4), 370–382. PubMed doi:10.1177/0020764008091659

Willis, S. L., Tennstedt, S. L., Marsiske, M., Ball, K., Elias, J., Koepke, K. M....Wright, E., for the ACTIVE Study Group. (2006). Long-term effects of cognitive training on everyday functional outcomes in older adults. *Journal of the American Medical Association, 296*(23), 2805–2814. PubMed doi:10.1001/jama.296.23.2805

Yong Joo, L., Soon Yin, T., Xu, D., Thia, E., Pei Fen, C., Kuah, C. W. K., & Kong, K. H. (2010). A feasibility study using interactive commercial off-the-shelf computer gaming in upper limb rehabilitation in patients after stroke. *Journal of Rehabilitation Medicine, 42*(5), 437–441. PubMed doi:10.2340/16501977-0528

ADDITIONAL READING

Beale, I. L., Kato, P. M., Marín-Bowling, V. M., Guthrie, N., & Cole, S. W. (2007). Improvement in cancer-related knowledge following use of a psychoeducational video game for adolescents and young adults with cancer. *The Journal of Adolescent Health, 41*, 263–270. PubMed doi:10.1016/j.jadohealth.2007.04.006

Bogost, I. (2007). *Persuasive games: The expressive power of videogames*. Cambridge, MA: MIT Press.

Buiza, C., Gonzalez, M. F., Facal, D., Martinez, V., Diaz, U., & Etxaniz, A. et al. (2009). Efficacy of cognitive training experiences in the elderly: Can technology help? *Universal Access in Human-Computer Interaction. Addressing Diversity. Lecture Notes in Computer Science, 5614*, 324–333. doi:10.1007/978-3-642-02707-9_37

Cardullo, S., Seraglia, B., Bordin, A., & Gamberini, L. (2011). Cognitive training with Nintendo Wii® for the elderly: An evaluation. *Journal of Cybertherapy and Rehabilitation, 4*(2), 159–161.

Cheok, A. D., Lee, S., Kodagoda, S., Tat, K. E., & Thang, L. N. (2005). A social and physical inter-generational computer game for the elderly and children: Age invaders. In *Proceedings of the 2005 Ninth IEEE International Symposium on Wearable Computers (ISWC'05)*.

de Kort, Y. A. W., & IJsselsteijn, W. A. (2008). People, places, and play: A research framework for digital game experience in a socio-spatial context. *Computers in Entertainment, 6*(2), 18:1-18:11.

Fitzgerald, M. M., Kirk, G. D., & Bristow, C. A. (2011). Description and evaluation of a serious game intervention to engage low secure service users with serious mental illness in the design and refurbishment of their environment. *Journal of Psychiatric and Mental Health Nursing, 18*(4), 316–322. PubMed doi:10.1111/j.1365-2850.2010.01668.x

Gamberini, L., Barresi, G., Maier, A., & Scarpetta, F. (2008). A game a day keeps the doctor away: A short review of computer games in mental healthcare. *Journal of Cybertherapy and Rehabilitation, 1*(2), 127–146.

Gee, J. P. (2003). *What video games have to teach us about learning and literacy*. New York, NY: Palgrave Macmillan. doi:10.1145/950566.950595

Huizinga, J. (1949). *Homo Ludens: A study of the play-element in culture*. London, UK: Routledge.

IJsselsteijn, W. A. (2004). *Presence in depth*. (Unpublished doctoral dissertation). Eindhoven University of Technology, Eindhoven, The Netherlands.

Kalapanidas, E., Davarakis, C., Fernández-Aranda, F., Jiménez-Murcia, S., Kocsis, O., & Ganchev, T., Hannes Kaufmann, Lam T., & Konstantas, D. (2009). PlayMancer: Games for health with accessibility in mind. *Communications & Strategies, 73*, 105–120.

Labat, J. M. (2008). Designing virtual players for Game Simulations in a pedagogical environment: A case study. In *Proceedings of Edutainment '08 the 3rd international conference on Technologies for E-Learning and Digital Entertainment*, (pp 487-496).

Lyons, E. J., Tate, D. F., Ward, D. S., Ribisl, K. M., Bowling, J. M., & Kalyanaraman, S. (2012). Do motion controllers make action video games less sedentary? [PubMed]. *Journal of Obesity, 2012*. doi:10.1155/2012/852147 PMID:22028959

McGaffey, A. L., Abatemarco, D. J., Jewell, I. K., Fidler, S. K., & Hughes, K. (2011). Fitwits MD&trade: An office-based tool and games for conversations about obesity with 9- to 12-year-old children. *Journal of the American Board of Family Medicine, 24*(6), 768–771. PubMed doi:10.3122/jabfm.2011.06.100278

Melzer, E., & Moffit, K. (1997). *Head mounted displays: Designing for the user*. New York, NY: McGraw-Hill.

Michael, D. R., & Chen, S. L. (2005). *Serious games: Games that educate, train, and inform*. New York, NY: Muska & Lipman/Premier-Trade.

Nap, H. H. (2008). *Stress in senior computer interaction*. (Unpublished doctoral dissertation). Eindhoven University of Technology, Eindhoven, The Netherlands.

Optale, G., Urgesi, C., Busato, V., Marin, S., Piron, L., Priftis, K., et al. (2010). Controlling memory impairment in the elderly using virtual reality memory training: A randomized controlled, pilot study. *Neurorehabilitation and Neural Repair, 24*(4), 348–357. PubMed doi:10.1177/1545968309353328

Perry, J. C., Andureu, J., Cavallaro, F. I., Veneman, J., Carmien, S., & Keller, T. (2011). Effective game use in neurorehabilitation: User-centered perspectives. In P. Felicia (Ed.), *Handbook of research on improving learning and motivation through educational games: Multidisciplinary approaches* (pp. 683–725). doi:10.4018/978-1-60960-495-0.ch032

Raessens, J., & Goldstein, J. (2005). *Handbook of computer game studies*. Cambridge, MA: MIT Press.

Ritterfeld, U., Cody, M., & Vorderer, P. (2009). *Serious games: Mechanisms and effects*. New York, NY: Routledge, Taylor and Francis.

Schultheis, M. T., & Rizzo, A. A. (2001). The application of virtual reality technology for rehabilitation. *Rehabilitation Psychology, 46*(3), 296–311. doi:10.1037/0090-5550.46.3.296

Sherry, J. L., Lucas, K., Greenberg, B., & Lachlan, K. (2006). Video game uses and gratifications as predictors of use and game preference. In P. Varderer, & J. Bryant (Eds.), *Playing video games: Motives, responses, and consequences* (pp. 213–224).

Tate, R., Haritatos, J., & Cole, S. (2009). HopeLab's approach to re-mission. *International Journal of Learning and Media, 1*(1), 29–35. doi:10.1162/ijlm.2009.0003

Thompson, D. J., Baranowski, T., Buday, R., Baranowski, J., Thompson, V., Jago, R., & Juliano Griffith, M. (2010). Serious video games for health: How behavioral science guided the development of a serious video game. *Simulation & Gaming, 41*(4), 587–606. PubMed doi:10.1177/1046878108328087

Vanden Abeele, V., & De Schutter, B. (2010). Designing intergenerational play via enactive interaction, competition and acceleration. *Personal and Ubiquitous Computing, 14*(5), 425–433. doi:10.1007/s00779-009-0262-3

Vorderer, P., Hartmann, T., & Klimmt, C. (2003). Explaining the enjoyment of playing video games: The role of competition. In D. Marinelli (Ed.), In *Proceedings of ICEC '03 the second international conference on Entertainment computing*, (pp. 1-9).

Williams, D. (2003). The video game lightning rod. *Information Communication and Society, 6*(4), 523–550. doi:10.1080/1369118032000163240

KEY TERMS AND DEFINITIONS

BCIs: Brain Computer Interfaces, which can use a variety of information types from the user's brain activity to execute actions and tasks.

Exergames: Games that are developed for the purpose of physical training.

HMD: Head Mounted Display, a binocular or monocular display that is worn on the head or integrated in a helmet and can track the position and angle of the head.

Rehabilitation Gaming: The use of digital games for motoric, cognitive, and social rehabilitation and recovery.

Self-Efficacy: Personal expectations about the ability to succeed in specific situations.

Serious Games: Games that are developed for the purpose of training, learning, and skill acquisition.

Virtual Reality: An interactive simulated environment that can induce physical presence and can be accessed by means of a head mounted display and data gloves.

This work was previously published in Serious Games for Healthcare: Applications and Implications, edited by Sylvester Arnab, Ian Dunwell, and Kurt Debattista, pp. 50-75, copyright 2013 by Medical Information Science Reference (an imprint of IGI Global).

Chapter 9
The Use of Motion Tracking Technologies in Serious Games to Enhance Rehabilitation in Stroke Patients

Andrew M. Burton
Nottingham Trent University, UK

David Brown
Nottingham Trent University, UK

Hao Liu
Nottingham Trent University, UK

Nasser Sherkat
Nottingham Trent University, UK

Steven Battersby
Nottingham Trent University, UK

Penny Standen
University of Nottingham, UK

Marion Walker
University of Nottingham, UK

ABSTRACT

Stroke is the main cause of long term disability worldwide. Of those surviving, more than half will fail to regain functional usage of their impaired upper limb. Typically stroke upper limb rehabilitation exercises consist of repeated movements, which when tracked can form the basis of inputs to games. This paper discusses two systems utilizing Wii™ technology, and thermal and visual tracking respectively to capture motions. The captured motions are used as inputs to specially designed games, which encourage the users to perform repeated rehabilitation movements. This paper discusses the implementation of the two systems, the developed games, and their relative advantages and disadvantages. It also describes the upcoming testing phase of the project.

INTRODUCTION

Stroke is the main cause of long term disability (Mackay & Mensah, 2004). Of those surviving, more than half will fail to regain functional usage of their impaired upper limb (Feys et al., 1998). UK National Clinical Guidelines recommend participation focused rehabilitation - substituting medical rehabilitation services with suitable social and leisure activities (Royal College of Physicians, 2004).

DOI: 10.4018/978-1-4666-5071-8.ch009

Community stroke rehabilitation can be expensive to provide due to therapy contact time and new ways are being explored to provide patients with alternative opportunities to practice upper limb tasks that will enhance recovery.

Typically stroke upper limb rehabilitation exercises consist of repeated movements, which when tracked can form the basis of inputs to games. In light of modern advances from the gaming industry of human motion tracking devices, such as the Nintendo™ Wii Remote, the Wii Balance Board (Clarka, Bryanta, Puab, Bennella, & Hunta, 2010), and optical cameras, the cost of accurate movement tracking systems has reached a level where systems could be deployed to the homes of patients. This increase in availability, combined with a suite of games that encourage participation could have a major impact on the successful rehabilitation of stroke patients.

In this paper, we introduce three 3D games which will encourage the sorts of repeated movements which have been considered effective in the reacquisition of post-stroke motor skills. We introduce two different motion tracking systems to enable stroke patients to control the games by hand and finger movements:

- The IRGlove system, uses Nintendo™'s Wiimote technology. Two Wiimotes mounted either side of a PC monitor track infra-red Light Emitting Diodes (LEDs) on the finger tips of a patient wearing a specially designed glove.
- The markerless motion capture (mocap) system uses dual cameras (both optical and thermal) to track hand movements and identify hand gestures, without any special deployment on the patients.

Both systems are able to capture grab/release motions, rolling motions of the wrist, as well as the 3D or 2D position of the hand in space.

This paper will outline the following: Section 2 introduces the three serious games specially developed for stroke rehabilitation. Section 3 introduces the IRGlove system. Section 4 introduces the markerless mocap system. Section 5 compares the two systems. And finally in Section 6 potential future work is discussed.

SERIOUS GAMES

Frequently stroke survivors are left with partial paralysis on one side of the body and movement can be severely restricted in the affected hand and arm. Effective rehabilitation must be early, intensive and repetitive, which leads to the challenge of how to maintain motivation for people undergoing therapy. Burke et al. (2009) and Brown et al. (2009) demonstrated that games may be an effective way of addressing the problem of engagement in therapy.

The Computing and Technology team at Nottingham Trent University have developed three serious games using C# and Microsoft XNA (Microsoft, 2010): The games are each designed to encourage particular hand exercises for stroke patients, and are described in the following section.

Slingshot Game

A screenshot of the Slingshot game is shown in Figure 1a. This was designed to encourage movement of the hand in the XY plane and pinching movements. The user moves their hand forward and then uses a pinching/hand-close action to load a ball. Force is applied to the arrow either by moving the hand backwards or via the period the pinch is held depending upon the user's ability. Once primed, the ball is aimed at the target via movement in the X and Y planes. The arrow is fired by releasing the pinch. Complexity is added to further levels of the game by introducing effects from both gravity and wind on the shots. The user fires three sets of three balls from an increasing distance away from the target. The score is calculated from their accuracy over the total nine shots.

SpaceRace Game

A screenshot of the SpaceRace game is shown in Figure 1b. This was designed to encourage movement of the hand in the XY plane and rotation of the wrist. User's hand represents a ship flying in 3D space. As the user moves their hand in the X and Y planes the ship responds accordingly. Rotation of the hand is also used to bank the ship. The user must navigate a course by flying through different shaped obstacles. The ship gets faster as the game progresses in order to increase difficulty. Complexity is variable dependent on user ability. In 'Easy' mode, the user only controls the banking of the spaceship by rolling their wrist, and the position is automatic. In 'Normal' mode, the user controls both banking and position, and in 'Hard' mode, the gaps the ship must fly through become smaller. The game ends when a preset number of collisions has occurred.

BalloonPop Game

A screenshot of the BalloonPop game is shown in Figure 1c. This was designed to encourage reach and grab movements. The user controls a grab which moves in 3D space with their hand motion. They must grab balloons positioned randomly in space and pop them on a pin at the front centre of the play volume. Moving the hand in the X and Y plane lines up the grab with the balloon. The user then has to grab the balloon via movement in the Z plane and a corresponding pinch action. The user then holds the pinch and drags the balloon to the pin to pop it. To increase complexity of the game it has three levels. 'Easy' where balloon is stationary and the grab is smooth and so only knocks the balloons position on collision, 'Normal' where the balloons are stationary but grab has spikes on the outside of it requiring careful grabbing to avoid bursting the balloon, and 'Hard' where the balloons move and must be grabbed with the spiky grab.

System Calibration

The games are accompanied by a calibration system which allows the user's capability to be accounted for when playing the games. All of the input motions can be calibrated including physical play area and depth, rotation and pinch. This means for example that if a particular user is unable to bring their fingers completely together in a pinch motion, that the game can trigger the pinch at a wider setting. The system of calibration is intended to be used to set up the games so that they can be tailored to an individual user's ability, and also may be used as a means of testing for and logging changes in a user's ability. The calibration settings can be captured using a suite of on-screen tests which guide the user to perform the required motions and capture the extents of their ability. These settings can be easily updated when a user's ability changes by simply re-running the calibration exercises.

IR GLOVE

An initial goal of the project was identifying low-cost technologies in order to create a system cheap enough to allow it to be used in the home of the patient throughout their rehabilitation - thus removing the restrictive need to travel to a central rehabilitation unit. The Wii™ Remote (or Wiimote) was a device that was quickly recognized as having potential in this area, as it can connect to an ordinary PC via Bluetooth giving an ability to track infrared diodes, is available off the shelf and costs only a few pounds.

In order to accurately track the positions of the fingers in 3D space, it is necessary to use a two camera system, or in this case a two Wiimote system. The Wiimote fields of view must overlap, and so a system was devised where the two Wiimotes are positioned either side of the viewing monitor pointing inwards at approximately 22° (as shown in Figure 2). The system uses only the infrared

Figure 1. Captures from the games: a) SlingShot, b) SpaceRace, and c) BalloonPop

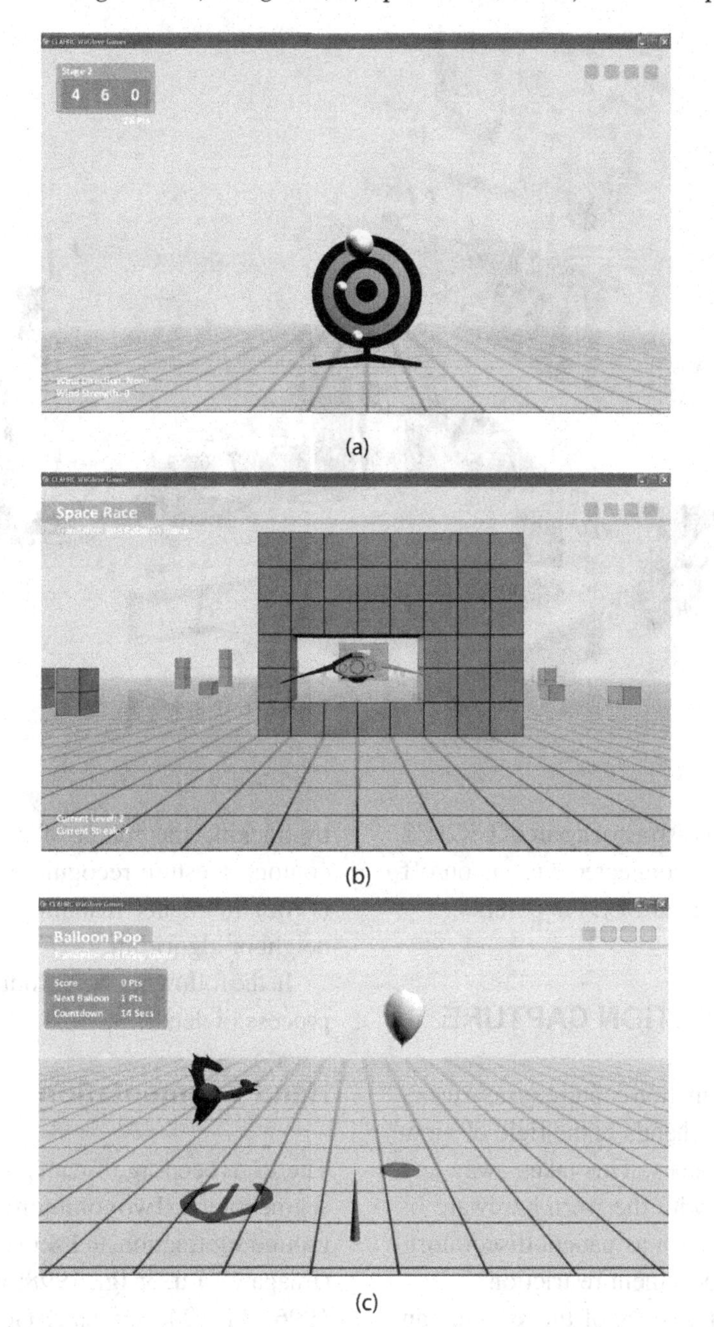

camera outputs from the Wiimote, disregarding all button press and accelerometer data.

The 'IRGlove' (close-ups photographs are shown in Figures 3a and 3b) was developed in order to allow the Wiimotes to track the hand, and must be worn by the patient. Each Wiimote™mote is capable of tracking 4 individual LEDs- and so the design of the glove must reflect this. At the end of each finger of the glove - excepting the ring finger, which is deemed to mimic the middle and little finger in its motion, an LED is mounted. On the back of the hand a power supply is

Figure 2. A typical home installation of the IRGlove system

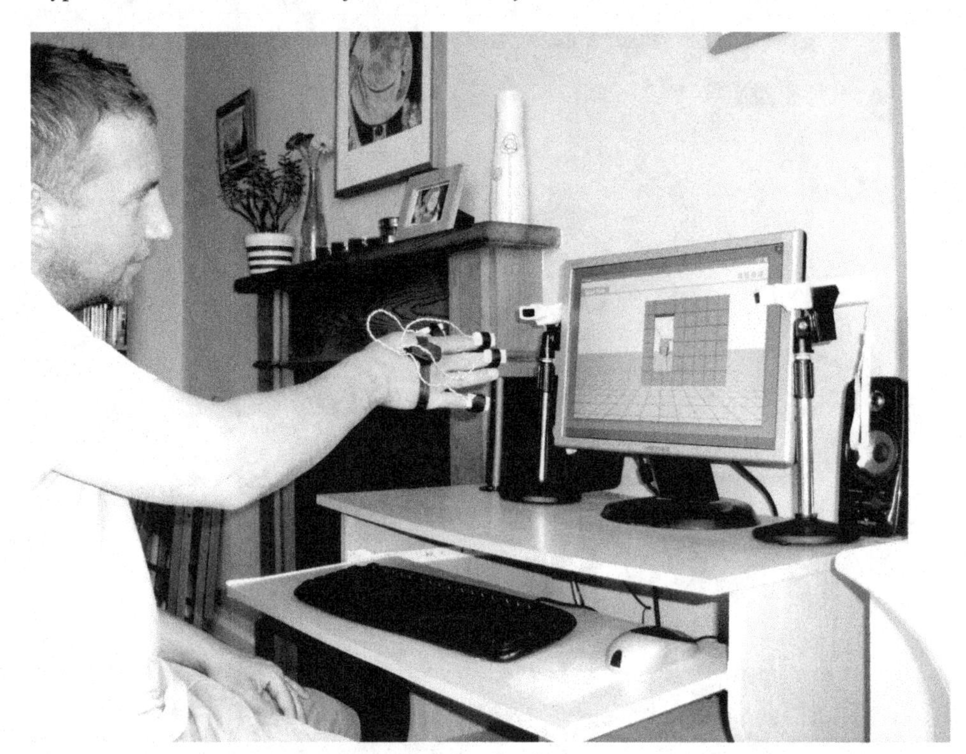

mounted currently consisting of a pair of CR2032 coin cells. These are connected via an on/off switch to the four infrared LEDs in parallel.

MARKERLESS MOTION CAPTURE

Using the markerless motion capture system users can simply place their hands in the field of view of cameras to play games. This takes away any associated problems with the worn hardware of the IRGlove system, such as patient discomfort, hygiene issues and movement restriction.

The development process of the system can be divided into three stages: hand segmentation, hand position tracking and gesture recognition. Segmentation is innovatively done by dual cameras based on skin colour histogram and skin temperature threshold. This approach can provide robust performance in uncontrolled environments. After segmentation, position tracking is performed by tracking the center of moments of the hand contour. Gesture recognition is achieved by Hu (1962) moments matching with the K-nearest neighbor algorithm.

In the following we introduce the development process in detail.

Hand Segmentation

There has been significant previous work on hand segmentation. Two common approaches are background subtraction and skin colour segmentation (Imagawa, Lu, & Igi, 1998; Kjeldsen & Kender, 1996; Raja, McKenna, & Gong, 1998). The fact the system is being designed to be implemented in the patient's home rules out the possible use of background modeling. Colour segmentation is more suitable in our case. However since skin colour may partially overlap with background colour, we add another filter, skin temperature, to get more robust performance.

Figure 3. Views of the IRGlove: a) showing the infrared diodes mounted on the fingertips, and b) showing the view from the back of the hand where the power supply and on/off switch are mounted

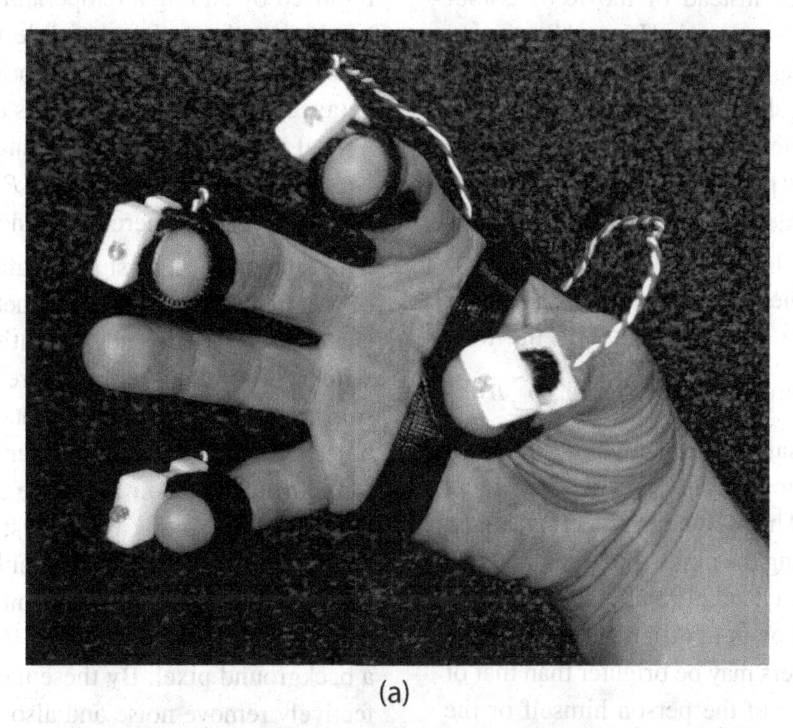

(a)

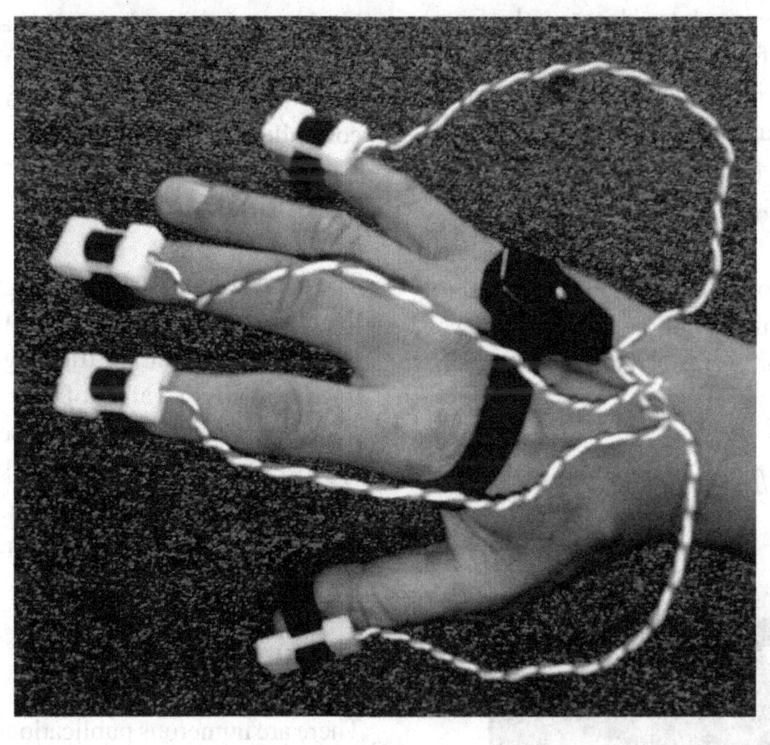

(b)

Skin-colour Based Segmentation: We use the HSV colour space instead of the RGB colour space. Moreover, we use only Hue and Saturation and ignore V (brightness) in order to minimize the influence of shadow and uneven lighting. Given a pixel which has a colour x, we want to classify it as a hand pixel if $P(hand \mid x) > \theta$, where $\theta(0 \sim 1)$ is a predefined probability threshold. $P(hand \mid x) = H(c) / H(C), x \in c$, where $H(c)$ is the frequency of the colour bin c in the skin colour histogram sample H, while $H(C)$ is the total frequency of all the colour bins in H, i.e., $\sum H(c_i)$. The histogram sample needs to be customized before segmentation because users' skin colours can be different. This is done by selecting a hand region and counting the number of pixel for each colour bin. Since a hand normally does not have exactly the same colour in different regions (e.g., the colour of fingers may be brighter than that of the palm) because of the person himself or the light reason, the user needs to select samples from different regions of his hand to get a comprehensive colour histogram) (Figure 4). For each selection s, we can get a histogram h_s. The histogram sample H is the accumulative value $\sum h_s$. Figure 5 shows a skin colour histogram sample.

It can be seen in Figure 4 that the desk is similar in colour to the hand. This could lead to false positive segmentation, as Figure 6 shows. To counter this problem, a thermal camera is introduced, and the extraneous pixels can then be removed by adding a temperature threshold. By combining thermal and visible images, a compound image where each pixel has two dimension data: colour and temperature, is achieved. Given a pixel which has a colour x and temperature t, we classify it as a hand pixel if $P(hand \mid x) > \theta$ & $T_{bottom} < t < T_{top}$, where T_{bottom} and T_{top} are respectively the preset lowest and highest temperature of a human hand. These parameters can also be customized during segmentation processing. After colour and temperature segmentation, sometimes the image still contains small noise objects which have similar colour and temperature to the hand. Since in the game playing context, the hand is normally the largest object, we can further add a noise filter by only selecting the biggest contour in the segmented image. Any pixel outside that contour then is classified to be a background pixel. By these means, we can effectively remove noise and also fill the holes or black dots left inside of the hand, as shown in Figure 6, and Figure 7 show the segmentation result. Note the background environment is uncontrolled in this experiment.

Figure 4. Select different hand regions

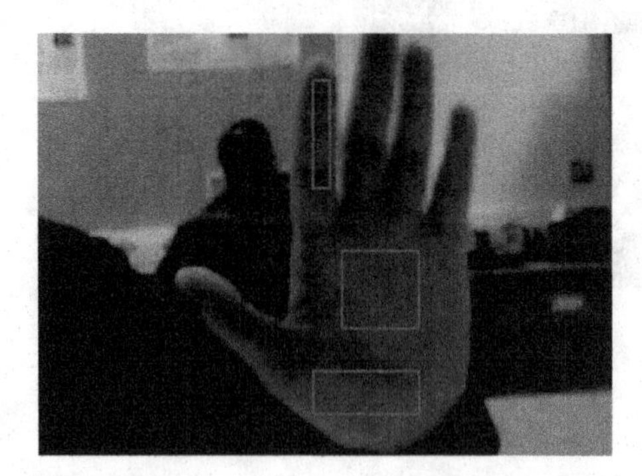

Position Tracking

The hand contour is filled with white colour and then the center of moments of the filled contour is found. The position tracking is then done by tracking the X, Y values of the center point. Relying on accurate preprocessed segmentation, this approach can achieve better tracking performance especially for fast and random movements than the well-known mean shift algorithm (Comaniciu & Meer, 1999)

Gesture Recognition

There are numerous publications on hand gesture recognition. Well-known approaches are Viola and Jones (Viola & Jones, 2001), Hidden Markov

Figure 5. A skin colour histogram H

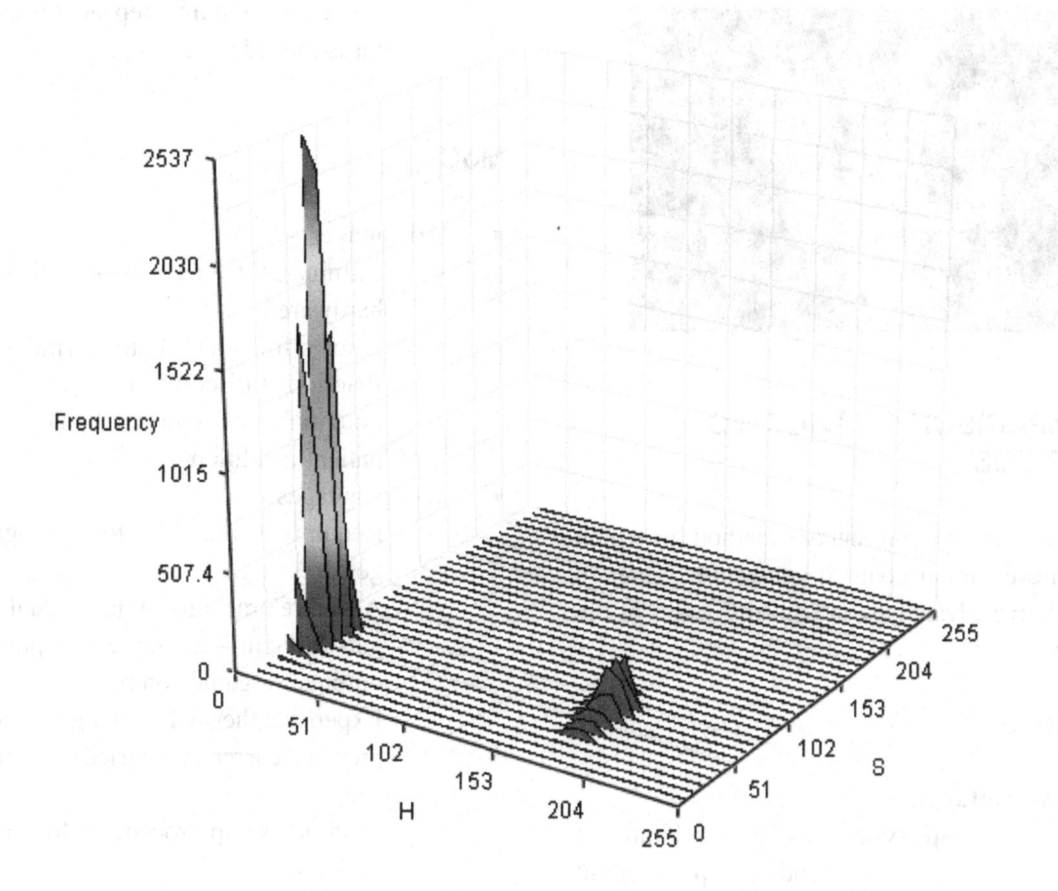

Model (Wilson, Member, Society, Bobick, & Society, 1999) and motion templates (Bobick & Davis, 1996). These approaches require training, and are computationally very expensive, and therefore not feasible for the real-time game scenarios. On other hand, there are many other recognition approaches relying on geometric features, such as convexity defects (Homma & Takenaka, 1985), Hu moments matching (Hu, 1962) and customized geometric detections.

Hu moments matching can achieve reliable performance no matter the size, rotation and location of the image. Here the system is set to detect whether the hand is open and closed. Users need to record 10 ~ 20 templates respectively for hand open and hand closed in advance. During the game the system compares the current hand gesture with those templates and uses the K-nearest neighbor algorithm to find the best matched gesture. This approach can be easily extended to support more gestures in the future.

The rolling of the wrist is detected by different means since a hand does not change its shape when rolling. It is done by detecting the ratio between length and width of the hand contour.

After image processing the motion capture result is sent to the XNA games in real time. Figure 8 and Figure 9 show two of the CLAHRC games controlled by user's hand. The inset images are of the hand as viewed in real-time by the system camera.

Figure 6. Colour-based segmentation

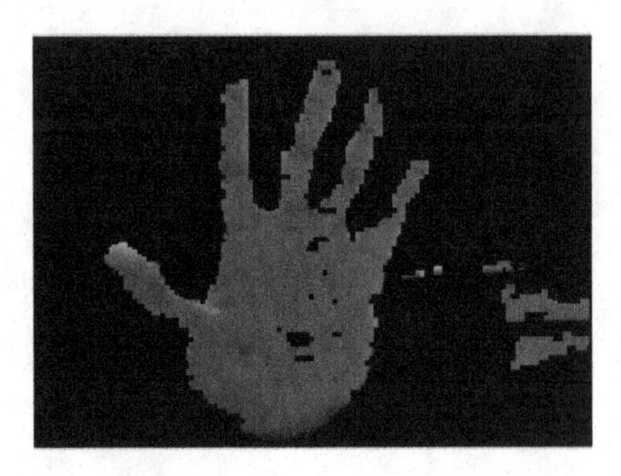

COMPARISON OF THE TWO SYSTEMS

The two systems represent two methods by which to achieve similar goals. It is useful to compare the relative advantages and disadvantages of the two systems:

IRGlove

- **Advantages:**
    - Cheap system using a mixture of off the shelf and cheap to build technology
    - Allows accurate tracking of 3D position of individual fingers giving the ability to detect hand motions in detail
    - Accurate tracking can give more accurate and detailed input to games making them potentially more responsive
- **Disadvantages:**
    - Potential hygiene issues
    - Potential restriction of the user hand movement
    - Potential to cause discomfort to the patient
    - Potentially difficult for patients to fit the glove without assistance
    - Requires maintenance, power supply needs regular recharging
    - Setup and calibration maybe complicated and required repeatedly if the kit is moved after setup.

MoCap

- Advantages
    - Eliminates hygiene issues of worn hardware
    - Allows free hand motion without restriction or discomfort
    - Potential to use thermal data to assess patient condition
- Disadvantages
    - Requires training to recognize gestures
    - Accurate tracking of individual fingers difficult – giving less responsive outputs for game control
    - Expensive thermal tracking technologies are currently restrictive to mass use
    - Complex setup procedures for the kit in the home

Figure 7. Colour + Temperature + Contour segmentation

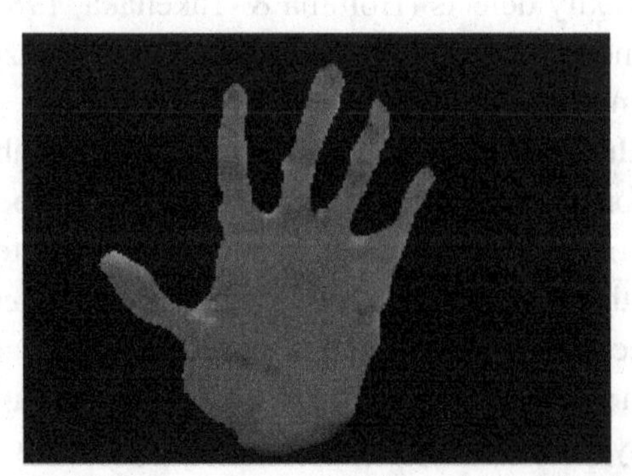

Figure 8. Playing the SlingShot game

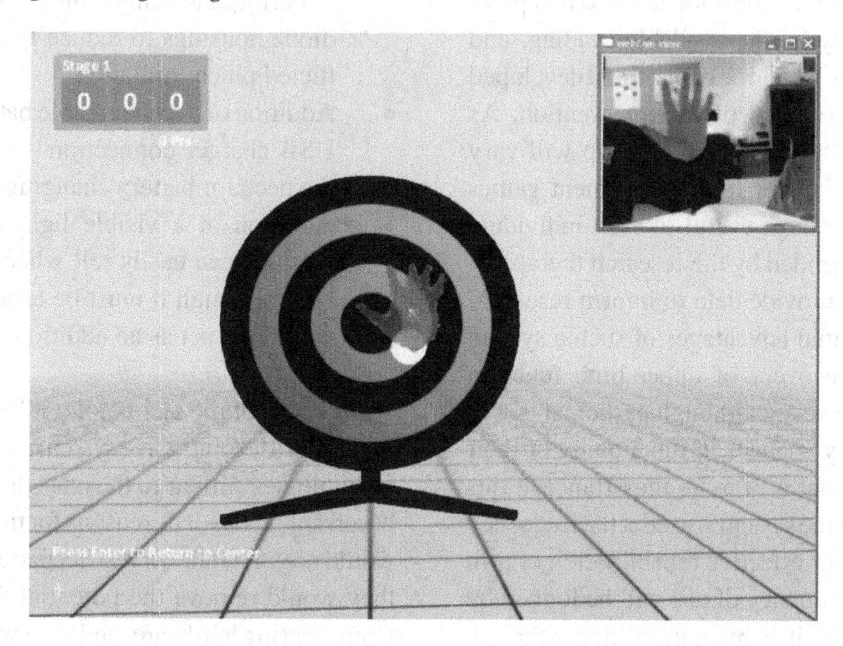

FUTURE WORK

A feasibility trial of the IRGlove system is currently underway. The intention of the trial is to evaluate the system in its current form in the homes of patients recovering from a stroke. Sixty patients are being randomly allocated to either the intervention or control group, the latter receiving only usual care which, with the present level of service provision, may be no rehabilitation at all.

Figure 9. Playing the SpaceRace game

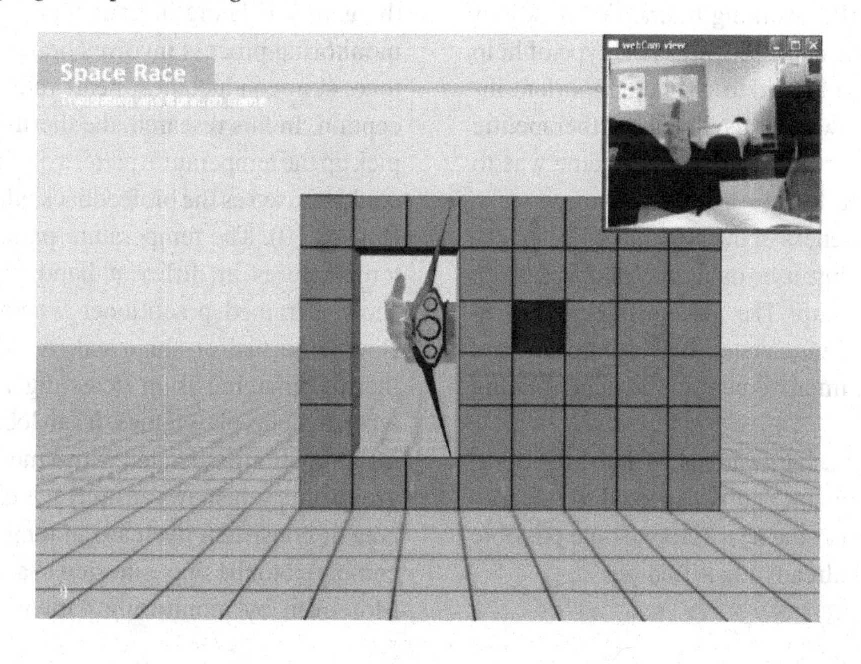

The three games suite developed up to this point was limited only by the available funding, and many more games could be devised and developed to add to the potential of the intervention. As participants in the intervention group will vary in their level of upper limb impairment games which would be most useful to each individual will be recommended by the research therapists. This work will provide data to inform researchers of the potential advantages of such a system in aiding the recovery of upper limb function after stroke. However, although participants will be tested for any changes in functioning of their affected arm, what is of more importance at this stage is whether the system is used at the frequency recommended for effective rehabilitation of arm functioning. Frequency of use will be logged by the computer but it is recognized that although long-term participation of stroke survivors in exercise programmes is universally poor, one reason for any lack of compliance in this study might be technical problems both failure of the system as well as participants' apprehension and unfamiliarity with such technology. For this reason, the study includes face to face support in terms of frequent visits from the research team, including both therapists and computer scientists, and also a help line during working hours. Frequency of visits and phone calls, as well as the type of help required will be logged to provide important information on how much technical and therapeutic help would be required if such a scheme was to be implemented fully.

The effectiveness of the system will be evaluated by analyzing user data and feedback from the trial usergroup. The system is expected to be a useful tool in encouraging, monitoring and evaluating the improvement in the rehabilitating patient.

The IRGlove hardware is in its third iteration of development and is expected to undergo another change in the near future, some possible enhancements already identified are:

- Tapering and lengthening the ends of the diode housings to reduce the hardware inflicted pinch limitation.
- Addition of a rechargeable battery unit with USB charger connection – thus removing the need for battery changing maintenance.
- Addition of a visible light diode so that the user can easily tell when the device is on – although it must be tested as one that would not act as an additional IR source.

In terms of the technology it has been recognized that a less intrusive system using markerless technologies similar to the ones discussed would be advantageous. The reasons for this are that they would not encumber or restrict patient movement, they would remove the potential for discomfort from wearing hardware, and they would be easier to administer (in terms of both removing the requirement for battery changing/maintenance, and removing the hygiene issues associated with worn hardware).

The markerless mocap system is still in its early development stage. Besides further software reliability enhancement, the possibility of using biofeedback detection to monitor stroke rehabilitation patients' physical status when playing the games is being investigated. The traditional monitoring process involves brainwaves, muscle tone, skin conductance, heart rate and pain perception. In this research, the thermal camera can pick up the temperature pattern of the hand, which could be used as the biofeedback when exercising (Figure 10). The temperature pattern represents temperatures in different hand regions. It can show a trained practitioner whether the blood flow is normal or abnormal. A valuable use of thermal imaging is in detecting muscle injury when patients play games. It can locate the area of inflammation associated with a muscle or muscle group injury. It shows atrophy which is seen as an area of consistent decrease in temperature when compared to the opposite side (Sandham, 2005). Moreover, by monitoring patient's temperature pattern we can monitor the current exercise status. With more hand exercises, the temperature pattern will change gradually. For example, when a hand open/close exercise is performed, the temperature of the middle of the palm will rise while the temperature of the finger region will drop because of a change of blood flow. It is proposed that each type of hand exercise corresponds to a rule of change of temperature pattern.

CONCLUSION

It is expected that the trial stages of this project will yield data fuelling future research and investigations into the use of the described technologies in the field of stroke rehabilitation. It is also likely that the testing stage will highlight and instigate further research in both the technologies employed and the scope for their use in other medical rehabilitation fields.

The games were developed at Nottingham Trent University under the CLAHRC-NDL (Collaboration for Leadership in Applied Health Research and Care - Nottinghamshire, Derbyshire and Lincolnshire), an applied health research partnership, funded by the National Institute for Health Research.

REFERENCES

Bobick, A., & Davis, J. W. (1996). Real-time recognition of activity using temporal templates. In *Proceedings of the 3rd IEEE Workshop on Computer Vision*, Sarasota, FL (pp. 39-42).

Brown, D., Standen, P., Barker, M., Battersby, S., Lewis, J., & Walker, M. (2009, October). Designing serious games using Nintendo™'s Wiimote controller for upper limb stroke rehabilitation. In *Proceedings of the 3rd European Conference on Games Based Learning*, Graz, Austria (pp. 68-77).

Burke, J. W., McNeill, M., Charles, D., Morrow, P., Crosbie, J., & McDonough, S. (2009, March 23-24). Serious games for upper limb rehabilitation following stroke. In *Proceedings of the Conference in Games and Virtual Worlds for Serious Applications*, Coventry, UK (pp. 103-110).

Figure 10. Left: temperature pattern. Right: temperature histogram

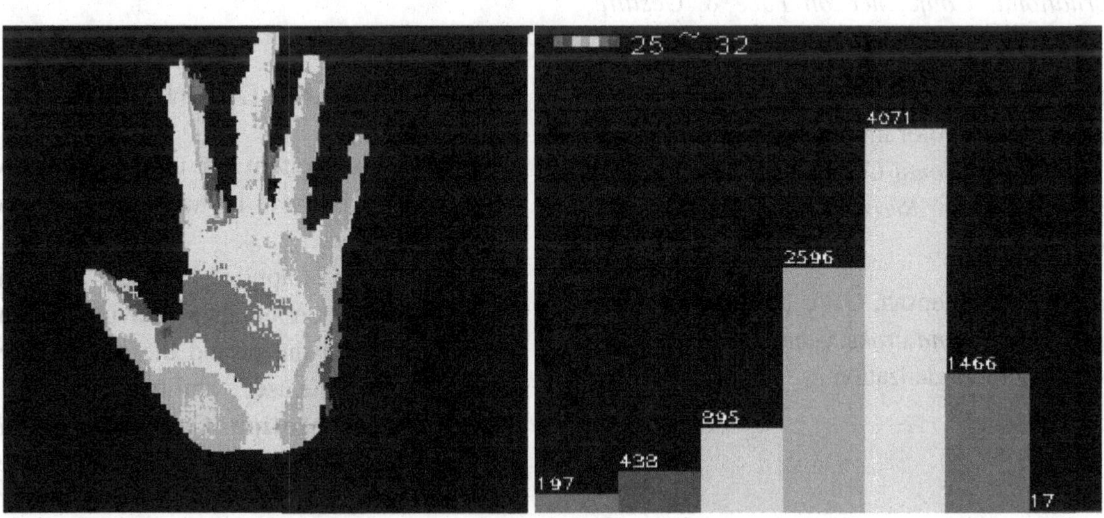

Clarka, R., Bryanta, A., Puab, Y., Bennella, P. M. K., & Hunta, M. (2010). Validity and reliability of the Nintendo™ Wii mote balance board for assessment of standing balance. *Gait & Posture*, *31*(3), 307–310. doi: doi:10.1016/j.gaitpost.2009.11.012

Comaniciu, D., & Meer, P. (1999). Mean shift analysis and applications. In. *Proceedings of the International Conference on Computer Vision, 2*, 1197–1203. doi:10.1109/ICCV.1999.790416

Feys, H. M., De Weerdt, W. J., Selz, B. E., Cox Steck, G. A., Spichiger, R., & Vereeck, L. E. et al. (1998). Effect of a therapeutic intervention for the hemiplegic upper limb in the acute phase after stroke: A single-blind, randomized, controlled multicenter trial. *Stroke*, *29*, 785–792. doi: doi:10.1161/01.STR.29.4.785

Homma, K., & Takenaka, E. (1985). An image processing method for feature extraction of space-occupying lesions. *Journal of Nuclear Medicine*, *26*(12), 1472–1477.

Hu, M. K. (1962). Visual pattern recognition by moment invariants. *I.R.E. Transactions on Information Theory*, *8*, 179–187. doi: doi:10.1109/TIT.1962.1057692

Imagawa, K., Lu, S., & Igi, S. (1998, April 14-16). Color-based hands tracking system for sign language recognition. In *Proceedings of the 3rd International Conference on Face & Gesture Recognition*, Nara, Japan (p. 462).

Kjeldsen, R., & Kender, J. (1996, October 14-16). Finding skin in color images. In *Proceedings of the 2nd International Conference on Automatic Face and Gesture Recognition*, Killington, VT (pp. 312).

Mackay, J., & Mensah, G. A. (2004). *The atlas of heart disease and stroke*. Geneva, Switzerland: World Health Organization.

Microsoft. (2010). *XNA game studio*. Retrieved from http://msdn.microsoft.com/ en-us/library/bb200104.aspx

Raja, Y., McKenna, S. J., & Gong, S. (1998, April 14-16). Tracking and segmenting people in varying lighting conditions using colour. In *Proceedings of the Third IEEE International Conference on Automatic Face and Gesture Recognition*, Nara, Japan (pp. 228-233).

Royal College of Physicians. (2004). *National clinical guidelines for stroke* (2nd ed.). London, UK: Royal College of Physicians.

Sandham, J. (2005). *Medical thermography*. Retrieved from http://www.ebme.co.uk/arts/thermog/

Viola, P., & Jones, M. (2002). *Robust real-time object detection*. Paper presented at the Second International Workshop on Statistical and Computational Theories of Vision – Modeling, Learning, Computing, and Sampling, Vancouver, BC, Canada.

Wilson, A. D., Member, S., Society, I. C., Bobick, A. F., & Society, I. C. (1999). Parametric hidden Markov models for gesture recognition. *IEEE Transactions on Pattern Analysis and Machine Intelligence*, *21*, 884–900. doi: doi:10.1109/34.790429

KEY TERMS AND DEFINITIONS

Calibration System: A calibration allows the user's capability to be accounted for when playing the games.

Force: Force is applied to in the SlingShot rehabilitation system either by moving the hand backwards or via the period the pinch is held depending upon the user's ability.

Markerless Motion Capture Systems: Markerless motion capture systems allow users to simply place their hands in the field of view of cameras to play games. This takes away any associated problems with the worn hardware, such as patient discomfort, hygiene issues and movement restriction.

Serious Games: The application of gamification to traditional environments where entertainment is otherwise secondary to the task at hand. In healthcare this can include more interesting rehabilitation systems.

Wii™ Remote (or Wiimote): A device that was quickly recognized as having potential in the area of rehabilitation, as it can connect to an ordinary PC via Bluetooth giving an ability to track infrared diodes, is available off the shelf and costs only a few pounds.

This work was previously published in the International Journal of Game-Based Learning, Volume 1, Issue 4, edited by Patrick Felicia, pp. 60-73, copyright 2011 by IGI Publishing (an imprint of IGI Global).

Chapter 10
The Psychology of Trolling and Lurking:
The Role of Defriending and Gamification for Increasing Participation in Online Communities Using Seductive Narratives

Jonathan Bishop
Centre for Research into Online Communities and E-Learning Sytems, European Parliament, Belgium

ABSTRACT

The rise of social networking services have furthered the proliferation of online communities, transferring the power of controlling access to content from often one person who operates a system (sysop), which they would normally rely on, to them personally. With increased participation in social networking and services come new problems and issues, such as trolling, where unconstructive messages are posted to incite a reaction, and lurking, where persons refuse to participate. Methods of dealing with these abuses included defriending, which can include blocking strangers. The Gamified Flow of Persuasion model is proposed, building on work in ecological cognition and the participation continuum, the chapter shows how all of these models can collectively be used with gamification principles to increase participation in online communities through effective management of lurking, trolling, and defriending.

INTRODUCTION

The study of online communities has led to such colourful expressions as trolling, flaming, spamming, and flooding being developed in order to describe behaviours that benefit some people while disrupting others (Lampe & Resnick, 2004). Since the proliferation of technologies like the '*circle-of-friends*' (COF) for managing friends lists in online communities (Romm & Setzekom, 2008), the use of the Internet to build online communities, especially using social networking services

DOI: 10.4018/978-1-4666-5071-8.ch010

has grown – but so has the amount of Internet abuse on these platforms. Facebook is currently one of the more popular COF-based websites (Davis, 2008). In addition to this, microblogging, such as Twitter, have 'status updates', which are as important a part of social networks Facebook and Google+, as the circle of friends is. These technologies have made possible the instantaneous expression of and access to opinion into memes that others can access quickly, creating what is called, 'The public square' (Tapscott & Williams, 2010) . The public square is the ability to publish and control editorial policy, and is currently available to all with access to and competency in using the Internet and online social networking services.

It is clear in today's age that there are a lot of demands on people's time, and they have to prioritise which social networking services, or other media or activity they use. This is often based on which is most gratifying and least discomforting. It has become apparent that introducing gaming elements into such environments, where they would not usually be – a concept called 'gamification' – can increase interest and retention in them. Such systems can promote positive activities by members and reduce the number of people not taking part, called 'lurkers' (Bishop, 2009c; Efimova, 2009). It can also promote activities like 'trolling' where content is created for the 'lulz' of it – that is for the fun of it. These can have upsides and downsides, but it is clear gamification can play a part in managing it.

The Problem of Lurking and Trolling Behaviour

Besides social software, gamification and consumerisation have been identified as the big themes for cloud applications (Kil, 2010). Gamification offers online community managers, also known as systems operators (sysops), the opportunity for a structured system that allows for equitable distribution of resources and fair treatment among members. Finding new ways to makes ones' website grow is a challenge for any sysop, so gamification may be the key. Often this is looked on in a technical way, where such platforms are encouraged to move from simple resource archives toward adding new ways of communicating and functioning (Maxwell & Miller, 2008). It is known that if an online community has the right technology, the right policies, the right content, pays attention to the strata it seeks to attract, and knows its purpose and values then it can grow almost organically (Bishop, 2009c). A potential problem stalling the growth of an online community is lack of participation of members in posting content, as even with the right technology there is often still a large number of 'lurkers' who are not participating (Bishop, 2007b). Lurkers are defined as online community members who visit and use an online community but who do not post messages, who unlike posters, are not enhancing the community in any way in a give and take relationship and do not have any direct social interaction with the community (Beike & Wirth-Beaumont, 2005). Lurking is the normal behaviour of the most online community members and reflects the level of participation, either as no posting at all or as some minimal level of posting (Efimova, 2009). Lurkers may have once posted, but remain on the periphery due to a negative experience.

Indeed, it has been shown that lurkers are often less enthusiastic about the benefits of community membership (Howard, 2010). Lurkers may become socially isolated, where they isolate themselves from the peer group (i.e. social withdrawal), or are isolated by the peer group (i.e. social rejection) (Chen, Harper, Konstan, & Li, 2009). Trolling is known to amplify this type of social exclusion, as being a form of baiting, trolling often involved the Troller seeking out people who don't share a particular opinion and trying to irritate them into a response (Poor, 2005).

The Practice of Defriending in Online Communities

While the Circle of Friends allows the different techno-cultures that use online communities to add people as friends, it also gives them the power to remove or delete the person from their social network. This has been termed in the United States of America as 'unfriending' or in the United Kingdom as 'defriending'. Defriending is done for a number of reasons, from the innocent to the malicious to the necessary. For instance, a user can innocently suspend their account or want to 'tidy-up' their Circle of Friends, so that only people they actually know or speak to are in it. There can be malicious and ruthless acts of 'cutting someone dead' or permanently 'sending them to Coventry' so that they are no longer in one's network or able to communicate with oneself (Thelwall, 2009). And users can do it, through a 'blocking' feature to cut out undesirable people who are flame trolling them so much that it impairs their ability to have a normal discourse. Being able to 'block' the people they don't want to associate with, this means that it is impossible for them to reconnect without 'unblocking'. Such practice on social networking sites can lead to users missing out on the context of discussions because they are not able to see hidden posts from the person they blocked or who blocked them, to them seeing ghost-like posts from people whose identities are hidden but whose comments are visible for the same reason. Any form of defriending, whether intended innocently or otherwise, can lead to the user that has been defriended feeling angry and violated, particularly if the rules for killing a community proposed by Powazek (2002) haven't been followed. This can turn the user into an E-Venger, where by the user will seek to get vengeance against the person that defriended them through all means possible. If they're a famous person then this could mean posting less than flattering content on their Wikipedia page or writing negative comments about them in other online communities. If they're a close friend whose personal details they have to hand, then it could mean adding their address to mailing lists, or sending them abusive emails.

Gamification

As of the end of 2010, the Facebook game, Farmville, had more than 60 million users worldwide, or 1 per cent of the world's population with an average of 70 minutes played weekly (Hurley, 2000). Concepts like "Gamification", which try to bring video game elements in non-gaming systems to improve user experience and user engagement (Yukawa, 2005) are therefore going to be an important part in current and future online communities in order to increase participation of constructive users and reduce that of unconstructive users. It seems however the gaming elements of online communities need not be 'designed' by the **sysops**, but developed independently by the users, in some cases unintentionally or unknowingly.

For instance, it has become a game on Twitter for celebrities to try and outdo one another by exploiting the 'trending' feature which was designed to tell users what was popular. Celebrities like the interviewing broadcaster, Piers Morgan, and reality TV personality Alan Sugar talked up in the press their programmes which went head to head, and Ms Morgan claimed victory because he and his guest, Peter Andre, on his Life Stories programme appeared higher in the most mentioned topics on Twitter. Also, consumers joined in this activity which could be called 'ethno-gamification' by agreeing to prefix 'RIP' to various celebrities names in order to get that term to appear in the trending column. In the same way 'hypermiling' has become a term to describe ethno-gamification where people try to compete with one another on how can use the least amount of fuel in their vehicles, so this could be called 'hypertrending' as people seek to try to get certain terms to trend higher than others. Examples of both of these are in Figure 1.

So it seems that gaming is essential to the way humans use computer systems, and is something that needs to be exploited in order to increase participation in online communities, which may not have the membership or status of established platforms like Facebook and Google+. Table 1, presents a restructuring of the extrinsic motivators and mechanical tasks in gamification identified by (Wilkinson, 2006) as interface cues, which are 'credibility markers' which act as mediating artefacts when attached to a user's cognitive artefacts (Bishop, 2005; Norman, 1991; Weiler, 2002). These are categorised according to whether they are 'authority cues,' signalling expertise, or 'bandwagon cues,' which serve as 'social proof' by allowing someone to reply on their peers. These are followed by and inclusion of the UK health authority's guidance on communities and behaviour change (Esposito, 2010; Smith, 1996).

These stimuli and post types will need to be tailored to individuals dependent on their 'player type' and 'character type'. The dictionary, NetLingo identified four types of player type used by trollers; playtime, tactical, strategic, and domination trollers (Leung, 2010). Playtime Trollers are actors who play a simple, short game. Such trollers are relatively easy to spot because their attack or provocation is fairly blatant, and the persona is fairly two-dimensional. Tactical Trollers are those who take trolling more seriously, creating a credible persona to gain confidence of others, and provokes strife in a subtle and invidious way. Strategic Trollers take trolling very seriously, and work on developing an overall strategy, which can take months or years to realise. It can also involve a number of people acting together in order to invade a list. Domination Trollers conversely extend their strategy to the creation and running of apparently bona-fide mailing lists.

UNDERSTANDING ONLINE COMMUNITY PARTICIPATION

Increasing participation in online communities is a concern of most sysops. In order to do this it is important they understand how the behaviour of those who take part in their community affects others' willingness to join and remain on their website.

The Lurker Profile

Lurkers often do not initially post to an online community for a variety of reasons, but it is clear that whatever the specifics of why a lurker is not participating the overall reason is because of the dissonance of their cognitions that they have experienced when presented with a hook into a conversation. Cognitions include goals, plans, values, beliefs and interests (Bishop, 2007b), and may also include 'detachments'. These may include that they think they don't need or shouldn't post or don't like the group dynamics (Preece, Nonnecke, & Andrews, 2004). In addition some of the plans of lurkers causing dissonance has been identified (Preece et al., 2004), including needing to find out more about the group before participating and usability difficulties. The cognitions of 'goals' and 'plans' could be considered to be stored in 'procedural memory, and the 'values' and beliefs could be considered to be stored in 'declarative memory'. The remaining cognitions, 'interest' and 'detachment' may exist in something which the author calls, 'dunbar memory', after Robin Dunbar, who hypothesised that people are only able to hold in memory 150 people at a time. It may be that lurkers don't construct other members as individuals, and don't therefore create an 'interest' causing their detachment cognitions to be dominant. The profile of a reluctant lurker therefore is that of a socially detached actor, fearing consequences of their actions, feeling socially

Figure 1. Piers Morgan's Twitter page and 'RIP Adele' search results showing 'ethno-gamification' in the form of 'hypertrending'

Table 1. Examples of interface cues and guidance for gamification use

Stimulus type (Post type)	Examples of interface cue	Guidance for use as mediating artefacts
Social (Snacking)	'group identity[1]', 'fun[2]', 'love[2]'	Users do perform snacking offer short bursts of content and consume a lot too. To take advantage of this, one should utilise local people's experiential knowledge to design or improve services, leading to more appropriate, effective, cost-effective and sustainable services. In other words allow the community to interact without fear of reprisals
Emotional (Mobiling)	'punishments[2]', 'rewards[2]'	Mobiling is where users use emotions to either become closer to others or make a distance from them. This can be taken advantage of to empower people, through for example, giving them the chance to increase participation, so as to also increase confidence, self-esteem and self-efficacy. This can be done through using leaders and elders to encourage newer members to take part.
Cognitive (Trolling)	'levels[1]', 'learning[2]', 'points[2]'	Trolling as a more generic pursuit seeks to provoke others, sometimes affect their kudos-points with others users. Such users should contribute to developing and sustaining social capital, in order that people see a material benefit of taking part.
Physical (Flooding)	'power[1]', 'mastery[2]'	Flooding is where users get heavily involved with others uses by intensive posting that aims to use the person for some form of gratification. Sysops should encourage health-enhancing attitudes and behaviour, such as encouraging members to abuse the influence they have.
Visual (Spamming)	'leader-boards[1]', 'badges[2]'	Spamming, often associated with unsolicited mail, is in general the practices of making available ones creative works or changing others to increase the success of meetings one's goals. Interventions to manage this should be based on a proper assessment of the target group, where they are located and the behaviour which is to be changed and that careful planning is the cornerstone of success. Designing visual incentives can be effective at reinforcing the message.
Relaxational (Lurking)	'meaning[2]', 'autonomy[1]'	Lurking is enacted by those on the periphery of a community. Their judgements for not taking part often relate to a lack of purpose or control. It is essential to build on the skills and knowledge that already exist in the community, for example, by encouraging networks of people who can support each other. Designing the community around allowing people to both see what others are up to, as well as allowing them to have a break from one another can build strong relationships. A 'd0 not bite the newbies' policy should be enforced.

isolated or excluded, trapped in a state of low flow but high involvement. Lurkers, it has been argued are no more "tied" to an online community than viewers of broadcast television are "tied" to the stations they view (Beenen et al., 2004). However, it can be seen that some more determined lurkers are engaged in a state of flow with low involvement in doubting non-participation. Some have suggested lurkers lack commitment Building and sustaining community in asynchronous learning networks, but they are almost twice as likely to return to the site after an alert (Rashid et al., 2006). Indeed, lurkers belong to the community, and while they decide not to post in it, they are attracted to it for reasons similar to others (Heron, 2009). It has been argued that most lurkers are either shy, feel inadequate regarding a given topic, or are uncomfortable expressing their thoughts in written form (Jennings & Gersie, 1987), but others suggest lurking is not always an ability issue (Sherwin, 2006).

Some researchers characterised lurkers as against hasty conversation rather than a problem for the community (Woodfill, 2009). Often lurkers are afraid of flame wars and potential scrutinising of their comments (Zhang, Ma, Pan, Li, & Xie, 2010). Marked and excessive fear of social interactions or performance in which the person is exposed to potential scrutiny is a core feature of social phobia (Simmons & Clayton, 2010), which has similar facets to lurking (Bishop, 2009d). Perhaps one of the most effective means to change the beliefs of lurkers so that they become novices is for regulars, leaders and elders to nurture novices in the community (Bishop, 2007b). It is known that therapist intervention can help overcome social phobia (Scholing & Emmelkamp, 1993). It could be that through 'private messaging' features that a leader could speak to a registered member who is yet to post. After all, a community is a network of actors where their commonality is their dependence on one another, so feeling a need to be present is essential.

Feelings of uncertainty over the use of posted messages is common to lurkers All social situations carry some uncertainty, which people with social phobia find challenging (Waiton, 2009). Lurking can potentially lead to social isolation, such as not naming anyone outside of their home as a discussion partner (Pino-Silva & Mayora, 2010). Lurkers are less likely to report receiving social support and useful information and often have lower satisfaction levels with group participation sessions (Page, 1999). Leaders can post more messages to encourage all members to post messages (Liu, 2007). Uncertainty caused by poor usability leads to non-participation by lurkers (Preece et al., 2004), and this can be tackled by having the right technology and policies (Bishop, 2009c). Developing trust involves overcoming, particularly in trading communities (Mook, 1987). Such trust was evident in The WELL (Whole Earth 'Lectronic Link), where members use their real names rather than pseudonyms (Rheingold, 2000). Requiring actors to use their real names could help a lurker overcome their uncertainties about others' true intentions.

The Troller Profile

A generic definition of trolling by '*Trollers*' could be 'A phenomenon online where an individual baits and provokes other group members, often with the result of drawing them into fruitless argument and diverting attention from the stated purposes of the group' (Moran, 2007). As can be seen from Table 2, it is possible to map the types of character in online communities identified by (Bishop, 2009b) against different trolling practices. Also included is a set of hypnotised narrator types which affect the approach a particular character can take to influence the undesirable behaviour of others without resorting to defriending, which is explored in the empirical investigation later.

This makes it possibly to clearly see the difference between those who take part in trolling to harm, who could be called 'flame trollers' from those who post constructively to help others, called 'kudos trollers'. A flame is a nasty or insulting message that is directed at those in online communities (Leung, 2010). Message in this context could be seem to be any form of electronic communication, whether text based or based on rich media, providing in this case it is designed to harm or be disruptive. A 'kudos' on the hand can be seen to be a message that is posted in good faith, intended to be constructive.

The Effect of Gamification and Defriending on Online Community Participation

In 2007, as Facebook was emerging, (Bishop, 2007b) presented the ecological cognition framework (see Figure 2). The 'ECF' was able to show the different plans that actors make in online communities based on their different dispositional forces, which created 'neuro-responses' driving them to act, such as 'desires'. Four years earlier in 2003, research was pointing out that there were unique characteristics among those people forming part of the *net generation* (i.e. those born between

Table 2. Troller character types and counter-trolling strengths as narrators

Troller Character Type	Hypothesised Narrator types	Description
Lurker	Stranger	Silent calls by accident, etc., clicking on adverts or 'like' buttons, using 'referrer spoofers', modifying opinion polls or user kudos scores.
Elder	Catalyst	An elder is an out-bound member of the community, often engaging in 'trolling for newbies', where they wind up the newer members often without question from other members.
Troll	Cynic	A Troll takes part in trolling to entertain others and bring some entertainment to an online community.
Big Man	Sceptic	A Big Man does trolling by posting something pleasing to others in order to support their world view.
Flirt	Follower	A Flirt takes part in trolling to help others be sociable, including through light 'teasing'
Snert	Antagonist	A Snert takes part in trolling to harm others for their own sick entertainment
MHBFY Jenny	Pacifist	A MHBFY Jenny takes part in trolling to help people see the lighter side of life and to help others come to terms with their concerns
E-Venger	Fascist	An E-Venger does trolling in order to trip someone up so that their 'true colours' are revealed.
Wizard	Enthusiast	A wizard does trolling through making up and sharing content that has humorous effect.
Iconoclast	Detractor	An Iconoclast takes part in trolling to help others discover 'the truth', often by telling them things completely factual, but which may drive them into a state of consternation. They may post links to content that contradicts the worldview of their target.
Ripper	Rejector	A Ripper takes part in self-deprecating trolling in order to build a false sense of empathy from others.
Chatroom Bob	Striver	A chatroom bob takes part in trolling to gain the trust of others members in order to exploit them.

1977 and 1997). These included having dispositional forces with preference for *surveillance* and *escape*, factors which were not part of the ECF.

These online social networking services have shown that the ties that used to bring people to form online communities are different than what they used to be prior to 2007. The personal homepage genre of online community (Bishop, 2009a) is now the most dominant model of online community enabled through these services. Through actors forming *profiles,* linked together with the circle of friends and microblogging content, they can control the visibility of objects such as actors (e.g. their friends) and artefacts (e.g. the content they want to see). They are in effect creating their own online community dedicated to the people they consider friends.

The Participation Continuum

One of the most important concepts in creating online communities that can harness gamification is the relationship between 'flow' and 'involvement'. When an actor is engaged in a state of flow their concentration is so intense that they forget about their fears and become fully immersed and completely involved in what they are doing (Csikszentmihalyi, 1990). Decision-making in such a state becomes more fluid and actors respond almost without thought for the consequences of their actions. In a high state of flow, Snerts will have low involvement cognitively and post flames with little restraint, often trolling for their own benefit, which then deters lurkers from becoming posters. A structure based on the ecological cognition framework for decision making in human-centred computer systems has been

Figure 2. The ecological cognition framework

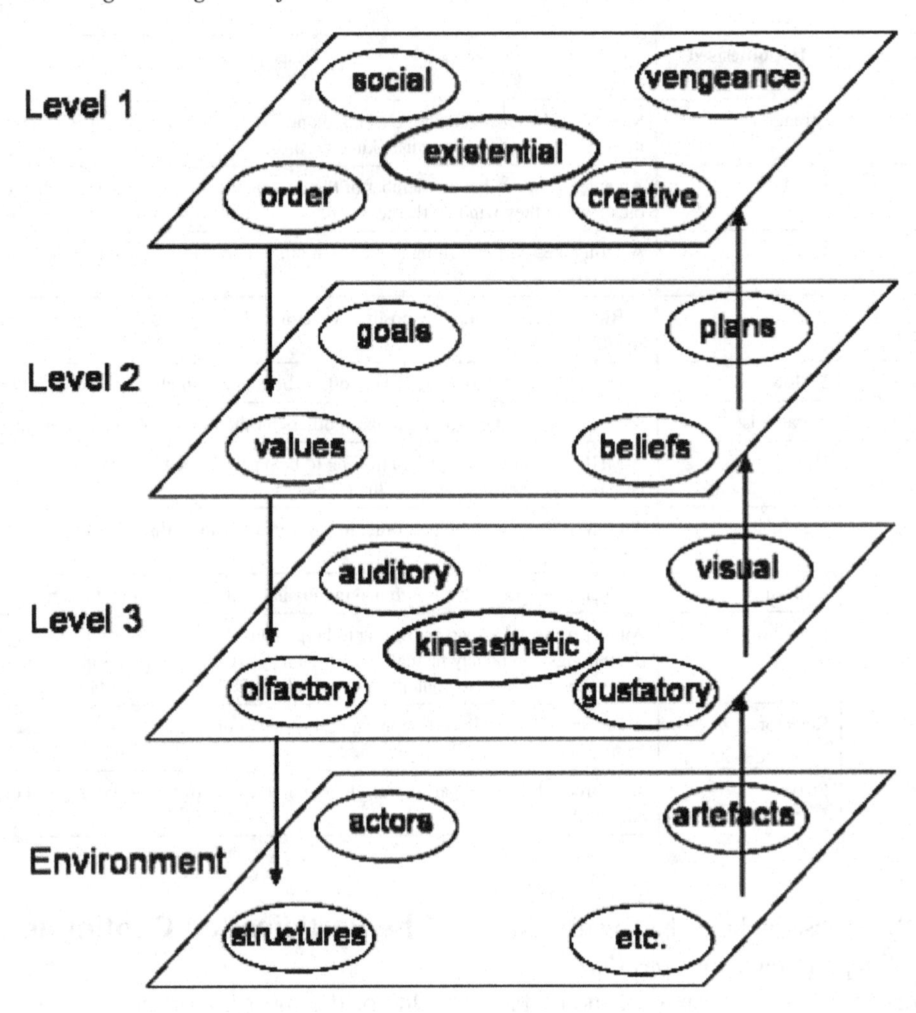

proposed (Bishop, 2007a), which introduced the concepts of deference, intemperance, reticence, temperance and ignorance. This was extended through the participation continuum, to suggest that these cognitive states will lead to empression, regidepression, depression, suppression and repression respectively in the case of the original five judgements (Bishop, 2011b). A six cognitive state, proposed in that paper, reflects the dilemma that lurkers go through, which is compression when they experience incongruence due to congruence when trying to avoid cognitions which are not compatible with their ideal self. Decompression on the other hand is when they start to break this down. These concepts are presented in the model in Figure 3, called the participation continuum.

There appears to be a 'zone of participation dissonance', between the level at which an actor is currently participating and what they could achieve if there was greater support for usability and sociability. This distance between fully 'mediating' their transfer to enhancement of participation could be called the 'Preece Gap', after Jenny Preece, who set out how to design for usability and support sociability (Preece, 2001). As can be seen from the participation continuum in Figure 1, the higher the state of flow for a lurker, the more likely they are to be 'dismediating' from enhancement towards preservation by not to posting due to low involvement. Equally, the higher the state of flow for a poster the more likely they are to keep mediating towards enhancement and

Figure 3. The participation continuum

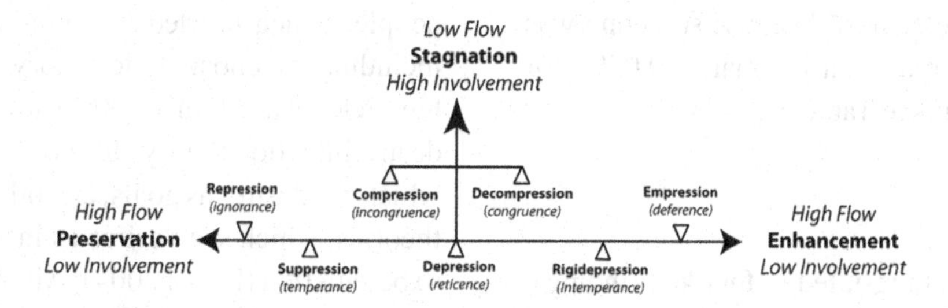

away from preservation within the community with little effort (i.e. involvement). The process in the middle resembles the visitor-novice barrier in the membership lifecycle (Kim, 2000). A lurker who has had bad experiences may be sucked into stagnation through rationalisation of non-participation, going from minimal posting (Efimova, 2009) to lurking (i.e. where they give up posting) and back out again after the intellectualisation process. This resembles a 'battering' cycle (Bishop, 2010), where the actor will be under a barrage of flaming abuse, then be told all is forgiven and they can come back as in (Bishop, 2009b).

AN INVESTIGATION INTO DEFRIENDING IN ONLINE COMMUNITIES

A study was designed to use a narrative analysis to analyse defriending activity and extend the understanding the ECF brings to online community research. Narrative analysis is a tool researchers can use to explore the intersection between the individual and society (Kil, 2010). Narrative analysis in Internet studies essentially uses both text and online "talk" to construct a holistic view of the online interactions, looking at cognition as well as affect (Yukawa, 2005). Narrative analysis is the most prevalent approach that has emphasized alternatives to categorising analysis, but much of narrative research, broadly defined, involves categorising as well as connecting analysis, and the distinction has not been clearly defined (Maxwell & Miller, 2008). Narratives were selected from Google's Blog Search by searching for the terms, "I deleted him as a friend", "he deleted me as a friend". "I deleted her as a friend" and "She deleted me as a friends". The ethnomethodological narrative analysis approach of (Bishop, 2011a) was then used to code the text in the blog posts to identify the different 'Methods', 'Memes', 'Amities', 'Rules' and 'Strategies' that impact on the decision to defriend someone or why someone was defriended.

Descriptives

The difficulties of a romantic relationship accounted for just over 2,700 (13.4%) of the cases where a female was defriended compared to less than 50 (0.47%) for men, suggesting that when a romantic relationship doesn't work out women are more likely to be defriended than men, or at least, people are more likely to disclose on a blog that they defriended a female because of relationship problems than they would males. Less than 20 males were defriended for a sex related issue compared to over 9,500 females. This may be because as Thelwall (2008) suggests, men use online social networking more for dating and women more for other forms of friendship. It became clear in the discourses there were often other people involved in the event leading to a person being defriended. In around 65 per cent of cases where males were defriended and 90 per cent where females were defriended there was another person involved. Over 3,000 females (16.4%) were defriended because someone was offended compared to only 4 males (0.08%) for the same reason (see Table 3).

Results

Analysing the data resulted in four key findings. Firstly, actors are provoked into responding to a state of disequilibrium, such as being defriended. Second, actors need to develop an awareness of the change in the environment before they are able to realise its impact on them. Thirdly, actors will first have a reaction to a state of disequilibrium before organising a response that causes them least dissonance. Fourthly and finally, actors will testify their experiences to others as a way of expressing their understanding in order to restore a state of equilibrium.

Finding 1: Actors are Provoked into Responding to a State of Disequilibrium

Understanding what drives actors to act is crucial to developing human-computer systems that adapt to and influence them. There has been extensive research into discovering what drives people, which has led to a number of theories, including psychoanalytic theory (Freud, 1933), hierarchical needs theory (Maslow, 1943), belief-desire-intention theory (Rao & Georgeff, 1998), which see desires as goals, and other desire-based theories, which see desires as instincts that have to be satisfied (Reiss, 2004). All of these theories suggest that actors are trying to satisfy some internal entity. This assumption ignores the role of the environment in shaping the behaviour of an actor and suggests that actors are selfish beings that only do things for shallow reasons.

There seemed from most of the narratives that there was something in the environment that provoked the actor to write about their defriending action. For instance, Era talking about a male she had known since the age of 12 who "made lots of sexual innuendos and jokes i.e. wolf whistles/ comments about my make up, perfume etc." ended her narrative saying, "I told him goodbye and removed him as a friend on FB. I wished him all the best in his life. Then he replies and says he only likes me as a friend. He denied that he ever flirted with me and said I was crazy and that I over-analyse things," suggesting that recognition

Table 3. Role of different factors in defriending narratives

Defriending discourse type	Males Defriended	Females Defriended
Effect of male on female friend	3,315	19,226
Effect of female on male friend	3,249	18,359
Employment mentioned	2,167	12,951
Sex	11	9,665
Break-ups and Dating	24	2,759
Offence	4	3,372
Little in common	3	1,835
Email related	25	1,386
Text message related	7	0
Application related	1	0
Total	*5,084*	*20,572*

of her experience was important and writing in the blogosphere might be a way she saw to achieve it.

Finding 2: Actors Need to Develop an Awareness of the Change in the Environment before they Are Able to Realise its Impact on Them

It was apparent in the data that those writing their narratives needed to gain an awareness of how the stimulus that provoked them affects them, so that they can understand its impact more appropriately. In one of the weblog narratives, a blogger, Julie, said; "I deleted her as a friend on Facebook because after waiting six months for her to have time to tell me why she was upset with me I got sick of seeing her constant updates (chronic posting I call it)". This supports the view accepted among many psychologists that perception and action are linked and that what is in the environment has an impact on an actor's behaviour. Perceptual psychologists have introduced a new dimension to the understanding of perception and action, which is that artefacts suggest action through offering affordances, which are visual properties of an artefact that determines or indicates how that artefact can be used and are independent of the perceiver (Gibson, 1986). This suggests that when an actor responds to a visual stimulus that they are doing so not as the result of an internal reflex, but because of what the artefact offers.

Finding 3: Actors Will First Have a Reaction to a State of Disequilibrium before Organising a Response that Causes them Least Dissonance

According to Festinger (1957) cognitive dissonance is what an actor experiences when their cognitions are not consonant with each other. For example if an actor had a plan to be social, but a belief that it would be inappropriate they would experience dissonance as a result of their plan not being consonant with their belief. Resolving this dissonance would achieve a state of consonance that would result in either temperance or intemperance. If this actor held a value that stated that they must never be social if it is inappropriate they could achieve consonance by abandoning the plan to be social which results in temperance. If the same actor had an interest in being social and a belief that it was more important to be social than not be social they might resolve to disregard their belief resulting in intemperance. If an actor experiences a desire without experiencing any dissonance they experience deference, as they will act out the desire immediately.

It became quite apparent early on in the analysis that those writing narratives would do to in such a way to cause least dissonance. For instance, one female blogger (Angie) when writing about a relationship breakdown with her friend, said, "I'm not sure if anything I write tonight will make any sense, but it's not as if anyone else reads these anyway so I guess it doesn't really matter how organized I keep it."

Finding 4: Actors Will Testify their Experiences to Others as a Way of Expressing their Understanding in Order to Restore a State of Equilibrium

It became apparent from looking at the weblog entries that bloggers got some sort of closure from writing the narratives. For instance, closing one of her blogs, Angie said, "As you can see, my brain is a ridiculously tangled ball of yarn at the moment and my thoughts are all over the place. Maybe some good old REM's sleep will massage the knots out. Until next time." Psychological closure, it is argued, is influenced by the internal world of cognition as well as the external world of (finished or unfinished) actions and (challenging or unchallenging) life events. Weblogs, according to some, serve similar roles to that of papers on someone's office desk, for example allowing them to deal with emerging insights and difficult to categorise ideas, while at the same time creating opportunities for accidental feedback and impressing those who drop by (Efimova, 2009).

REVIEW OF FINDINGS

The findings when mapped on to the ECF suggest several things. The first is that in online communities a stimulus is presented that provokes an actor into realising that an opportunity exists to post. For instance, a person may read something on an online news website which they disagree with so much that it provokes them into blogging about it. The next stage of the ECF, the impetus is governed by understanding and at is at this stage the actor beings to gain an awareness of how the stimulus affects them. The next stage is the realisation of its relevance to them and where they gain the intention to respond to it. In reference to the earlier example, it may be that the news article is disparaging about a particular cultural group they belong to, and it reignites old memories of discrimination that they want to respond to. The next narrative stage is where the reaction to this knowledge, where they may form a plan to do something about giving them a sense of aspiration. The next stage of the ECF, Judgement, would be where the actor organises their responses to their reaction and weighs up the positives and negatives to acting on it. For example, their head may be flooded with emotions about how they responded to previous situations that were similar, which they may want to write down to contextualise the current situation. Once they have taken the bold step to write the post, they will then testify their opinions at the response stage and may cycle through their thoughts until they have given the response they are comfortable with. Table 4 presents the stages of the ECF and how these related to the findings of this study.

Towards the Gamification Flow of Persuasion Model

The constructivism proposed by Lev Vygotsky in *Mind in Society* (Vygotsky, 1930) says there is a gap between what someone can achieve by themself and what they can achieve with a more competent peer. Vygotsky called this the zone of proximal development, and suggesting that through mediating with artefacts, which the author interprets to include signs such as language or tools such as software, an actor can have help to achieve their potential, in this case in learning. The preeminent *Oxford Dictionary of Law,* which defines a mediator as someone who assists two parties in resolving a conflict but has no decision-making powers, and the process and mediation, supports this conceptualisation proposed by Vygotsky, the author accepts. Equally the term 'dismediation' is the process where an actor, either through reflection or the intervention of another actor returns to a former state of preserving their original status quo. The example given in some texts on cognitive dissonance is where a consumer orders a car from a dealer and then experiences doubt over whether they made the right decision. It has been argued that a courtesy call can help an actor feel more confidence in their decision and reduce the experience, which I call reticence, as an intervention to create mediation towards enhancement, which in this case is the benefit from a new car, which acts as the 'seduction mechanism'. The seduction mechanism in this context refers to an intervention that stimulates substantial change in an actor's goals, plans, values, beliefs, interests and detachments. An example, which can be found in the existing literature (Bishop, 2007c), is where someone who has been lurking is presented with a post that provokes them so much they feel compelled to reply. However, it is clear that not everyone reacts the same way to a seduction mechanism, as some may take longer to fully change their behaviour than others. A framework is therefore needed to explain these differences, and an extension to the participation continuum is presented in Figure 4.

DISCUSSION

Encouraging participation is one of the greatest challenges for any e-community provider. Attracting new members is often a concern of many small

Table 4. Description of stages of the ECF with reference to narrative stages

ECF Stage	Narrative Stage	Description
Stimuli	Provocation	There is a spark that makes someone want to post to an online community. This stimulus provokes an actor into seizing the opportunity to make a contribution.
Impetus	Awareness	Once someone has been given an incentive to post the next stage is to get an understanding of what they can do through gaining an awareness of what has happened.
Intent	Realisation	Once someone has an awareness of how an opportunity affects them the next stage is for them to realise how relevant it is to them to give them the intention to go further.
Neuroresponse	Reaction	Once someone has realised the relevance of a particular action to them they react to it without knowing the consequences giving them a feeling of aspiration.
Judgement	Organisation	Once someone has aspired to a particular course of action they may experience dissonance through organising the proposed action in line with their thoughts. They or their nervous system will then make the choice to take a particular action.
Response	Testimony	Once someone has made the judgement to take a particular action the next stage is to express that choice. In terms of narratives this is their testimony, which may encompass the various aspects of the previous stages.

online communities, but in larger e-communities which are based on networks of practice, the concern is often retaining those members who make worthwhile contributions. These communities still have their 'classical lurkers' who have never participated, but they also appear to what could be called 'outbound lurkers', referred to as elders, who used to participate frequently, but now no longer do as much. One reason for this is that the actors have lost their ties through being 'defriended' by other actors in the network. Some of the reasons for this defriending behaviour has been explored in this chapter. They vary from issues in the workplace to difficulties in romantic relationships, whether romantic partners or strangers who take part in flame trolling. What is clear that defriending has an impact on those affected by them and are explained in the narratives they produce on weblogs. This suggests that while defriending can have an impact in one community, such as causing 'outbound lurking', it can increase participation in another. Actors will always have a desire to share their experiences, and as has been shown through this chapter they follow a clear six-part cycle in expressing themselves, and their narratives take on 10 different personas based on their individual differences. It could therefore be concluded that one online communities loss is another's gain, as participation in these environments has now become so pervasive that if a person is forced not

Figure 4. The gamification flow of persuasion model

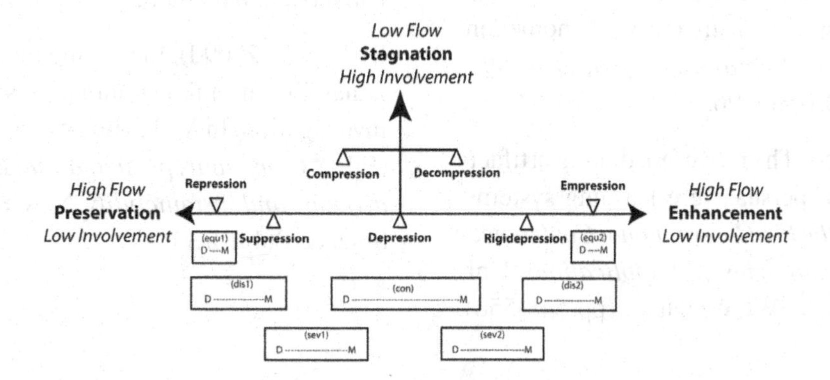

to participate in them and therefore become an 'outbound lurker,' or elder, they can always find another to meet their desires to express themselves.

This chapter has argued that essential to ensuring 'responsible trolling' is the use of gamification techniques. Gamification introduces elements from video gaming, such as points and leader-boards in order to incentivise positive behaviours and disincentivise negative ones. A model, called the 'gamified flow of persuasion' is presented, which builds on the earlier participation continuum. This explains how gamification based systems can be designed so as to help users transfer from one level of participation to another.

ACKNOWLEDGMENT

The author would like to thank those who provided feedback to earlier versions of this paper. In particular he would like to thank Jean Bishop for the thorough proof reading and for suggesting the names of the narrator types.

REFERENCES

Beenen, G., Ling, K., Wang, X., Chang, K., Frankowski, D., Resnick, P., & Kraut, R. E. (2004). Using social psychology to motivate contributions to online communities. *Proceedings of the 2004 ACM Conference on Computer Supported Cooperative Work,* (p. 221).

Beike, D., & Wirth-Beaumont, E. (2005). Psychological closure as a memory phenomenon. *Memory (Hove, England), 13*(6), 574–593. PubMed doi:10.1080/09658210444000241

Bishop, J. (2005). The role of mediating artifacts in the design of persuasive e-learning systems. *Proceedings of the First International Conferences on Internet Technologies and Applications,* University of Wales, NEWI, Wrexham, (pp. 548-558).

Bishop, J. (2007a). Ecological cognition: A new dynamic for human-computer interaction. In B. Wallace, A. Ross, J. Davies, & T. Anderson (Eds.), *The mind, the body and the world: Psychology after cognitivism* (pp. 327–345). Exeter, UK: Imprint Academic.

Bishop, J. (2007b). Increasing participation in online communities: A framework for human–computer interaction. *Computers in Human Behavior, 23*(4), 1881–1893. doi:10.1016/j.chb.2005.11.004

Bishop, J. (2007c). Increasing participation in online communities: A framework for human–computer interaction. *Computers in Human Behavior, 23*(4), 1881–1893. doi:10.1016/j.chb.2005.11.004

Bishop, J. (2009a). Enhancing the understanding of genres of web-based communities: The role of the ecological cognition framework. *International Journal of Web Based Communities, 5*(1), 4–17. doi:10.1504/IJWBC.2009.021558

Bishop, J. (2009b). Increasing capital revenue in social networking communities: Building social and economic relationships through avatars and characters. In S. Dasgupta (Ed.), *Social computing: Concepts, methodologies, tools, and applications* (pp. 1987–2004). Hershey, PA: IGI Global. doi:10.4018/978-1-60566-984-7.ch131

Bishop, J. (2009c). Increasing membership in online communities: The five principles of managing virtual club economies. *Proceedings of the 3rd International Conference on Internet Technologies and Applications - ITA09,* Glyndwr University, Wrexham.

Bishop, J. (2009d). Increasing the economic sustainability of online communities: An empirical investigation. In M. F. Hindsworth, & T. B. Lang (Eds.), *Community participation: Empowerment, diversity and sustainability*. New York, NY: Nova Science Publishers.

Bishop, J. (2010). *Multiculturalism in intergenerational contexts: Implications for the design of virtual worlds.* Paper Presented to the Reconstructing Multiculturalism Conference, Cardiff, UK.

Bishop, J. (2011a). *The equatrics of intergenerational knowledge transformation in techno-cultures: Towards a model for enhancing information management in virtual worlds. Unpublished MScEcon.* Aberystwyth, UK: Aberystwyth University.

Bishop, J. (2011b). Transforming lurkers into posters: The role of the participation continuum. *Proceedings of the Fourth International Conference on Internet Technologies and Applications (ITA11),* Glyndwr University.

Chen, Y., Harper, F. M., Konstan, J., & Li, S. X. (2009). Group identity and social preferences. *The American Economic Review, 99*(1). doi:10.1257/aer.99.1.431

Csikszentmihalyi, M. (1990). *Flow: The psychology of optimal experience.* New York, NY: Harper & Row.

Davis, S. (2008). With a little help from my online friends: The health benefits of internet community participation. *The Journal of Education, Community and Values, 8*(3).

Efimova, L. (2009). Weblog as a personal thinking space. *Proceedings of the 20th ACM Conference on Hypertext and Hypermedia,* (pp. 289-298).

Esposito, J. J. (2010). Creating a consolidated online catalogue for the university press community. *Journal of Scholarly Publishing, 41*(4), 385–427. doi:10.3138/jsp.41.4.385

Festinger, L. (1957). *A theory of cognitive dissonance.* Evanston, IL: Row, Peterson.

Freud, S. (1933). *New introductory lectures on psycho-analysis.* New York, NY: W.W. Norton & Company, Inc.

Gibson, J. J. (1986). *The ecological approach to visual perception.* Lawrence Erlbaum Associates.

Heron, S. (2009). Online privacy and browser security. *Network Security,* (6): 4–7. doi:10.1016/S1353-4858(09)70061-3

Howard, T. W. (2010). *Design to thrive: Creating social networks and online communities that last.* Morgan Kaufmann.

Hurley, P. J. (2000). *A concise introduction to logic.* Belmont, CA: Wadsworth.

Jennings, S., & Gersie, A. (1987). *Drama therapy with disturbed adolescents* (pp. 162–182).

Kil, S. H. (2010). Telling stories: The use of personal narratives in the social sciences and history. *Journal of Ethnic and Migration Studies, 36*(3), 539–540. doi:10.1080/13691831003651754

Kim, A. J. (2000). *Community building on the web: Secret strategies for successful online communities.* Berkeley, CA: Peachpit Press.

Lampe, C., & Resnick, P. (2004). Slash (dot) and burn: Distributed moderation in a large online conversation space. *Proceedings of the SIGCHI Conference on Human Factors in Computing Systems,* (pp. 543-550).

Leung, C. H. (2010). Critical factors of implementing knowledge management in school environment: A qualitative study in Hong Kong. *Research Journal of Information Technology, 2*(2), 66–80. doi:10.3923/rjit.2010.66.80

Maslow, A. H. (1943). A theory of motivation. *Psychological Review, 50*(4), 370–396. doi:10.1037/h0054346

Maxwell, J. A., & Miller, B. A. (2008). Categorizing and connecting strategies in qualitative data analysis. In P. Leavy, & S. Hesse-Biber (Eds.), *Handbook of emergent methods* (pp. 461–477).

Mook, D. G. (1987). *Motivation: The organization of action*. London, UK: W.W. Norton & Company Ltd.

Moran, J. (2007). Generating more heat than light? Debates on civil liberties in the UK. *Policing*, *1*(1), 80. doi:10.1093/police/pam009

Norman, D. A. (1991). Cognitive artifacts. In J. M. Carroll (Ed.), *Designing interaction: Psychology at the human-computer interface* (pp. 17–38). New York, NY: Cambridge University Press.

Page, S. E. (1999). Computational models from A to Z. *Complexity*, *5*(1), 35–41. doi:10.1002/(SICI)1099-0526(199909/10)5:1<35::AID-CPLX5>3.0.CO;2-B

Pino-Silva, J., & Mayora, C. A. (2010). English teachers' moderating and participating in OCPs. *System*, *38*(2). doi:10.1016/j.system.2010.01.002

Poor, N. (2005). Mechanisms of an online public sphere: The website slashdot. *Journal of Computer-Mediated Communication*, *10*(2).

Powazek, D. M. (2002). *Design for community: The art of connecting real people in virtual places*. New Riders.

Preece, J. (2001). *Online communities: Designing usability, supporting sociability*. Chichester, UK: John Wiley & Sons.

Preece, J., Nonnecke, B., & Andrews, D. (2004). The top 5 reasons for lurking: Improving community experiences for everyone. *Computers in Human Behavior*, *2*(1), 42.

Rao, A. S., & Georgeff, M. P. (1998). Decision procedures for BDI logics. *Journal of Logic and Computation*, *8*(3), 293. doi:10.1093/logcom/8.3.293

Rashid, A. M., Ling, K., Tassone, R. D., Resnick, P., Kraut, R., & Riedl, J. (2006). Motivating participation by displaying the value of contribution. *Proceedings of the SIGCHI Conference on Human Factors in Computing Systems*, (p. 958).

Reiss, S. (2004). Multifaceted nature of intrinsic motivation: The theory of 16 basic desires. *Review of General Psychology*, *8*(3), 179–193. doi:10.1037/1089-2680.8.3.179

Rheingold, H. (2000). *The virtual community: Homesteading on the electronic frontier* (2nd ed.). London, UK: MIT Press.

Romm, C. T., & Setzekom, K. (2008). *Social network communities and E-dating services: Concepts and implications*. London, UK: Information Science Reference. doi:10.4018/978-1-60566-104-9

Scholing, A., & Emmelkamp, P. M. G. (1993). Exposure with and without cognitive therapy for generalized social phobia: Effects of individual and group treatment. *Behaviour Research and Therapy, 31*(7), 667–681. PubMed doi:10.1016/0005-7967(93)90120-J

Sherwin, A. (2006, April 3). A family of Welsh sheep - The new stars of Al-Jazeera. *Times (London, England)*, (n.d), 7.

Simmons, L. L., & Clayton, R. W. (2010). The impact of small business B2B virtual community commitment on brand loyalty. *International Journal of Business and Systems Research*, *4*(4), 451–468. doi:10.1504/IJBSR.2010.033423

Smith, G. J. H. (1996). Building the lawyer-proof web site. Paper presented at the *Aslib Proceedings, 48,* (pp. 161-168).

Tapscott, D., & Williams, A. D. (2010). *Macrowikinomics: Rebooting business and the world*. Canada: Penguin Group.

Thelwall, M. (2008). Social networks, gender, and friending: An analysis of MySpace member profiles. *Journal of the American Society for Information Science and Technology*, *59*(8), 1321–1330. doi:10.1002/asi.20835

Thelwall, M. (2009). Social network sites: Users and uses. In M. Zelkowitz (Ed.), *Advances in computers: Social networking and the web* (p. 19). London, UK: Academic Press. doi:10.1016/S0065-2458(09)01002-X

Vygotsky, L. S. (1930). *Mind in society*. Cambridge, MA: Waiton, S. (2009). Policing after the crisis: Crime, safety and the vulnerable public. *Punishment and Society*, *11*(3), 359.

Weiler, J. H. H. (2002). A constitution for Europe? Some hard choices. *Journal of Common Market Studies*, *40*(4), 563–580. doi:10.1111/1468-5965.00388

Wilkinson, G. (2006). Commercial breaks: An overview of corporate opportunities for commercializing education in US and English schools. *London Review of Education*, *4*(3), 253–269. doi:10.1080/14748460601043932

Woodfill, W. (2009, October 1). The transporters: Discover the world of emotions. *School Library Journal Reviews,* 59.

Yukawa, J. (2005). Story-lines: A case study of online learning using narrative analysis. *Proceedings of the 2005 Conference on Computer Support for Collaborative Learning: Learning 2005: The Next 10 Years!* (p. 736).

Zhang, P., Ma, X., Pan, Z., Li, X., & Xie, K. (2010). Multi-agent cooperative reinforcement learning in 3D virtual world. In *Advances in swarm intelligence* (pp. 731–739). London, UK: Springer. doi:10.1007/978-3-642-13495-1_90

KEY TERMS AND DEFINITIONS

Elder: An elder is a person who is out-bound from participating in an online community, who whilst having made lots of contributions are no longer interesting in doing so. Where they cause mischief with newer members it is called 'trolling for newbies.'

Eyeballers: The collective term for Elders, Trolls and Lurkers is eyeballer, which refers to people who participate mainly at the opportune moment for them.

Flame Trolling: Flame trolling is the posting of messages to the Internet in order to get a negative response. Where this is to make oneself laugh while harming others it is known as 'trolling for the lulz.'

Kudos Trolling: Kudos trolling is the posting of messages to the Internet in order to get a positive response. Where this is to encourage others to laugh it is known as 'trolling for the lolz.'

Lurker: A lurker is a person whose only participation in an online community is browsing its contents.

Lurking: Lurking is a persistent state of non-participation in posting to an online community.

Troll: A Troll is a person who posts provocative messages on the Internet for humorous effect. It contrasts with a troll (i.e. with a lower-case 't') who posts messages to offend others.

Troller: A troller is person who posts to a website in order to entice others into respond.

Trolling: Trolling is the posting of provocative or offensive messages on the Internet.

This work was previously published in Virtual Community Participation and Motivation: Cross-Disciplinary Theories, edited by Honglei Li, pp. 160-176, copyright 2012 by Information Science Reference (an imprint of IGI Global).

Section 3
Pedagogical Issues

Chapter 11
Designing Educational Games: A Pedagogical Approach

Stephen Tang
Liverpool John Moores University, UK

Martin Hanneghan
Liverpool John Moores University, UK

ABSTRACT

Play has been an informal approach to teach young ones the skills of survival for centuries. With advancements in computing technology, many researchers believe that computer games[1] can be used as a viable teaching and learning tool to enhance a student's learning. It is important that the educational content of these games is well designed with meaningful game-play based on pedagogically sound theories to ensure constructive learning. This chapter features theoretical aspects of game design from a pedagogical perspective. It serves as a useful guide for educational game designers to design better educational games for use in game-based learning. The chapter provides a brief overview of educational games and game-based learning before highlighting theories of learning that are relevant to educational games. Selected theories of learning are then integrated into conventional game design practices to produce a set of guidelines for educational games design.

INTRODUCTION

Computer gaming is an extremely popular trend among youth in the 21st century (Pearce, 2006) yet is often seen as a concern by the general public with the potential harm it may introduce based on studies of video gaming effects in the 1980's and 1990's. But should those concerns neglect the educational potential of computer games? Computer games are able to generate enormous levels of motivational drive for game players as opposed to formal classes which are perceived as

DOI: 10.4018/978-1-4666-5071-8.ch011

"boring" or rather "dry" (BECTa, 2006; Prensky, 2002). The energy that game players often invest for computer games is phenomenal. Though some may comment that aspects of learning in computer games may not be suitable for academic learning, e.g. (Adams, 2005), nevertheless exploiting such technology to aid learning is still possible when used appropriately.

Educational games, also known as *instructional games*, take advantage of gaming principles and technologies to create educational content. Early versions of educational games were often incarnations of interactive multimedia courseware that incorporated simple mini-games such as puzzles and memory games as rewards attempting to inject fun into learning (albeit often developed by inexperienced educational game designers). Most educational games were developed for children who have lower expectations of interactive content as compared to teenagers and adults. Financial and technological constraints presented major barriers to production of high quality educational games that could meet teenage and adult expectations, and such constraints still exist today. Hence there is a common misconception that educational games are simply for children.

In actual fact, there are a number of educational games for adults aimed mainly in medical (Moreno-Ger, Blesius, Currier, Sierra, & Fernández-Manjón, 2008) and business education (Faria, 1998). *Training simulators* (a term more familiar to the adult population) simulate real-world experience intended for development of skills where the challenges presented accurately replicate real-world scenarios requiring the user to overcome problems using realistic procedural acts defined through hardware interfaces. Training simulators are most popular in the fields of aviation (Telfer, 1993), medicine (Colt, Crawford, & III, 2001) and military applications (Nieborg, 2004). *Serious games* is a more recent term used for representing software applications that employ gaming principles and technologies for non-entertainment purposes including education and training (Sawyer & Smith, 2008; Zyda, 2005).

Designing games with good game-play is not a science or an art, but often quoted as a 'craft' requiring skills to engage and immerse game players in a realistic setting while also encouraging replayability. Game designers are brilliant at creating "hooks" to engage gamers, but in the context of game-based learning it is important to emphasize the aspects of academic value that can develop skills that are useful to the learner. This chapter presents a general model and guidelines for designing educational games by incorporating theories of learning into games design practices. Variables influencing learning and a selection of theories of learning related to educational games design are described before being mapped to elements of game design to form guidelines for designing educational games. Some conclusions from this work are presented at the end of this chapter.

VARIABLES INFLUENCING LEARNING

Learning is generally perceived as the process of acquiring new knowledge which often takes place in a formal classroom setting. However, learning can also take place informally after school hours through interactions with peers and the surrounding environment making learning a constant process. The ability to learn and adapt are crucial in our daily lives and has a direct relationship with human performance when executing tasks. Learning is enriched us with knowledge and skills gained through experience from direct and indirect interaction with the subject matter which proves useful in future similar events and scenarios.

Learning as a cognitive process is affected by a number of psychological factors which can be categorised as internal or external. Internal factors are factors originated by the learner them self and are closely related to the functioning of the human mind and emotions (Bransford, L., & Crocking, 1999). External factors can be those that are sourced from teachers, peers or the environment for example (Hattie, 2005). Some internal factors can be controlled using appropriate pedagogy to ensure learners achieve the learning objectives associated to a lesson, while others require discipline, dedication and effort from learners. Learning requires full commitment from participants as learners themselves account for 50% variance of achievement, whereas the teacher (amongst the external factors) contribute to a learner's achievement with a variance of 30% (Hattie, 2005).

It is important to recognise that teachers and effective teaching strategies can impact learning. Creemers (1994) identified nine variables relating to teaching instruction as *advance organisers*; *evaluation*; *feedback*; *corrective instruction*; *mastery learning*; *ability grouping*; *homework*; *clarity of presentation* and *questioning*. Meta-analysis on teaching and learning can also provide additional pointers on variables that influence learning. Hattie (2003) in his study identified 33 variables. Amongst the additional variables that provide a positive effect on learning are *quality of instruction*; *class environment; challenge of goals*; *peer tutoring*; *teaching style; peer effects*; *simulation and games*; *computer-assisted instruction*; *testing*; *instructional media*; *programmed instruction*; *audio-visual aids*; *individualisation* and *behavioural objective*. These influential variables are essential inputs for designing educational games that can maximise the positive effect on learners.

In a recent quantitative meta-analysis on computer games as learning tools, 35 of the 65 studies reported that computer games have significant positive effects on learning, 17 with mixed results, 12 indicated similar effects to conventional instruction and one study reported otherwise (Ke, 2009). It is not surprising that the correlation between computer games and learning has increased the effect size from 0.34 (Hattie, 2003) to 2.87 (Marzano, 2009) as computer games have emerged as popular culture among the multimedia generation. These findings affirm the claims by supporters of game-based learning that computer games have tremendous potential for positive learning. However, the extent of this impact is still very much dependent on the abilities of educational game designers.

THEORIES OF LEARNING

Applications of theories of learning often relate to instructional design concerned with research and theory about instructional strategies and the process of developing and implementing those strategies. In the context of designing game-based learning content, it is important that educational games are developed using pedagogically sound theories that are relevant to instructional design. For brevity we will therefore focus only on theories of learning which are deemed useful in designing educational games.

SHAPING LEARNERS' BEHAVIOUR

A crucial part of learning is to shape the learner's behaviour by encoding knowledge of cause and effect. *Behaviourism* is built upon the belief that learners can be conditioned to make a response in which learning is perceived as the result of association forming between stimuli and responses. Thorndike's 'Law of Effect' (Thorndike, 1933) suggests that stimulus and response can be strengthened through rewards and can, over time, become habitual. Conversely, the connection between stimulus and response is weakened when discomfort is experienced. Skinner's 'Operant Conditioning theory' (Skinner, 1935) states that learners can be conditioned to respond to stimulus through reinforcement, and over time continue to behave as such even when the stimuli is not present.

Hull's 'Drive Reduction theory' (1951) focuses more on human behaviour and argues that motivation is essential in order for responses to occur. It suggests that responses can become habitual when the stimuli and response cycle is reinforced reducing the latency of such responses. His framework can be closely tied with Maslow's 'Hierarchy of Needs theory' (Maslow, 1946) and the work of Thorndike and Skinner.

Weiner's 'Attribution theory' (1979) extends the concepts of motivation and his findings on attributions to achievement; ability, effort, task difficulty and luck (luck is interpreted as being at the right place in right time and the right execution of action) is also worthy of mentioning in this chapter.

ADAPTING TO LEARNERS' COGNITIVE NEEDS

Human cognition is a combination of short-term and long-term memory. Short-term memory provides the working memory for cognitive processing, while long-term memory stores cognitive constructs which are composed of multiple elements of information grouped into one single representation. Miller's 'Information Processing theory' (2003) presents an interesting principle on the limitation of a human's cognitive capacity reporting that an average human can remember between five to nine elements of information in his infamous 'seven ± two' theory. It explains why humans cluster elements of information to address the capacity of short-term memory before encoding it permanently into the long-term memory.

Sweller's 'Cognitive Load theory' (1994) focuses on the interaction between information structure and working memory in relation to learning. It suggests that cognitive load should be kept to a minimum in the learning process for optimal learning through elimination of working memory load for tasks which can be completed physically, usage of worked examples of problem solving methods and usage of audio and visual components to increase working memory capacity.

In designing learner-oriented instructional material for computer-based training, Carroll's 'Minimalism theory' (1998) suggests that all learning tasks should be meaningful, interactive, contain error recognition and be relevant to the real-world scenario.

AN ENVIRONMENT FOR KNOWLEDGE CONSTRUCTION

Constructivists view learning as a process of knowledge construction through active participation. Bruner's 'Constructivist theory' (1960) presents a general framework to many other constructivists' work proposing that a theory of instruction should address four major aspects: predisposition towards learning; approaches to structure knowledge for ease of understanding; methods of presenting learning material; and nature of rewards and punishment. This was the basis for 'Discovery Learning' which instructs learners to solve situational problems based on their experience and knowledge (Bruner, 1961). The approach encourages learners to discover the facts and relationships themselves through interactions with the environment in search for solutions for the given problems.

Gagne's 'Conditions of Learning theory' (1970) advocates that effective learning takes place when learners are exposed to the right conditions. The theory proposes nine instructional events and corresponding cognitive processes: (i) *gaining attention* (reception); (ii) *informing learners of the objective* (expectancy); (iii) *stimulate recall of prior learning* (retrieval); (iv) *presenting the stimulus* (selective perception); (v) *providing learning guidance* (semantic encoding); (vi) *eliciting performance* (responding); (vii) *providing feedback* (reinforcement); (viii) *assessing performance* (retrieval); and (ix) *enhancing retention and transfer* (generalization) to facilitate learning at various levels.

This work has a similar objective to Reigluth's 'Elaboration theory' (1980) which postulates that instructions should be organised with increasing complexity for optimal learning. Sequencing instructions promotes semantic development allowing subsequent ideas to be incorporated. The 'Elaboration theory' proposes seven major strategy components: (i) *an elaborative sequence*; (ii) *learning prerequisite sequences*; (iii) *summary*; (iv) *synthesis*; (v) *analogies*; (vi) *cognitive strategies* and (vii) *learner control*.

Bloom's 'Taxonomy of Educational Objectives' (1956) categorises cognitive development into six major components: (i) *knowledge*; (ii) *comprehension*; (iii) *application*; (iv) *analysis*; (v) *synthesis* and (vi) *evaluation*. Learning can also be clustered into affective and psychomotor domains. Learning in the affective domain focuses on development of emotions and can be categorised into (i) *receiving phenomena*; (ii) *responding to phenomena*; (iii) *valuing*; (iv) *organization* and (v) *internalising values* (Krathwohl, Bloom, & Masia, 1964). The psychomotor domain focuses on development of physical movement and coordination which can be measured in quality of execution. Learning in such a domain can be categorised in increasing behaviour complexity: (i) *perception*; (ii) *readiness*; (iii) *guided response*; (iv) *mechanism*; (v) *complex overt response*; (vi) *adaptation* and (vii) *origination* (Simpson, 1972).

Anderson & Krathwohl et. al. (2000) redefine Bloom's Taxonomy by renaming and reorganizing the cognitive processes as (i) *remember*; (ii) *understand*; (iii) *apply*; (iv) *analyse*; (v) *evaluate* and (vi) *create* which are simpler to understand and apply. In addition, the revision also includes dimensions of knowledge categorised into *factual*, *conceptual*, *procedural* and *meta-cognitive* which are used with cognitive processes to categorise knowledge comprehensively.

Marzano & Kendall's 'New Taxonomy of Educational Objectives' (2006) is a comprehensive and more recent revision of Bloom's work. Based on Bloom's work, the new taxonomy organises the six levels of processing as: (i) *retrieval*; (ii) *comprehension*; (iii) *analysis*; (iv) *knowledge utilization*; (v) *meta-cognitive system*; (vi) *self system*. These are grouped into three systems of thought: the cognitive system (level 1 to level 4), the meta-cognitive system (level 5) and the self system (level 6). Each level correlates with different knowledge domains – *information*, *mental procedures* and *psychomotor procedures*.

LEARNING IN ADULTHOOD

Adult learning is mostly supported by theories on experiential learning. Knowles' 'Andragogy theory' (1996) notes that adults should be informed of the learning objectives and that learning should be problem-centred reflecting current issues. More importantly, he suggested that adults should learn experientially.

Kolb's 'Learning Cycle theory' (1984) presents a useful descriptive model of the adult experiential learning process known to many educators at the tertiary level. The model suggests that the adult learning process is a cycle of four stages better described as (i) *Concrete Experience*; (ii) *Reflective Observation*; (iii) *Abstract Conceptualization*; and (iv) *Active Experimentation*. The model was then adopted by Honey and Mumford (1982) in their proposal of a typology of learners that distinctively describes the learning styles in each stage.

EDUCATIONAL GAMES DESIGN

Designing computer games is a creative and innovative process of imagining and describing the 'game world' in a detailed manner. There are two approaches employed for designing educational games: *instructor* where the designer places education as a high priority; and *entertainer* where the designer places entertainment as a high priority. There have been several proposals on designing educational games from the *instructor* perspective focusing on various aspects of educational games:

- Malone (1980) in his early thoughts of designing educational games presented interesting guidelines through his study of fun by identifying the essential characteristics of a good computer game as *challenge*, *fantasy* and *curiosity*.

- Prensky (2001) in his well-received book "Digital Game-based Learning" shares his research findings on types of learning and possible game styles.
- Pivec, Dziabenko and Schinnerl (2003) proposed six steps of educational game design as: (i) *determine the pedagogical approach*; (ii) *situate the task in a model world*; (iii) *elaborate the details*; (iv) *incorporate underlying pedagogical support*; (v) *map learning activities to interface actions*; and (vi) *map learning concepts to interface objects*.
- Paras and Bizzocchi's (2005) propose an integrated model for educational games design promoting Csikszentmihalyi's (1991) 'Flow theory' as the bridge for understanding and implementing motivation through play.
- Denis and Jouvelot's (2005) motivation-driven educational game design principles namely (i) *reify values into rules*; (ii) *give power*; (iii) *tune usability*; (iv) *derail the game-play*; and (v) *favour communication* centred on Ryan and Deci's (2000) 'Self Determination Theory'.
- Fisch (2005) emphasises the importance of *making educational content an integral part of game-play, relating game challenges to learning, providing feedback* and *creating linkages for offline learning* as considerations for making educational games educational.

These proposals are certainly useful in designing educational games. Building upon the existing knowledge in this domain, this section compiles the best practices on educational games design and presents insights on educational game design based on the theories of learning introduced earlier from both entertainer and instructor perspectives.

ELEMENTS OF PEDAGOGY IN EDUCATIONAL GAMES

Before delving into models and guidelines for designing educational games, it is important to identify and understand the elements of pedagogy within educational games. Studies on the various aspects of computer games have provided enough information to identify the pedagogic elements in educational games that could be used as learning subjects. These pedagogic elements exist in educational games in the form of (i) *the properties and behaviour of in-game components*; (ii) *the relationships between in-game components*; and (iii) *the solving of problems in the scenario defined* (Tang, Hanneghan, & El-Rhalibi, 2007). The following subsections describe and map each pedagogy element previously mentioned to the 'Taxonomy of Educational Objectives' in each learning domain. This is offered as an aid to teachers for defining assessable learning objectives and designing suitable learning activities in educational games. In this chapter, Anderson & Krathwohl *et. al.*'s taxonomy for cognitive domain is favoured over Marzano & Kendell's for its simplicity and relation to the widely practiced Bloom's taxonomy (See Figure 1).

Properties and Behaviours of In-game Components

Actors and objects, as in-game components represented in educational games, are valid subjects for learning. Learners can learn by simply observing the properties of these in-game components that are presented visually and aurally. Learners can also interact with these in-game components to learn about the physical and cognitive behaviour possessed by these in-game components. The amount of knowledge extracted from these in-game components depends on the detail provided within the actors' and objects' identity. Some classes of actors can be programmed with the ability to converse with other actors through dialogue to guide or direct learners. The

Figure 1. Pedagogy elements in educational games organised in taxonomy of educational objectives in cognitive, affective and psychomotor domains

Elements of Pedagogy in Educational Games	Cognitive Domain	Affective Domain	Psychomotor Domain
Tasks and Problems in Scenario	Create Evaluate Analyse Apply	Internalising Values Organization	Origination Adaptation Complex Overt Response Mechanism Guided Response
Relationship of in-game components	Understand	Valuing Responding to Phenomena	Set
Properties of in-game components	Remember	Receiving Phenomena	Perception

Difficulty ↑

content of this dialogue can be presented aurally or visually by using user interface components. Learning about the properties and behaviours of actors and objects allows learners to develop an understanding toward these in-game components and relationships between the encoded properties and behaviours. The properties and behaviour of in-game components are pedagogy elements can be assessed using the *'remember'* cognitive process. It can also be assessed in the affective and psychomotor domains as *'receiving phenomena'* and *'perception'* respectively.

Relationship between In-Game Components

Learning about the properties and behaviours of in-game actors and objects allows learners to classify these into their distinctive classes and subsequently develop knowledge of understanding of the relationships that are defined among actors and objects. These relationships can be learned by observing the cause and effect of an interaction. Every action taken by the learner is associated with a meaningful response that can be represented visually or aurally (or both) to foster construction of knowledge. Interacting with in-game components can help learners to develop knowledge and promote understanding of the usage of the real-world equivalents of these in-game components in solving problems. The relationship between in-game components can also help learners to develop greater understanding toward actors and objects observable via emergent properties that are introduced through interactions. Emergent properties are noticeable once learners have developed an understanding of the relationships and applications of the collection of parts as a combined whole. Problem semantics defined through relationships in educational games are assessable via the *understanding*, *responding to phenomena* and *valuing* categories in the affective domain, and *set* category in the psychomotor domain.

Tasks and Problems in a Given Scenario

Performing tasks and solving problems interactively in a scenario staged by appropriate domain experts provides learners with the cognitive, affective and psychomotor challenges at a higher level in the taxonomy of educational objectives. Tasks are direct interactions with in-game components that serve a specific purpose. Performing a task in an educational game requires the learner to have knowledge of the properties and behaviour of the in-game components involved and the relationships that exist between these components. Defining and arranging a set of tasks to cognitively challenge learners is the essence of designing problems in educational games. These problems can be designed to reflect educational objectives that require learners to perform *analysis*, *application*, *evaluation* or *creation* in order to complete the tasks assigned. Tasks can be combined to form complex problems in a defined scenario. Tasks are measurable interactions and therefore problems presented to learners are also assessable. Performing tasks and solving problems are learning activities that help in building learners' experience and cognitive skills when approaching similar problems in the real world. An example of a real-world task might be demonstration of knowledge of safe handling of a certain chemical or operating a machine correctly in the virtual environment. Social problems in the scenario can be assessed also through the affective domain requiring learners to organise values in priority when resolving conflicts and demonstration of internalised values when similar social problems arise. In the psychomotor domain learners can be assessed on *mechanism*, *complex overt response*, *adaptation* and *origination* of actions to perform the given tasks or solve a problem. Some of these categories are often best assessed on specific hardware interfaces such as a steering wheel, joystick or other specialised controllers.

EDUCATIONAL GAME DESIGN METHODOLOGY

Designing computer games is a process of (1) imagining the game; (2) defining the way it works; (3) describing the elements that construct the game; and (4) communicating the information to the development team. Computer games often begin life either as a genuinely novel game idea, technology-driven concept or based upon existing intellectual property (IP) from movies, comics and novels. Designing educational games differs greatly from conventional game design process because it principally involves pedagogy. Teachers often begin by defining learning objectives or learning outcomes (depending on the adopted instructional design model) when designing lessons rather than applying mechanisms that promote fun (Kelly, et al., 2007). Topics are then identified and structured, and relevant learning activities are devised to complement.

When designing educational games, there is a need to adopt a hybrid game design methodology that infuses activities from instructional design beginning with the definition of learning objectives. Rules and game-play are then designed to support the learning objectives but yet remain interesting to captivate learner, and hence the need to place education first and entertainment second. These differences in approach require close collaboration between teachers and the game design team during the process of educational game design and often require a number of iterations (involving various stages of design, rapid prototyping, play-testing and revision) before it is released for use (Kelly, et al., 2007). These measures are taken to ensure that actual learning takes place within educational games. Building on Morrison, Ross and Kemp's (2006) model of instructional design plan for designing learning software, an educational game design methodology is presented in Figure 2.

The methodology consists of thirteen activities that are grouped into three phases; *plan*, *prototype* and *finalise*. In the planning phase, teachers are expected to define the learning objectives and design goals, understand the learners, identify the learning activities, sequence it and design a story to set the scene for the educational game and to link the learning activities defined. Planning activities are similar to early phases of instructional planning and are carried out by the teacher with the game designer (although this may in fact be the same person!) and the development team in building working prototypes of the game levels (where each level may form a specific problem for the end-user). During the prototyping phase, the development team will be involved in designing details of the game level, prototyping,

Figure 2. Educational game design methodology

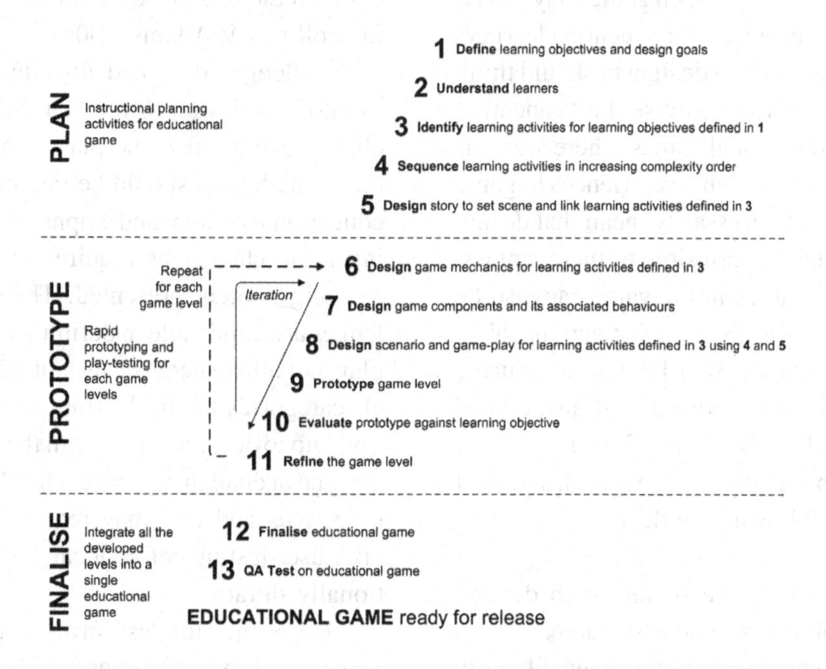

evaluating and refining the game level. If required, they may go through the design activities again to make necessary adjustments so as to meet the learning objectives and design goals defined or to further satisfy the learners' needs. It is best to have the prototyped game level play-tested by a focus group as part of the evaluation process. Once all game levels are completed, the complete game can then be finalised and tested against quality parameters before its final release.

DESIGN GUIDELINES FOR EDUCATIONAL GAMES

Complementing the educational game design methodology is a set of design guidelines to help design better educational games with pedagogy in mind while still maintaining the element of entertainment. These guidelines focus on four core areas in educational game design namely: *game-play and challenges*; *game structure*; *embedding pedagogic content*; and *motivational design*.

Game-Play with Embedded Pedagogy

Play can be regarded as an activity having its rules constantly under negotiation. In game design, designers often regard game-play as series of goal-directed activities with interesting choices designed for the purpose of enjoyment. Educational game designers however should design game-play as meaningful activities derived from the practice of a particular knowledge domain. The semantics of such activities can be derived from the relationships between the available actions and responses present within the game and should be discernible and integrated into the larger context of a game (Salen & Zimmerman, 2003). Such a proposal is not meant to restrict the designer's creativity, but rather issues a plea to them not to take short-cuts in designing such activities, for example by blending "hack and slash" or "shoot-em up" genres into educational games. Although studies shows that game players are aware of the consequences of violent and irresponsible acts in real-life (Dawson, Cragg,

Taylor, & Toombs, 2007), such game-play activities may not be appropriate for educating learners in many domains. Instead designers should think of innovative ways to maximise the "education factor" within educational games. There are various ways to introduce "fun" experiences to game players; it does not necessarily mean that design choices should be irresponsible or mere fantasy. Each interesting choice in the game can also be designed with responsibilities that actions taken within the game world will be able to educate game players about knowledge and also ethical values in the real world (Sicart, 2005).

To create game-play adhering to such a context we propose the following guidelines:

- Activities should be in line with defined learning objectives and assessable.
- Activities should be of increasing difficulty order and achievable.
- Activities should be a form of intellectual exercise (or psychomotor challenge if challenges are meant for assessing the psychomotor domain) with minimal abstraction.
- Activities should be applicable and readily transferable to a real world scenario.
- Activities should be carefully balanced and achievement should be based on Weiner's 'Attribution Theory' rather than predestined winning.
- Learners should be given feedback (either in the form of positive or negative reinforcement or rewards) to assist in success and error recognition.

Challenge in computer games is important as it invites participants to be involved in an action or series of actions that can distinctively justify their superiority in mastering it. As challenges are part of game-playing, challenges should be tied closely with the defined activities. Challenges can be either *pure* or *applied*, each requiring different approaches to solve the problems expressed therein (see Table 1 below). A more detailed description of the challenges presented in Table 1 is described in (Rollings & Adams, 2003).

Challenges designed for educational games should be relevant to the learning objectives to elicit meaningful game-play. More importantly these challenges should be tied closely with the educational content and emphasize that overcoming these challenges requires mastery over the learning content presented. Though most challenges are applicable in defining engaging game-play, not all are necessarily suitable for assessing all categories of the learning domain. It is entirely up to designers to design the most appropriate type of challenge to assess the desired learning objectives and this may require some revision after user-testing before it can be deemed educationally suitable.

One of the simplest forms of cognitive challenge, used to test learners' ability to remember factual information embedded within storytelling or game-play, is the memory-based challenge. For example, a challenge can be designed to be a progression barrier (i.e. a roadblock or assessment point) that requires learners to collect a list of items in which the answer is presented in the cut-scene prior to the game-play. A challenge can also incorporate more than one applied or pure challenge to add complexity and depth. For example, the task can be made more complex and interesting by requiring the learner to interact with various non-player characters in order to obtain each item while possibly incorporating trading of existing items to obtain further items.

For challenges that require a higher level of cognition, educational game designers can incorporate tasks that require learners to devise a solution for a given purpose. Mind Rover[2], an intelligent robot simulation, is an example of a computer game that presents lateral-thinking challenges. The learner takes the role of a researcher to program the intelligence of robotic vehicles to race around tracks and battle against other robotic vehicles.

The affective domain can also be tested through moral challenges and conflict. These different

Table 1. Forms of Pure and Applied Challenges (Rollings & Adams, 2003)

Pure Challenges	Applied Challenges
Logic and Inference Challenges	Races
Lateral Thinking Challenges	Puzzles
Memory Challenges	Exploration
Intelligence-Based Challenges	Conflict
Knowledge-Based Challenges	Economies
Pattern Recognition Challenges	Conceptual Challenges
Moral Challenges	
Spatial Awareness Challenges	
Coordination Challenges	
Reflex Challenges	
Physical Challenges	

levels may not be directly assessable in all games but quite often present themselves in multiplayer modes which invite collaboration between game players and promote development of social values during post-game sessions.

As for the psychomotor domains reflex challenges are the simplest form of challenges available for designers to stimulate a game player's sensory mechanism. Races and coordination challenges are suitable for exercising higher levels of the psychomotor domain that involve programmed responses, responding to different circumstances and creation of a new regime for optimal response. Challenges involving physical movements are now becoming increasingly popular thanks to innovative input technologies such as the Nintendo WiiMote and WiiFit, the Apple iPhone, the Sony Sixaxis controller and PlayStation Eye camera system.

A Structure for Integrated Learning

Game structure represents the main construct of a computer game and is usually organised in the form of levels which further broken down into achievable objectives within the game scenario. Such a structure is relevant, but it should be presented in an environment that can assist knowledge construction. In many commercial computer games, it is noticeable that levels are ordered in increasing difficulty, organised in a hierarchical or linear structure and often driven by storytelling. An ideal structure for educational games should depict the characteristics from well-designed commercial computer games to include spaces for pedagogies such as *active learning*, *experiential learning*, *problem-based learning* and *situated learning*. Based on the theories of learning by Gagne, Reigluth and Kolb, an ideal educational game structure is proposed in Figure 3.

The proposed structure closely resembles that of well-designed computer games but aims at the presentation of learning material in a guided and elaborative manner closely integrated with meaningful activities introduced as a form of play. It organises game events into a series of instructional events for learning to take place within educational games. These events are grouped as segments of a game, similar in concept to chapters in a book. Events are made up of game elements such as the game screen and cut scene (or animated information segment) to attract the learner's attention, inform of learning objectives, link learners to previous lessons and provide summative feedback on the learner's performance. The game scenario, the core element in a computer game, provides a stage for other instructional events such as presentation of stimuli to learners, assessment of a learner's performance and providing learners with guidance and feedback during game-play. Tutorials and levels each represent a different form of learning activity within a game

Figure 3. An integrated learning structure for educational games

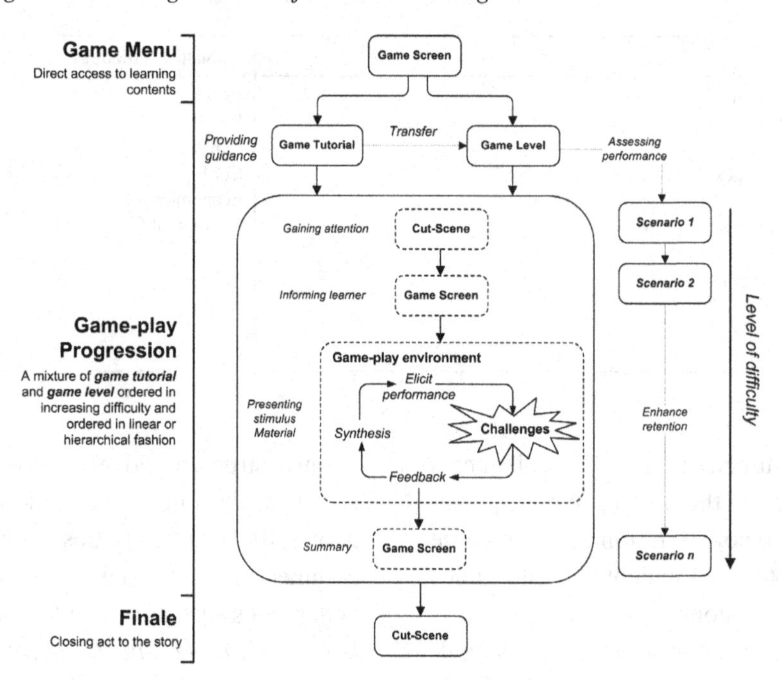

scenario. A tutorial provides the learner with guidance while a level is used for assessment of and transfer of knowledge or skills. Different forms of challenge can be used to assess the different forms of learning outcome associated with the scenario. Similar challenges can be assessed in subsequent levels to enhance retention of knowledge and fine-tune skill acquisition. Scenarios presented during a tutorial or level should be organised in increasing difficulty order as an attempt to introduce 'scaffolding' or a learning growth mechanism. As the design process continues, game-play spaces can be identified and coupled with relevant learning material making learning more integrated within educational games rather than isolated using a so-called 'blended learning' approach.

Embedding Pedagogic Content

One of the goals of exploiting computer games as a learning tool is to present educational materials within the familiar gaming environment. Factual information, descriptive information and scenarios are common forms of knowledge that need to be conveyed to learners. Learning content in educational games should be congruent with any lesson given in the event that the educational game is used as a support mechanism either in a classroom setting or during private study. In this context, game-play is used as an exercise to revise and practice the knowledge in a simulated environment to enhance understanding or to provide reinforcement. Rather than embed an entire 'textbook' into an educational game, Fisch (2005) proposes that designers should attempt to create linkages for offline learning by means of testing concepts or by requiring game players to search for information which is not provided within the game itself. Although such an approach may reduce the degree of immersion in educational games, it is necessary to have game players apply what they have learnt in the classroom within educational games to enhance their understanding in the subjects taught. This can then help to achieve the primary objective of using computer games as a medium to educate learners with fun and meaningful activities. Alternatively, educational games can embed all the necessary information by linking game players to appropriate web resources where this information is readily accessible.

To reiterate this we propose the following guidelines as part of our efforts to make learning material more accessible to learners:

- Learning content should be embedded within the storytelling and narrative components of the game whenever possible.
- Learning content should use multiple representations where appropriate (i.e. use auditory and visual information) and be concise but avoid oversimplification.
- Learning content should contain challenges allowing learners to apply acquired knowledge in the present (and any future unseen) scenario and thus increase retention of the concepts therein.
- Learning content should not have more than seven key concepts in order to aid information recall.

Extending the discussion on pedagogic content in educational games is the issue pertaining to the quality of graphics both in art and in technology suitable for educational games. The "graphics versus game-play" debate is a long-standing argument. A game with stunning graphics attracts game players but poor game-play is often equated with poor replayability. In addition, poor graphics quality may affect the learner's first impressions of the whole offering thereby discouraging their commitment to invest time in the game. In the context of educational games however, game-play should be given higher priority and graphics should be used merely for visualisation purposes. This is often born out of necessity since the development budgets for educational games are often a fraction of that spent on commercial entertainment software.

Motivating Learners

Motivation is the 'drive' that individuals experience either for personal reasons or to receive external rewards (or avoid external punishments). Highly motivated learners tend to spend a greater proportion of time devoted to learning a subject area. It follows then that we should aim to better motivate learners within the context of an educational game to promote deeper learning of the concepts within.

Bixler (2006), in his studies of motivation in relation to educational games, compared four motivational frameworks, namely Keller's ARCS Model; Wlodkowski's Time Continuum Model of Motivation and Motivational Framework for Culturally Responsive Teaching; and Malone and Lepper's Taxonomy of Intrinsic Motivations for Learning. His findings reveal common motivational constructs such as: (i) *obtain and sustain the learners' attention*; (ii) *relate learning objectives to learners*; (iii) *develop the learners' competency in subject area*; and (iv) *provide control to learners in achieving the learning objectives*. Further evidence can be found in the motivational frameworks within the theories of learning proposed by Gagne and Reigluth. Malone and Lepper (1987) in their framework state individuals can be motivated by providing the right *challenge*, arousing their *curiosity*, providing the *controls* and creating a *fantasy* to aid learners in experiencing well-deserved satisfaction over achievements within the virtual world. *Competition*, *cooperation* and *recognition* are parameters affecting motivation on the intrapersonal level defined in the framework.

In practice game designers introduce events that hold the game player's interest through creative storytelling or artwork and attempt to create psychological proximity to immerse game players into the game (Schell, 2005). This can be achieved via the conceptual design of the virtual world, the user interface, the chosen avatar (or player and non-player characters within the game world) and the storyline. Story and narrative provides access to degrees of awareness by manipulating a game player's abilities such as imagination, empathy and focus in the virtual world. Immersion can be divided into three stages; *engagement*, *engrossment* and *total immersion* (Brown & Cairns, 2004). When game players are totally immersed, they tend to ignore their physical surroundings and focus only on the game-play. Well-designed and balanced game-play often provides game players the optimal experience described by Csíkszentmihályi (1991) as *flow* which is the ultimate goal of motivated game-play. A game player who is experiencing flow is focussed, energised and committed to complete the challenges presented.

Educational game designers should consider the following further guidelines which are relevant in designing motivation in educational games.

- Complex activities designed should be made of smaller, achievable tasks to guide learners in achieving the main objective.
- The story and narrative used should be closely related to common cases in the real world.

The user interface also plays an important role in motivation. In addition to being thematic, the user interface should also be designed to be functional to provide learners the control required in the virtual environment to accomplish the tasks assigned.

- Controls should be made simple and natural to reduce the learning barrier in using computer games as a medium for learning.
- Graphical User Interface (GUI) components should be grouped logically with each group not exceeding seven items to aid learners in remembering.

CONCLUSION

Educational games are commonly used at home as learning aids to encourage children to learn or as a computer-aided learning tool in the case of adult learner. In the past, both these approaches have gained significant support from academia, government and industry bodies. With the current generation of 'digital natives' being exposed to interactive entertainment during childhood years (Rideout, Vandewater, & Wartella, 2003) and being brought up in the digital era, it is widely thought that game-based learning will become a preferred method of learning for many of this generation in the future. Although game-based learning has some of the accepted and desirable learning approaches embedded such as active learning, experiential learning, situated learning and problem-based learning (Tang, Martin Hanneghan, & El-Rhalibi, 2009), it should be pointed out that it is not a panacea. Educational games are a viable alternative to existing computer-aided learning technologies that can assist in persuading and encouraging digital natives to acquire knowledge. This new medium can make learning more effective since the approach is more relevant to this particular target group's lifestyle. There are still substantial barriers for teachers (technical, pedagogical and social) to convince this group to immerse themselves in actual game-based learning. Therefore it is important that educational games are designed with pedagogically sound theories to encourage further learning when disconnected from the virtual learning environment. The effect video gaming has on society should not be ignored; instead such knowledge should be used to help channel appropriate energy and resources to promote learning adaptable to the learning styles of today's entertainment-generation.

REFERENCES

Adams, E. (2005). *Educational Games Don't Have to Stink!* Retrieved 31 January, 2009, from http://www.gamasutra.com/features/20050126/adams_01.shtml

Anderson, L. W., Krathwohl, D. R., Airasian, P. W., Cruikshank, K. A., Mayer, R. E., & Pintrich, P. R. et al. (2000). *A Taxonomy for Learning, Teaching, and Assessing: A Revision of Bloom's Taxonomy of Educational Objectives* (2nd ed.). Boston: Allyn & Bacon.

BECTa. (2006). *Computer Games in Education: Findings Report.* Retrieved 19 February, 2008, from http://partners.becta.org.uk/index.php?section=rh&rid=13595

Bixler, B. (2006). *Games and Motivation: Implications for Instructional Design.* Paper presented at the 2006 NMC Summer Conference.

Bloom, B. S. (1956). *Taxonomy of Educational Objectives, Handbook I: The Cognitive Domain.* New York: David McKay Co. Inc.

Bransford, J. D. L. B. A., & Crocking, R. R. (1999). How People Learn: Brain, Mind, Experience, and School. Washington, DC: National Academic Press.

Brown, E., & Cairns, P. (2004). *A grounded investigation of game immersion.* Paper presented at the Conference on Human Factors in Computing Systems.

Bruner, J. S. (1960). *The Process of Education.* Cambridge, MA: Harvard University Press.

Bruner, J. S. (1961). The act of discovery. *Harvard Educational Review, 31*(1), 21–32.

Carroll, J. M. (Ed.). (1998). *Minimalism Beyond the Nurnberg Funnel (Technical Communication, Multimedia and Information Systems).* Cambridge, MA: MIT Press.

Colt, H. G., Crawford, S. W., & III, O. G. (2001). Virtual Reality Bronchoscopy Simulation*: A Revolution in Procedural Training. *Chest, 120*(4), 1333–1339. PubMed doi:10.1378/chest.120.4.1333

Creemers, B. P. M. (1994). *The Effective Classroom.* London: Cassell.

Csíkszentmihályi, M. (1991). *Flow: The Psychology of Optimal Experience.* New York: Harper Perennial.

Dawson, C. R., Cragg, A., Taylor, C., & Toombs, B. (2007). Video Games Research to improve understanding of what players enjoy about video games, and to explain their preferences for particular games. London: British Board of Film Classification (BBFC).

Denis, G., & Jouvelot, P. (2005). *Motivation-Driven Educational Game Design: Applying Best Practice to Music Education.* Paper presented at the Advances in Computer Entertainment (ACE) 2005.

Faria, A. J. (1998). Business Simulation Games: Current Usage Levels - An Update. *Simulation & Gaming, 29*(3), 295–308. doi:10.1177/1046878198293002

Fisch, S. M. (2005). *Making Educational Computer Games "Educational".* Paper presented at the 4th International Conference on Interaction Design and Children (IDC2005).

Gagne, R. M. (1970). *The Conditions of Learning and Theory of Instruction* (2nd ed.). New York: Holt, Rinehart & Winston.

Hattie, J. A. (2003). Teachers make a difference: What is the research evidence? 2003 Australian Council for Educational Research Conference. Melbourne.

Hattie, J. A. (2005). *What is the nature of evidence that makes a difference to learning?* Paper presented at the Research Conference 2005 VIC: Australian Council for Educational Research.

Honey, P., & Mumford, A. (1982). *Manual of Learning Styles*. London: P. Honey.

Hull, C. L. (1951). *Essentials of behavior*. New Haven, CT: Yale University Press.

Ke, F. (2009). A Qualitative Meta-Analysis of Computer Games as Learning Tools. In R. E. Ferdig (Ed.), *Handbook of Research on Effective Electronic Gaming in Education* (Vol. 1-32). Hershey, PA: Information Science Reference.

Kelly, H., Howell, K., Glinert, E., Holding, L., Swain, C., & Burrowbridge, A. et al. (2007). How to build serious games. *Communications of the ACM*, *50*(7), 44–49. doi:10.1145/1272516.1272538

Knowles, M. (1996). Androgogy: An emerging techology for adult learning. In R. Edwards, A. Hanson, & P. Raggatt (Eds.), *Boundaries of Adult Learning* (pp. 82–96). London: Routledge.

Kolb, D. A. (1984). *Experiential Learning*. Englewood Cliffs, NJ: Prentice-Hall.

Krathwohl, D. R., Bloom, B. S., & Masia, B. B. (1964). *Taxonomy of Educational Objectives: Classification of Educational Goals, Handbook II: Affective Domain*. New York: David McKay Co., Inc.

Malone, T. W. (1980). *What makes things fun to learn? heuristics for designing instructional computer games*. Paper presented at the Proceedings of the 3rd ACM SIGSMALL symposium and the first SIGPC symposium on Small systems.

Malone, T. W., & Lepper, M. R. (1987). Making Learning Fun: A Taxonomy of Intrinsic Motivations for Learning. In R. E. Snow, & M. J. Farr (Eds.), *Aptitude, Learning and Instruction: Conative amd affective process analyses* (Vol. 3, pp. 223–254). Hillsdale, NJ: Lawrence Erlbaum Associates.

Marzano, R. J. (2009). *MRL Meta-Analysis Database Summary*. Retrieved 5th July 2009: http://files.solution-tree.com/MRL/documents/strategy_summary_6_10_09.pdf

Marzano, R. J., & Kendall, J. S. (2006). *The New Taxonomy of Educational Objectives* (2nd ed.). Thousand Oaks, CA: Corwin Press.

Maslow, A. H. (1946). A Theory of Human Motivation. In P. L. Harriman (Ed.), *Twentieth Century Psychology: Recent Developments in Psychology* (pp. 22–48). New York: The Philosophical Library.

Miller, G. A. (2003). The Magical Number Seven, Plus or Minus Two: Some Limits on Our Capacity for Processing Information. In B. J. Baars, W. P. Banks, & J. B. Newman (Eds.), *Essential Sources in the Scientific Study of Consciousness* (pp. 357–372). Cambridge, MA: MIT Press.

Moreno-Ger, P., Blesius, C., Currier, P., Sierra, J. L., & Fernández-Manjón, B. (2008). Online Learning and Clinical Procedures: Rapid Development and Effective Deployment of Game-Like Interactive Simulations. In Z. Pan, A. D. Cheok, W. Müller, & A. E. Rhabili (Eds.), *Transactions on Edutainment I* (Vol. 5080, pp. 288–304). Berlin, Heidelberg: Springer Verlag. doi:10.1007/978-3-540-69744-2_22

Morrison, G. R., Ross, S. M., & Kemp, J. E. (2006). *Designing Effective Instruction* (5th ed.). London: Wiley.

Nieborg, D. B. (2004). *America's Army: More Than a Game*. Paper presented at the Transforming Knowledge into Action through Gaming and Simulation, Munchen: SAGSAGA.

Paras, B., & Bizzocchi, J. (2005). *Game, Motivation, and Effective Learning: An Integrated Model for Educational Game Design*. Paper presented at the DiGRA 2005 – the Digital Games Research Association's 2nd International Conference, Simon Fraser University, Burnaby, BC, Canada.

Pearce, C. (2006). Productive Play: Game Culture From the Bottom Up. *Games and Culture, 1*(1), 17–24. doi:10.1177/1555412005281418

Pivec, M., Dziabenko, O., & Schinnerl, I. (2003). *Aspects of Game-Based Learning.* Paper presented at the Third International Conference on Knowledge Management (IKNOW 03), Graz, Austria.

Prensky, M. (2001). *Digital Game-Based Learning.* New York: Paragon House.

Prensky, M. (2002). The Motivation of Gameplay or the REAL 21st century learning revolution. *Horizon, 10,* 1–14. doi:10.1108/10748120210431349

Reigeluth, C. M., Merrill, M. D., Wilson, B. G., & Spille, R. T. (1980). The elaboration theory of instruction: A model for sequencing and synthesizing instruction. *Instructional Science, 9*(3), 195–219. doi:10.1007/BF00177327

Rideout, V. J., Vandewater, E. A., & Wartella, E. A. (2003). *Zero to Six: Electronic Media in the Lives of Infants, Toddlers and Preschoolers.* Menlo Park, CA: Kaiser Family Foundation.

Rollings, A., & Adams, E. (2003). *Andrew Rollings and Ernest Adams on Game Design.* St. Carmel, IN: New Riders Publishing.

Ryan, R. M., & Deci, E. L. (2000). Self-determination theory and the facilitation of intrinsic motivation, social development and well-being. *The American Psychologist, 55*(1), 68–78. PubMed doi:10.1037/0003-066X.55.1.68

Salen, K., & Zimmerman, E. (2003). *Rules of Play: Game Design Fundamentals.* Cambridge, MA: The MIT Press.

Sawyer, B., & Smith, P. (2008). *Serious Games Taxonomy.* Retrieved 26 March, 2008, from http://www.dmill.com/presentations/serious-games-taxonomy-2008.pdf

Schell, J. (2005). Understanding entertainment: story and gameplay are one. [CIE]. *Computers in Entertainment, 3*(1), 6. doi:10.1145/1057270.1057284

Sicart, M. (2005, 16-20 June 2005). *The Ethics of Computer Game Design.* Paper presented at the DiGRA 2005 - the Digital Games Research Association's 2nd International Conference, Simon Fraser University, Burnaby, BC, Canada.

Simpson, E. J. (1972). *The classification of educational objectives in the psychomotor domain: The psychomotor domain* (Vol. 3). Washington, DC: Gryphin House.

Skinner, B. F. (1935). Two Types Of Conditioned Reflex And A Pseudo Type. *The Journal of General Psychology, 12,* 66–77. doi:10.1080/00221309.1935.9920088

Sweller, J. (1994). Cognitive load theory, learning difficulty, and instructional design. *Learning and Instruction, 4*(4), 295–312. doi:10.1016/0959-4752(94)90003-5

Tang, S. Martin Hanneghan, & El-Rhalibi, A. (2009). Introduction to Games-Based Learning. In T. M. Connolly, M. H. Stansfield & L. Boyle (Eds.), Games-Based Learning Advancements for Multi-Sensory Human Computer Interfaces: Techniques and Effective Practices (pp. 1-17). Hershey, PA: Idea-Group Publishing.

Tang, S., Hanneghan, M., & El-Rhalibi, A. (2007). *Pedagogy Elements, Components and Structures for Serious Games Authoring Environment.* Paper presented at the 5th International Game Design and Technology Workshop (GDTW 2007), Liverpool, UK.

Telfer, R. (1993). *Aviation Instruction and Training.* Aldershot, UK: Ashgate.

Thorndike, E. L. (1933). A proof of the law of effect. *Science, 77,* 173–175. PubMed doi:10.1126/science.77.1989.173-a

Weiner, B. (1979). A Theory of Motivation for Some Classroom Experiences. *Journal of Educational Psychology, 71*(1), 3–25. PubMed doi:10.1037/0022-0663.71.1.3

Zyda, M. (2005). From Visual Simulation to Virtual Reality to Games. *Computer, 38*, 25–32. doi:10.1109/MC.2005.297

KEY TERMS AND DEFINITIONS

Behaviourism: Behaviourism is built upon the belief that learners can be conditioned to make a response in which learning is perceived as the result of association forming between stimuli and responses. Many types of gamification make use of behaviourist ideas in a more structured and systematic context.

Conditions of Learning Theory: Conditions of Learning Theory advocates that effective learning takes place when learners are exposed to the right conditions.

Constructivism: Constructivism is an educational philosophy that has the view that learning is a process of knowledge construction through active participation.

Drive Reduction Theory.: Drive Reduction theory focuses on human behaviour and argues that motivation is essential in order for responses to occur. It suggests that responses can become habitual when the stimuli and response cycle is reinforced reducing the latency of such responses.

Educational Games: Educational games, also known as instructional games, take advantage of gaming principles to gamify technologies in order to create engaging educational content.

Learning: Learning is generally perceived as the process of acquiring new knowledge which often takes place in a formal classroom setting. However, learning can also take place informally after school hours through interactions with peers and the surrounding environment making learning a constant process.

Serious Games: Serious Games is a term used for representing software applications that apply gaming principles and technologies through the use of gamification for non-entertainment purposes including education and training.

ENDNOTES

1. The terms 'computer games' and 'video games' refer to digital games on a home computer and console platform respectively and are often used interchangeably. For clarity, this article uses the term 'computer games' to represent digital games on all platforms.
2. More details of Mind Rover can be found at http://www.lokigames.com/products/mindrover/.

This work was previously published in Design and Implementation of Educational Games: Theoretical and Practical Perspectives, edited by Pavel Zemliansky and Diane Wilcox, pp. 108-125, copyright 2010 by Information Science Reference (an imprint of IGI Global).

Chapter 12
Games and the Development of Students' Civic Engagement and Ecological Stewardship

Janice L. Anderson
University of North Carolina – Chapel Hill, USA

ABSTRACT

In recent years, researchers and classroom teachers have started to explore purposefully designed computer/video games in supporting student learning. This interest in video and computer games has arisen in part, because preliminary research on educational video and computer games indicates that leveraging this technology has the potential to improve student motivation, interest, and engagement in learning through the use of a familiar medium (Gee, 2005; Mayo, 2009; Squire, 2005; Shaffer, 2006). While most of this early research has focused on the impact of games on academic and social outcomes, relatively few studies have been conducted exploring the influence of games on civic engagement (Lenhart et al, 2008). This chapter will specifically look at how Quest Atlantis, a game designed for learning, can potentially be utilized to facilitate the development of ecological stewardship among its players/students, thereby contributing to a more informed democratic citizenry.

INTRODUCTION

Computer/video games and virtual worlds have emerged as a pervasive influence on American society and culture in a relatively short period of time (Mayo, 2009; Squire, 2006). Students of all ages engage these environments as much or more than they watch television (Buckley & Anderson, 2006; Entertainment Software Association, 2006; Mayo, 2009; Michigan State University, 2004) which has led researchers to examine motivational factors such as the desire

DOI: 10.4018/978-1-4666-5071-8.ch012

to play, focusing on how computer/video games can be utilized to facilitate student learning in the classroom (Squire, 2006). Computer/video games and virtual worlds have been developed as models for improving the learning environment of students by implementing the types of clear goals and challenges that are presented to students through the gaming platforms, allowing for and challenging students to collaborate creating the potential for transforming learning in all types of settings, including schools (Gee, 2003; Shaffer, 2006; Barab et al. 2008; Barab et al., 2007). These virtual environments make it plausible to immerse students within networks of interaction and back-stories which engages them in problem solving and reflection in both real and in-world relationships and identities (Barab, 2008). This type of virtual-engagement represents what Gee (2003) identifies as empathetic embodiment of complex systems, where students develop an understanding of and appreciation for one or more aspects of the context of the virtual worlds in which they are engaged.

Many of the massively multiplayer online games (MMOGs) and multi-user virtual environments (MUVEs) provide students with the opportunity to role play, engaging them in a collaborative processes that facilitates participation and leads to problem solving, hypothesis generation and identity construction (Barab, 2008). These environments allow student players to become engaged in an evolving discourse as members of a community of practice (Barab, 2008; Lave & Wenger, 1991; Squire, 2006). By creating experiences of legitimate peripheral participation (Lave & Wenger, 1991) which emphasize conceptual understanding as a means to address authentic situations (e.g. taking on the role of a scientist, a politician, engineer, etc.), students come to a new way of knowing different from the more traditional, didactic approaches to curriculum and instruction (Barab, Hay, Barnett, & Keating, 2000; Brown, Collins, & Duguid, 1989). By balancing academic content, legitimate peripheral participation, background narratives, and game rules, these virtual worlds can be utilized to support disciplinary-specific learning in content areas such as science, social studies, and civics (Barab, 2008).

Much of the current research has focused on how games and virtual worlds impact academic and social outcomes (e.g. aggression, violence), while relatively little research has been conducted exploring the impact of games and virtual worlds on civic engagement and society (Lenhart, Kahne, Middaugh, McGill, Evans, & Vitak, 2008). Citizenship, democracy and education are inextricably bound to the life of a nation (Bennett, Wells, & Rank, 2008). Schools work to prepare or help students build tools that enable them to play an active role in society as engaged, educated participants. An educated citizenry will contribute their skills and talents to preserve a democratic society. According to John Dewey (1916), community participation is a key to this maintenance. This community participation consists of individuals united through common interests, goals and ideas, but also allow for "free and full interplay" (Dewey, 1916, p. 83) with those who assert differing viewpoints and perspectives. This is the point of education, to nurture the development of individuals who can think and critically analyze, contributing to a democratic society.

In recent years, post-industrial democracies have acknowledged a crisis in student civic engagement, noting a lack of participation in elections and other traditional civic activities as students mature into adulthood (Bennett, Wells & Rank, 2008). However, while recognizing that students have become disengaged and disconnected from current political practices, most post-industrial democracies have continued to frame their conceptions of citizenship without regard to changing social identities and new and emerging ways of learning, (e.g. gaming, social networking and the internet) among young people (Bennett, Wells, & Rank, 2008). According to the Civic Mission for Schools (Gibson & Levine, 2008), while schools are the main source of civic education today, they fail to account for how students view citizenship roles differently from their parents. This disconnect suggests the need to extend educational methods for citizenship beyond traditional textbooks to include "critical engagement with issues and community involvement" (Bennett, Wells & Rank, 2008, p. 4) in order to fully engage students in becoming democratic citizens.

While technology has been "blamed" for fostering isolation among users, the reality is youth who participate in video/computer game and virtual environments are often actively, and sometimes unknowingly, engaged in new forms of civic life that differs from that of previous generations of non-digital natives like their parents and other adults (Bers, 2008). This type of virtual civic engagement stems from students participation and immersion in a digital culture based on experiential learning and online knowledge sharing with others in their own classrooms and across the globe associated with social networking and digital media (Jenkins, 2006; Bennett, Wells, & Rank, 2008) . The emergence and popularity of social networking and media sites such as Facebook©, LinkedIn©, MySpace© and YouTube© provides evidence of this type of virtual civic engagement. As a result, collaborative problem solving and the circular flow of ideas among peers (Jenkins, 2006) is often the norm within this digital native generation. Gaming, likewise, becomes a major component within the range of social activities with a large percentage of youth participation (Lenhart et al, 2008). Multi-user environments provide the opportunity for students to take on the role of producers, as opposed to consumers, of knowledge (Jenkins, 2006; Bers, 2008). Participation in digital environments allows students to explore civic identities by participating in events and discourse related to civic issues in new and unique ways across global communities. This allows students to begin to understand the difficulties that arise from globalization (Bers, 2008).

In recent years, the emergence and growing popularity of the green movement is putting a new emphasis on issues like recycling, composting, and community-based gardening (Biswas et al, 2000) and bringing attention to the size and impact of one's carbon footprint (Weber & Matthews, 2008). With this new interest in ecological stewardship, citizens are becoming increasingly concerned with their impact on the environment, not only in their local communities but across the planet. Through participation in local and digital/virtual environments, youth begin to see a juxtaposition of the worlds; the two worlds are mutually beneficial for learning new skills, so that students take their experiences within the digital environment and transfer them to their local situation where they can work toward long term solutions for environmental and social issues as actively engaged, real-world citizens.

THEORETICAL FRAMEWORK

Contemporary viewpoints of the nature and philosophy of science are rooted in the notion that science is not simply the accumulation of a myriad of facts about the world, but rather it involves the construction of ideas and theories about how the world may be. This view allows for challenges, conflict, and disputes as opposed to common agreement on the nature of science (Giere, 1991; Popper, 1959; Kuhn, 1962). Multi-user, digital environments work to develop these ideas by situating disciplinary content within broader contextual frameworks (Bers, 2008; Sadler, Barab & Scott, 2007; Barab, 2008). According to Papert (1980), the constructivist nature of these environments promotes higher-order learning because they engage the individual in creating personally meaningful artifacts that can benefit and be shared with others within a community. Through this reflection on external objects, internal knowledge is also developed (Papert, 1980).

Quest Atlantis, (QA), builds and expands upon these constructivist principles through the framework of socio-scientific inquiry (Barab, Sadler,

Heiselt, Hickey, & Zuiker, 2007). Socio-scientific inquiry engages students in "the process of using scientific methods to interrogate rich narratives about societal issues that have a scientific basis, yet whose solution claims with political, economic and social concerns" (Barab et al., 2007, p. 61). Based upon the three core concepts of narrative engagement ("context"), inscription construction/deconstruction ("resources") and scientific inquiry ("practice"), *Quest Atlantis* allows students to utilize these core constructs to create compelling solutions to "real world problems." The narrative of the virtual environment contextualizes the scientific content or problem, which, in this study, revolves around why the fish are declining in *Taiga.* Barab and colleagues (2007) see this contextualization as a mechanism to transform student learning from "facts or concepts to be memorized into useful tools to address significant issues" (p. 61).

The inscriptions, or resources, focus on the written or printed objects (e.g. charts, tables, graphs, schemes, diagrams, etc.). These allow the students to demonstrate and represent knowledge, as well as focus on data extrapolated from the larger narrative (Roth & McGinn, 1998). Scientific knowledge is often communicated through these types of simplified representations, or inscriptions, turning them into a conceptual tool that allows students to make sense of the world and creatively solve problems (Barab et al, 2007). The scientific inquiry practices that exist within these gaming and virtual world environments allow students to utilize a dynamic approach towards asking questions, making and testing hypothesis and discoveries, and considering the impact of all possible solutions within the context of the community. *Quest Atlantis* makes use of this socio-scientific inquiry framework to gain insight into how virtual world environments and other computer games are leveraged to address all types of academic content. The focus of this study was analyzing how students learned science content, and developed their own ecological stewardship and civic engagement while engaging with *Quest Atlantis.*

The framework of pedagogical praxis (Shaffer, 2004) utilizes the ideas of situated learning first developed by Brown, Collins, and Duguid (1989); this notion of situated learning was later expanded upon by Lave and Wenger (1991), who begin with the premise that "under the right conditions, computers and other information technologies can make it easier for students to become active participants in meaningful projects and practices in the life of their community" (p. 1401). Praxis-based educational models, such as digital environments, are designed to encourage experiences where students learn through engagement and participation (Bers, 2008; Shaffer, 2004) as opposed to the more traditional knowledge based models that focus solely on subject specific content. Pedagogical praxis further develops Lave and Wenger (1991) notion of communities of practice by incorporating legitimate peripheral participation (Shaffer, 2004) and Schon's model (1985, 1987) of reflective practice which suggests that one must "think in action" (p. 1402) through these experiences. According to Schon (1985), individuals who make a connection between knowing and doing through reflection are able to "combine reflection and action, on the spot, ... to examine understandings and appreciations while the train is running" (p. 27). Multi-user digital environments provide students with the tools to engage in this form of legitimate peripheral participation where conceptual understanding, in both disciplinary content and practices, in authentic situations is valued.

Citizen models, reflecting global civic and environmental engagement, demonstrate what is described by Westheimer and Kahne as participatory and justice-oriented forms of citizenship. A participatory citizen is actively engaged in his/her community and is working to solve social and environmental problems (e.g. pollution in the rivers). Citizens who take a justice-oriented perspective identify the various social, behavioral, and environmental problems, the structures that perpetuate these problems, and the actions that are needed to change the patterns that contribute to the replication of problems over time (Westheimer & Kahne, 2004).

While commercial games such as *World of Warcraft, SimCity,* and *Second Life* are environments that can potentially nurture the development of ecological stewardship and civic engagement, this study will focus on the use of an educational game, *Quest Atlantis,* developed by education and learning science researchers, to examine how students learn (or do not learn) science content and how interaction with one's environment impacts (or does not impact) the player's sense of ecological stewardship and civic engagement.

This work contributes to a developing body of research that examining the impact of using computer/video games and virtual worlds within educational settings (e.g. Nelson, Ketelhut, Clarke, Bowman & Dede, 2005; Barab et al., 2007; Squire, (2006); Neulight, N. Kafai, Y., Kao, L., Foley, B. & Galas, C. (2007)) It also contributes to the research base on using multi-user virtual environments (MUVEs) through the examination of how students learn science content within the *Quest Atlantis* environment and how they are able to translate these experiences to their own lives, becoming civic and ecological stewards of their communities.

QUEST ATLANTIS

Quest Atlantis (QA), developed at Indiana University, is a multi-user virtual environment that combines strategies used in commercial gaming environments while integrating lessons from educational research on learning and motivation. The immersive gaming environment is designed for students (ages 9-14) to engage in forms of play that allows them to explore social responsibility within the context of both fictional and non-fictional realities while promoting the civic engagement of its participants (www.questatlantis.org). Student engagement is accomplished through a compelling narrative. The back-story focuses on the problems of a mythical world called *Atlantis*, where students encounter issues similar to the challenges faced in their own experiences on Earth. *Atlantis* is described to the participants as a planet experiencing a myriad of social and environmental issues. Students are invited by a group of concerned *Atlantan* citizens to help them solve some of these crucial issues. The story plays out with the 3D world as well as in novels, comic books, and a global community of participants. The game platform provides an immersive context for the students to engage in real-world inquiry through the fictional world, *Atlantis*.

The 3D world provides the platform of engagement for students, who teleport to virtual locations within *Atlantis* to perform educational activities known as *quests*, talk with other students and mentors, and build virtual personae in real-time. The virtual personae, or avatars, can be customized to reflect the student's own identity through their choice of hair and skin color, clothes and other accessories such as hats, glasses and backpacks. Students, through their avatar, respond to *Quests* (developmentally appropriate activities with task descriptions and goals) in order to help the Council of Atlantis solve some of their problems and restore lost knowledge within *Atlantan* society.

Groups of *Atlantans*, known as the Council, are determined to restore *Atlantis* back to its previous magnificence by enlisting the help of student questers from Earth. *Questers* teleport to *Atlantis* via OTAK, the computer designed by the council. Upon entering the *Quest Atlantis* environment, students are free to visit a number of virtual worlds, each with their own unique theme, and council member supporting it (e.g. *Ecology World* with its environmental awareness focus headed by Council Member Lan). Each world is made up of several villages with their own *quests*. The *quests* within the worlds and villages are connected to academic standards, both at the national and state level, and to social commitments such as environmental awareness. Students are invited to bring their own experiences, families, and cultures to help them solve the problems of *Atlantis*. *Quests* within the world can vary, ranging from simple simulations to complex application problems.

Student *questers* navigate their avatars through the 3D world and interact with other players via a text-based chat window and respond to non-player characters (NPCs) with structured dialogues where they propose solutions and communicate ideas about the problem they are solving. As students complete each quest experience, their final responses/solutions can be typed directly into the game space, or they can upload up to four files, including word processing documents, spreadsheets, presentations, movies, or any other file type. *Quests* are generally assessed by the supervising teacher who assumes the role of a non-player character such as Ranger Bartle in *Taiga,* or Lan, the council member in *Ecology World*. All correspondence about the *quests* is generated by the non-player character/teacher. A teacher can also assign questers to conduct peer reviewing or edits of their fellow classmates. *Questers* access their work and feedback through the 2D window space, which becomes their homepage complete with their electronic portfolios.

While students interacted with their teacher and other non-player characters, there were no internal supports like an intelligent tutor to scaffold students' construction of knowledge. In this study, however, students participants utilized a field notebook, developed by the researcher, to guide their *quests* in *Taiga*. This field notebook provided the students with scaffolds for note-taking as they progressed through *Taiga* and encouraged them to connect their game play to extensions into "real world" experiences and situations.

The focus of this chapter is on *Taiga*, one of the many worlds within *Quest Atlantis. Taiga* is a park located along two water-ways and inhabited by a variety of non-player characters including loggers, tourists, and indigenous farmers. The world also includes a fishing resort and park administration. *Taiga* is designed to help students learn about environmental science concepts such as eutrophication, erosion, water quality indicators and hypothesis testing through their interaction with these virtual characters and data concerning a declining fish population within the park's rivers.

Student *questers* are invited to assume the role of a field investigator helping Ranger Bartle solve his dilemma – the decline of the fish in the river. They begin their narrative immersion by interviewing the various stakeholders/non-player characters found in the park and identifying the possible factors contributing to the decline of the fish population. The non-player characters provide a diverse set of perspectives on the problem of fish decline for the students to analyze.

Students collect water samples, analyze data and formulate a hypothesis that is based upon their understanding of the scientific evidence, along with their analysis of the stakeholder's perspectives of the problem. After proposing an initial solution, students are allowed to travel to the future to see the impact of their ideas on *Taiga*. Depending on their choices, the students will encounter different scenarios that are reflective of their choices. Upon returning to the present, student questers are given the opportunity to revise their solution into a nuanced argument which balances the scientific evidence with a greater understanding of both the political and economical impact on the community. The success of the student within *Taiga* is dependent upon their understanding of 1) water quality indicators such as pH, dissolved oxygen, nitrates and phosphates; 2) the processes of eutrophication and erosion; and 3) the dynamic relationship between the indicators, the processes and the outcomes within the *Taiga* water-ways.

Figure 1.

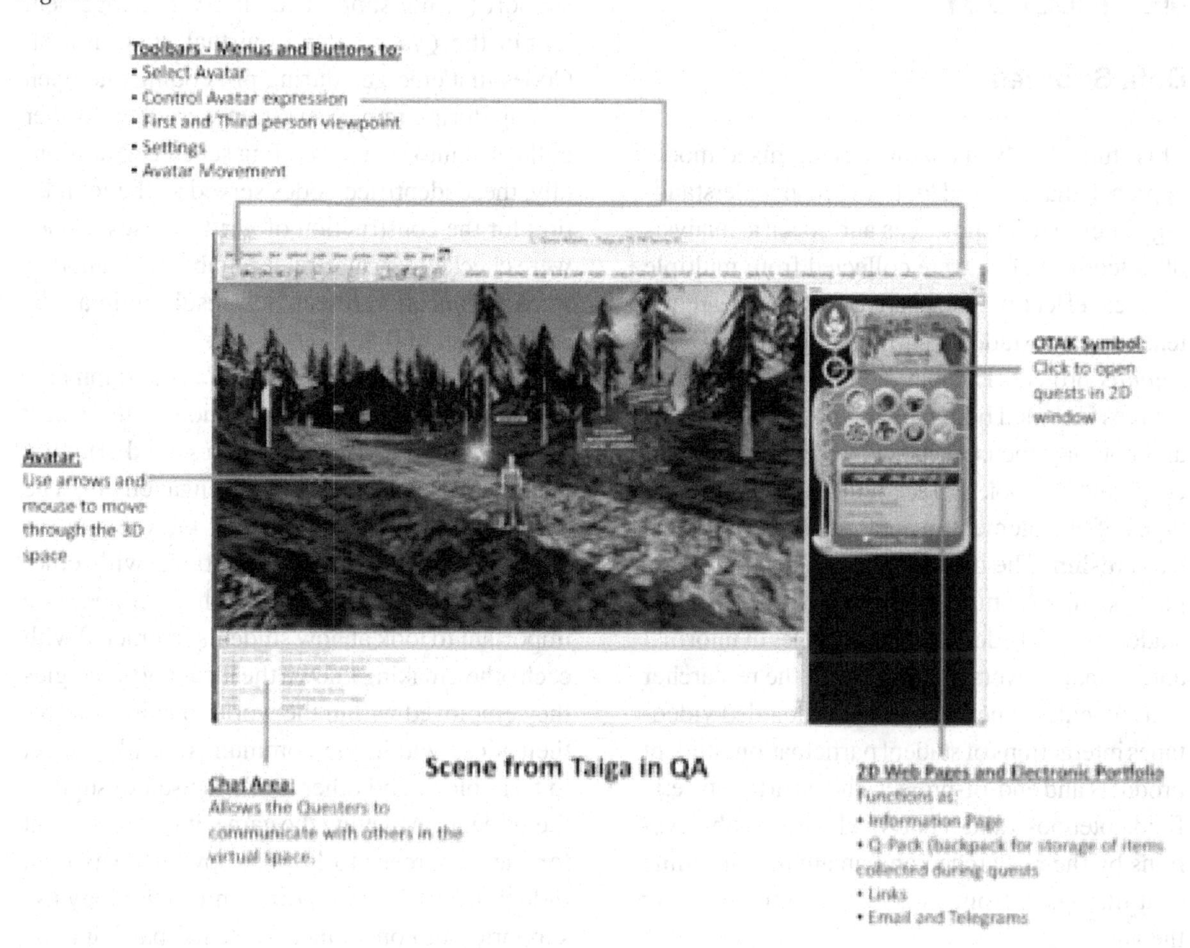

STUDY CONTEXT

Context: School and Students

This study occurred in three fifth grade classrooms (n=50) of two urban schools, Chamberlain Elementary and Edison Elementary, part of the *Northeast Public School System*[1] in the United States. Northeast Public Schools is a large urban district that faces many of the problems plaguing urban centers including poverty, low academic achievement, English Language Learner (ELL) issues, high-risk students, and lack of student engagement. Both schools struggle to meet their annual "Adequate Yearly Progress" (AYP) and are often categorized as being either failing or needs improvement schools. Both Chamberlain and Edison fall in the "Needs Improvement" category in English Language Arts and "No Status" for Mathematics in the statewide Assessment System used to determine AYP under the federal No Child Left Behind (NCLB) legislation. Neither school met its Annual Yearly Progress (AYP) in English Language Arts. However, Edison met AYP for Mathematics, while Chamberlain did not. Both of these schools have predominately Black and Latino(a) populations, which account for seventy to seventy-five percent of the school populations. Chamberlain is also linguistically diverse with over seventy languages spoken at home.

METHODOLOGY

Data Sources

This study involved a multi-tiered, mixed model approach that allowed for both broad understandings of classroom practices and specific analysis of outcomes. Data were collected from multiple sources reflecting perspectives of the researcher, teachers, and students participating in the *Quest Atlantis* project (Lincoln & Guba, 1985). These sources included both pre- and post- assessments and not only focused on science content, but incorporated or took into account process skills and types of engagement in areas such as ecological stewardship. The data sources included pre- and post- semi-structured interviews of a subset of students (n=20), detailed journal notes of informal and formal conversations between the researcher and students. Other data sources included videotapes interactions of student participation, student products and end-of-project student artifacts (e.g. field notebooks and reports), classroom observations by the researcher, and an archive of online chat dialogues from student interactions within the game.

Data Analysis

Data was triangulated in order to overcome any weakness or intrinsic bias arising from the use of a single data source. An interpretive approach was utilized to analyze the qualitative data sources (Denzin & Lincoln, 2000). The extensive data collected throughout the project provided an in-depth picture of how students re-conceptualized ecological stewardship and engagement over time. Through interaction analysis (Jordan& Henderson, 1995), data were coded by analyzing segments of video that focused on a specific topic of interest (e.g. science content knowledge, ecological stewardship). Further analysis specifically examined the student discourse and actions around ecological stewardship and environmental consciousness during game play, and how the game seemed to support (or not support) their civic engagement within the *Quest Atlantis* virtual environment. Codes that emerged during pre-coding and open coding during implementation were then further collapsed into an axial coding scheme. Additionally, these identified codes served as the foundation for the construction of written cases. Codes were checked for inter-rater reliability by another science education researcher, resulting in a reliability level of 0.8.

What was apparent in the data collection process was that student participation in the *Quest Atlantis* virtual environment represented a fluidity of mutual engagement and disengagement. The social construction of students' knowledge was observed through their interactions with other students and with the game. It therefore became important to look at how students interacted with each other, making known the ideas and strategies they employed within the world to gain entry for their ideas within the community. With respect to technology and other artifacts used to support the students in playing the game, it was important for the researcher to look at how artifacts (e.g. field notebooks and reports) and technology use supported or constrained students' participation in the activity. How were the students occupying space in-world and voicing their ideas? How were these ideas incorporated into problem-solving? Were their ideas translated into practice in the students' own real-world experiences or, if not, could they?

Students in each of the participating classrooms engaged in playing *Quest Atlantis* over 15-20 class periods of 45-60 minutes each. A purposeful sampling of the students was done to reflect the demographics of the school population. An attempt was made to select an equal number of male and female students. Of the students that were interviewed, eight identified as African American (four male, four female), eight were Latino(a) (three male, five female), one Caucasian (male), and two Asian/Pacific Islander (one male, one female).

While the data was extremely rich, this chapter will focus on two students, Rebecca and Keith, who most clearly articulated the impact of playing *Quest Atlantis* on the development of their own ecological stewardship and civic engagement in their communities. Keith was a ten year old Asian/Pacific Islander male student from Edison Elementary while Rebecca was an eleven year old African American female student from Chamberlain Elementary.

FINDINGS/ RESULTS

Through case studies, several themes surfaced that clearly illustrated the ways in which student discourse about notions of ecological stewardship emerged and evolved as students navigated the quests in *Taiga.* First, by playing the game students generally internalized and understood science content knowledge on topics of water quality, ecosystems and system dynamics. The students were also able to supply the researcher with well-developed science content answers while demonstrating a clear understanding of these concepts. While the primary focus of the *Taiga* experience was the acquisition of basic science content knowledge on water quality and ecosystems, the conversations and discourse that occurred between students, teachers and researchers demonstrated that the students were also able to offer a more nuanced understanding of how the content they encountered within *Quest Atlantis* was connected to real-world localized water environments.

Secondly, students were able translate the science content and nuanced explanations they learned by playing *Quest Atlantis* to their communities, both virtual and real, whereby giving voice to environmental issues impacting their areas allowing them to begin efforts to solve these problems. These ideas emerged from analysis of both video and audio data collected during the student interviews and in-world experiences. This became particularly evident during the student interviews. The following section will provide further detail about how Keith and Rebecca illustrate these themes.

SCIENCE CONTENT KNOWLEDGE

Analyzing the student responses during these interviews, there was a consistent trend of students moving from vague, non-scientific responses to more nuanced, data driven explanations using precise language and knowledge acquired from the in-world experiences. *Quest Atlantis*, and *Taiga* specifically, appeared to facilitate students' learning of specific science concepts involving water quality and ecosystems within a problem based environment at the fifth grade level. This water quality and ecosystem knowledge was demonstrated through the written assessments where a t-test showed a significance ($p < 0.005$) between the paired average ($n=50$) pre-assessment score of 25 + 8 and the post-assessment score of 29 + 8, where $r=0.830$.[2] More specifically, Keith and Rebecca's scores both improved ten points from pre- to post-assessment, with Keith improving from 24 to 34 and Rebecca improving from 29 to 39.

The water quality and ecosystem knowledge gained by the students from engaging in the virtual environment was then applied to scenarios of their local environments in the interviews. Students applied their new knowledge and demonstrated notions of ecological stewardship and civic responsibility in their responses. For example, during the pre-engagement interviews, students were asked to determine if a particular water source was safe for their families to drink. Nearly all of the students interviewed (18 of 20) indicated that they would "look at it to see if it were clean to drink," and offered no details as to *how* they would "look" to see if this were indeed true. Rebecca's response was indicative of these students:

Researcher: *Imagine that you found out that the river near your house, like the [winding] river was the source of your drinking water. What could you do to make sure the water was safe to drink for you and your family?*

Rebecca: *Well, I can look at the river and see if people are swimming in it...and then look to see if was polluted or something... and to look and see if there is mud and pollution and stuff... and people are swimming in it... then it probably wouldn't be very good to drink out of... but if it were a fresh river and no one was swimming....there wasn't any pollution and stuff it might be safer to drink...*

Researcher: *So is there anything you might do to see if the water was polluted? How would you know?*

Rebecca: *Like tests or something?*

Researcher: *Yes, are there any tests you might run?*

Rebecca: *You could take a cup and fill it with water to see if its muddy and then I would look to see if there were any things that clean out the water or if they just took the water out of the river....*

In her response, Rebecca focused on people swimming in the river. She talked about pollution, but did not account for the sources of the pollution, nor how she might have worked to eradicate the pollution. Additionally, she did not indicate *how* she would determine if there was pollution beyond taking a sample to analyze. When probed for specifics, the response remained similar - that the water was muddy and therefore polluted.

However, after Rebecca had completed the four quests within Taiga, her answers were different. In the post-engagement interview, Rebecca responded in a much more sophisticated manner with respect to the science content knowledge:

Researcher: *Ok, so let's say that you heard that the [winding] river near your house was where your drinking water came from. What might you do to see if the water was actually safe to drink?*

Rebecca: *Well, first I would get a water sample from the river and test it. The water could be really bad to drink because the water could have like, be like acid, and that wouldn't be very good. I would also test for oxygen and other stuff... it could be... hmmm.... Nitrogen or phosphate or something, you know... kind of like the Mulus [the indigenous stakeholder non-player characters in Taiga] when I played in QA... they had that fertilizer stuff that got in the water...you know, up by that farm area near Norbe.... then the water gets all green and stuff...*

Researcher: *So what kinds of tests do you think you might run?*

Rebecca: *Well.... I'd like to see what is in the water....and like what the pH level was to see if there was acid since that is bad... and turbidity... you know.. the mud and how clear the water is 'cause that makes a difference with the oxygen...*

Rebecca was able to articulate that water quality testing would be central for determining whether the water was safe to drink. Additionally, she was able to identify specific examples (e.g. dissolved oxygen and nitrogen/phosphate testing) and, in several instances, relate it back to her in-world experiences (e.g. the algal bloom from the fertilizer runoff). The types of water quality tests that she identified were the tests that she encountered, learned about and utilized in *Taiga*.

Researchers also asked the students to think about waterways in their own community; if they were polluted and how might this pollution impact living organisms. When questioned about how the students' believed the waterways became polluted, all of the interview participants (n=20) recognized the impact of human activity on the environment through pollutants like boating, factory waste, human pollution and littering. Student responses were consistent between the pre- and post-interviews, however, in post-interview responses, students also addressed the impact of the pollution on food webs and river ecosystems. In the following response, Keith discussed how human activity impacted the turbidity of the river and ultimately, the food chain:

Researcher: *How do you think the river became so polluted?*

Keith: *Well, people litter a lot in the city....and sometimes it gets in the water... and dirt and stuff, when ever they run or do other stuff... maybe cut down some trees or something... dirt gets in the water and then the turbidity changes...and you know... the dirt then makes the temperature go up...get hotter... and then the water gets warmer and that is not good...*

Researcher: *So, why is that not good? How does that affect the fish, plants and other animals along the river?*

Keith: *Well... I think that the temperature... it might sorta kill the ecosystem because of how the temperature is... you know the oxygen...it gets messed up and there isn't as much when the temperature is high...so then the fish, they like die...the small fish die, and then the big fish die because they don't have any food anymore...and the food chain is sorta messed up... other animals might not have anything to eat either...*

Like Rebecca before him, Keith's response reflected a nuanced scientific understanding of water quality concepts and, in this instance, dissolved oxygen. This understanding and experience was facilitated through his engagement with the *Taiga* community and reflection in action (Schon, 1985, 1987) on the problems faced by *Taiga*.

CIVIC ENGAGEMENT AND ECOLOGICAL STEWARDSHIP

Examining the context of the students' post-interview responses, what emerged was that students made specific connections between the problems encountered in the virtual world environment and those they saw within their own communities. They began to see themselves as being able to actively engage in community building based upon the in-world experiences in *Quest Atlantis*. For example, during the course of the post-interview, it became evident that Rebecca saw herself as an engaged citizen who wanted to work as a change agent in her local environment. Rebecca's responses reflected that she was not only situating herself as a participatory citizen (Westheimer & Kahne, 2004), but she was beginning to think about the long term impacts of pollution in her community and was therefore moving towards a justice-oriented citizen orientation (Westheimer & Kahne, 2004). She stated:

Rebecca:*.....and if there is no oxygen, then animals like the fish can't live and if there is too much fertilizer and stuff then the water will get all green, with... what is it called... is it algae?*

Researcher: *yes...*

Rebecca: *That would be bad..the water would be dirty and people shouldn't drink out of it at all or they might get sick... so I would want to make sure that I let people know that... and...* ***and maybe even get them to help me clean it up so we could use it again****....I like to fish, and if the fish are dead, I can't go fishing....so that's what I would do....*

It was apparent from this brief interlude that Rebecca recognized the need for community action, similar to what she encountered in her *Taiga* experiences, and which demonstrated the type of empathetic embodiment described by Gee (2003). Through her experiences in *Quest Atlantis*, Rebecca came to understand how her participation in a civic community, including virtual communities like *Taiga*, was impacted by her choices and was reflected in the greater global community (Bers, 2008; Shaffer, 2004). This type of participatory citizenship (Westheimer & Kahne, 2004) encouraged community engagement around social and environmental problems, in this instance, impacted the quality of a community waterway. *Taiga* had created an opportunity for pedagogical praxis (Shaffer, 2004) where students, like Rebecca and Keith, were able to participate in communities of practice (Lave & Wenger, 1991) that allowed them to simulate the types of civic and democratic dynamics (Dewey, 1900/1956) that created opportunities for them to both reflect (Schon, 1985, 1987) and act on social and environmental issues.

This participatory citizen perspective emerged again through the conversations with Keith. By understanding the basis of the problems through legitimate peripheral participation (Lave & Wenger, 1991) in the *Taiga* community, students like Keith can begin to examine the structural causes of social and environmental issues and seek solutions. The newly grasped, nuanced scientific knowledge can lead to actions rather than a exhibiting a "divorce between knowledge and action"(Dewey, 1909, 41). Keith through this response demonstrated an understanding of how he moved from this type of participatory citizenship to a more justice oriented citizenship in the example below:

Researcher: *So if the animals don't have anything to eat, what would you do?*

Keith: *Well... ummm...I think I might want to get people not to make pollution...like pick up their trash...and not throw stuff in the water...and maybe not cut down so many trees so that the dirt doesn't get in the water and that way there is oxygen for the fish to breath... and then they can all live too...*

Reseacher: *So how might you do that?*

Keith: *umm...maybe we could get everybody to go down to the river and clean it up....like maybe have a school activity where we do that....or maybe ask my family to help too...*

Keith's response demonstrated participatory citizenship because it reflected a desire to make changes in his environment. However, it also indicated the beginning of a type of justice oriented citizenship because he not only identified the types of behaviors that were contributing to the environmental problem, but he suggested mechanisms for change and solutions to the problem (Westheimer & Kahne, 2004). Both Rebecca and Keith's responses built upon the types of citizenry models that are played out with the *Quest Atlantis* environment. *Quest Atlantis* seeks to engage students within the world, through participation in the community and by proposing solutions that address major environmental issues in ways that can improve the quality of life in their communities over time. By seeing the impact that they made within the virtual environment, students began to make connections and understand how they can use these same types of skills within their own communities, fostering a new sense of ecological stewardship and civic engagement.

DISCUSSION

The underlying story of *Taiga* in *Quest Atlantis* described a society experiencing social and environmental problems. Within this context, students became part of a community of practice (Lave & Wenger, 1991) that was working to solve issues that were similar to the types of problems indigenous to our own society. Building on the Dewey's idea of linking school and society (Dewey, 1900/1956), learning environments such as *Quest Atlantis,* allow students to use the virtual world as a tool to support learning and engagement in epistemologically meaningful projects situated in legitimate "real world" experiences. The advantage of using pedagogical-based praxis models is that the open-ended nature of the environments allow students to have multiple types of experiences within the context of the same world (Bers, 2008; Shaffer, 2004).

In *Quest Atlantis,* and in particular *Taiga*, the underlying message was that students could all work towards solving environmental problems in order to create a better society for both current and future generations. By immersing the students in a high-tech virtual play-space, they acquired not only scientific knowledge, but also an understanding of civic responsibility through their participation in a virtual democratic community (Barab et al, 2007). Through their participation in *Taiga* and its communities of practice within the Atlantan society (Lave & Wenger, 1991), students began to understand the impact of their decisions on the virtual community. They could then apply those lessons to their own communities, recognizing how they could impact local change. This process reflected the framework of Westheimer and Kahne (2004), participatory and justice-oriented citizens.

The *Taiga* missions took on these citizenry models by engaging students with the citizens of *Atlantis*, in order to solve important environmental issues (participatory) and engaged them in proposing solutions that encouraged changes in the behavior and practices of community members. This virtual civic engagement ultimately meant that community problems (e.g. the fish dying in the rivers of *Taiga*) were not perpetuated, but rather eradicated over time (justice oriented). Within the virtual world structure and through the game narrative, students were able to see the impact of their decisions on the *Taiga* community. By participating in this type of decision making within the virtual community, students developed the needed skills including an understanding of the types of questions and action plans needed to create change within their own communities. This was evident in the interviews with Rebecca and Keith, where they not only applied the science content knowledge acquired through their play, but discussed how they could engage their own community to tackle problems of water pollution in local waterways.

Students developed a rich conceptual, perceptual and ethical understanding of the science of water quality through their participation in the narrative of *Taiga*. This allowed them to apply their knowledge to a real-world problem as opposed to simply acquiring scientific facts. Inquiry became the means by which the students engaged with science content and was the tool to solve the environmental problems of *Taiga*. This allowed for an appreciation of the underlying science content and the role that political and economic factors play in scientific decision-making (Barab et al, 2007).

Through the creation of empathetic embodiment (Gee, 2003), students came to understand the unique dynamics and complex systems that are found in within the context of *Taiga*, creating a mechanism for legitimate participation and an understanding of what it means to be an ecological steward within a community of practice (Lave & Wenger, 1991). Through the students' recognition of the need for community action to protect their waterways, they demonstrated the ability to transfer the knowledge gained from their participation to their own communities demonstrating both social responsibility and civic engagement. By creating opportunities for students to become active participants within their communities of practice (Shaffer, 2004; Lave & Wenger, 1991), *Quest Atlantis,* and other virtual worlds and games like it, build on the work of John Dewey (1900/1956;1915), who saw the classroom as a student-centered community of learners.

The collaborative nature of *Taiga* and *Quest Atlantis* is reflective of the type of curriculum that Dewey (1900/1956) envisioned. The larger, global communities afforded by technology, allow students to view the world from broader viewpoint, accounting for a variety of perspectives and solutions to common problems. The knowledge that students construct from these interactions and the proposed solutions as represented in their interviews, demonstrated what Dewey saw as experiential learning, constructed in a social and technological context that is different from traditional epistemologies (Dewey, 1900/1956; Jenkins, 2006; Bers, 2008; Shaffer, 2004). The types of learning and development of ecological stewardship and civic engagement demonstrated by the students in this study confirmed that this type of interactive, virtual world approach could be beneficial when applied to educating students about civic engagement. Part of educating for a democratic society is making sure that students reach their full potential while contributing to the life of a democracy (Martínez Alemán, 2001). It is possible that in contemporary times, being environmentally friendly or ecological stewards, particularly with the emergence of the green movement, is valuable for ensuring a clean environment for many generations. How *Quest Atlantis* and *Taiga* addressed issues of ecological stewardship impacted how students perceived their own ecological and civic responsibilities. The students participating in *Quest Atlantis* environments not only learned about water quality and ecosystems as was demonstrated by their gains in scientific knowledge between pre- and post-assessments, but were able to identify complex problems and phenomenon in their own local waterways suggesting mechanisms for implementing change and their development as an ecological stewards. This suggests that virtual worlds like *Quest Atlantis* can potentially provide inquiry experiences that allow students to engage in experiential learning (Dewey 1900/1956).

IMPLICATIONS AND CONCLUSIONS

There are definite limitations of this study. The study was conducted in three classrooms in just two schools in an urban setting and focused on two students. Despite this snapshot view, this study begins to reveal the potential that this curricular instantiation can have on inquiry based scientific pedagogy and its potential to develop ecological stewardship and civic engagement for participants. While *Quest Atlantis* and *Taiga* appeared to accomplish this for the two students, as demonstrated by the interviews and the assessments, it was also clear that they could have benefitted from internal scaffolds or supports to help them take on these perspectives and guide them through probing questions. Games or virtual environments that seek to engage students in this manner will need to develop intelligent tutors which prompt students to think about the application of scientific concepts learned during game play to issues in their own communities. Virtual environments, like *Quest Atlantis*, allow students to "do science" in an immersive environment that encourages scientific debate and looks at the broader impacts of the scientific process within communities. Purposefully embedded scaffolds will help to bridge content with community engagement.

The goal of environments such as *Quest Atlantis* is for students to gain an understanding about what it means to participate in a democratic society. By allowing students to gain a sense of civic responsibilities and knowledge about what it means to be a good citizen who cares about the world, the students begin to understand, beyond procedural aspects, what it means to be part of a larger global community. From a theoretical perspective, the pedagogical model of praxis allows one to begin to understand the relationship between activity and learning in context (Shaffer, 2004). By creating learning experiences that immerse students within legitimate science experiences, students begin to internalize scientific ways of knowing allowing them to, in turn, apply knowledge to new contexts and situations. Students participating in *Quest Atlantis* began to recognize their own role in being a voice for issues of the environment that impacted not only the ecosystems, but the larger world. Students acquired a clear understanding of science concepts around water quality, ecosystems and system dynamics. They were also awakened to the roles they played within their own communities and began to make a connection to how they could facilitate change as global citizens of virtual and real worlds.

REFERENCES

Barab, S. A., Hay, K., Barnett, M., & Keating, T. (2000). Virtual solar system: Building understanding through model building. *Journal of Research in Science Teaching*, *37*(7), 719–756. doi:10.1002/1098-2736(200009)37:7<719::AID-TEA6>3.0.CO;2-V

Barab, S. A., Ingram-Goble, A., & Warren, S. (2008). Conceptual Playspaces. In R. Ferdig (Ed.), *Handbook on Research on Effective Electronic Gaming in Education*. Hershey, PA: IGI Global. doi:10.4018/978-1-59904-808-6.ch057

Barab, S. A., Sadler, T. D., Heiselt, C., Hickey, D., & Zuiker, S. (2007). Relating narrative, inquiry and inscriptions: Supporting consequential play. *Journal of Science Education and Technology*, *16*(1), 59–82. doi:10.1007/s10956-006-9033-3

Bennett, W. L., Wells, C., & Rank, A. (2008). *Young citizens and civic learning: Two paradigms of citizenship in the digital age*. Seattle, WA: University of Washington.

Bers, M. U. (2008). Civic identities, online technologies: From designing civic curriculum to supporting civic engagement. In W. L. Bennett (Ed.), *Civic Life Online*. Cambridge, MA: MIT Press.

Biswas, A., Licata, J. W., McKee, D., Pullig, C., & Daughtridge, C. (2000). The Recycling Cycle. *Journal of Public Policy & Marketing*, *19*(1), 93–105. doi:10.1509/jppm.19.1.93.16950

Brown, J. S., Collins, A., & Duguid, P. (1989). Situated cognition and the culture of learning. *Educational Researcher*, *18*(1), 32–42. doi:10.3102/0013189X018001032

Buckley, K. E., & Anderson, C. A. (2006). A theoretical model of the effects and consequences of playing video games. In P. Vorderer, & J. Bryant (Eds.), *Playing video games: Motives, responses, and consequences*. Mahwah, NJ: Lawrence Erlbaum Associates.

Denzin, N. K., & Lincoln, Y. S. (2000). The discipline and practice of qualitative research. In N. K. Denzin, & Y. S. Lincoln (Eds.), *Handbook of Qualitative Research* (2nd ed.). Thousand Oaks, CA: Sage.

Dewey, J. (1909). *Moral Principles in Education*. New York: Houghton Mifflin.

Dewey, J. (1916). *Democracy and Education*. New York: Macmillan.

Dewey, J. (1956). *The Child and the Curriculum*. Chicago: University of Chicago Press. (Original work published 1900)

Entertainment Software Association. (2006). *Essential Facts about the computer and video game industry.*

Gee, J. P. (2003). *What video games have to teach us about learning and literacy*. New York: Palgrave Macmillan. doi:10.1145/950566.950595

Gee, J. P. (2005). *Game-like learning: An example of situated learning and implications for the opportunity to learn* [Electronic Version]. Retrieved September 6, 2006 from http://www.academiccolab.org/resources/documents/Game-Like%20Learning.rev.pdf

Gibson, C., & Levine, P. (2003). *The Civic Mission of Schools*. New York: Carnegie Corporation.

Giere, R. (1991). *Understanding Scientific Reasoning* (3rd ed.). Fort Worth, TX: Holt, Rinehardt and Winston.

Jenkins, H. (2006). *Confronting the challenges of participatory culture: Media education for the 21st Century*. Chicago, IL: John D. and Catherine A. MacArthur Foundation.

Jordan & Henderson. (1995). Interaction analysis: Foundations and practice. *Journal of the Learning Sciences*, *4*(1), 39–104. doi:10.1207/s15327809jls0401_2

Kuhn, T. E. (1962). *The structure of scientific revolutions*. Chicago: University of Chicago Press.

Lave, J., & Wenger, E. (1991). *Situated learning: Legitimate peripheral participation*. Cambridge, UK: Cambridge University Press. doi:10.1017/CBO9780511815355

Lenhart, A., Kahne, J., Middaugh, E., Macgill, A. R., Evans, C., & Vitak, J. (2008). *Teens, Video Games and Civics: Teens' gaming experiences are diverse and include significant social interaction and civic engagement*. Washington, DC: Pew Internet and American Life Project.

Lincoln, Y. S., & Guba, E. G. (1985). *Naturalistic Inquiry*. Newbury Park, CA: Sage.

Martínez Alemán, A. M. (2001). The ethics of democracy: Individuality and educational policy. *Educational Policy*, *15*(3), 379–403. doi:10.1177/0895904801015003003

Mayo, M. J. (2009). Video games: A route to large-scale STEM education? *Science*, *323*, 79–82. PubMed doi:10.1126/science.1166900

Michigan State University. (2004). *Children spend more time playing video games than watching TV*. Retrieved from http://www.newsroom.msu.edu/site/indexer/1943/content.htm

Nelson, B., Ketelhut, D., Clarke, J., Bowman, C., & Dede, C. (2005, November). Design based research strategies for developing a science inquiry curriculum in a multi-user virtual environment. *Educational Technology*.

Neulight, N., Kafai, Y. B., Kao, L., Foley, B., & Galas, G. (2007, February). Children's participation in a virtual epidemic in the science classroom: Making connections to natural infectious diseases. *Journal of Science Education and Technology*, *16*(1), 47–58. doi:10.1007/s10956-006-9029-z

Papert, S. (1980). *Mindstorms: Children, Computers, and Powerful Ideas*. New York: Basic Books.

Popper, K. (1959). *The logic of scientific discovery*. London: Hutchinson.

Roth, W. M., & McGinn, M. K. (1998). Inscriptions: Toward a theory of representing as social practice. *Review of Educational Research*, *68*(1), 35–59. doi:10.3102/00346543068001035

Sadler, T. D., Barab, S. A., & Scott, B. (2007). What do students gain by engaging in socioscientific inquiry? *Research in Science Education*, *37*, 371–391. doi:10.1007/s11165-006-9030-9

Schon, D. A. (1985). *The design studio: An exploration of its traditions and potentials*. London: RIBA Publications.

Schon, D. A. (1987). *Educating the reflective practitioner: Toward a new design for teaching and learning in the professions*. San Francisco: Jossey-Bass.

Shaffer, D. W. (2004). Pedagogical praxis: The professional models for post-industrial education. *Teachers College Record*, *106*(7). doi:10.1111/j.1467-9620.2004.00383.x

Shaffer, D. W. (2006). *How Computer Games Help Children Learn*. New York: Palgrave. doi:10.1057/9780230601994

Squire, K. (2005). Changing the game: What happens when video games enter the classroom? *Innovate: Journal of Online Education*, *1*(6). Retrieved from http://innovateonline.info.proxy.bc.edu/index.php?view=article&id=82

Squire, K. (2006). From content to context: Videogames as designed experience. *Educational Researcher*, *35*(8), 19–29. doi:10.3102/0013189X035008019

Weber, C., & Matthews, H. S. (2008). Quantifying the global and distributional aspects of American Household carbon footprints. *Ecological Economics*, *66*, 379–391. doi:10.1016/j.ecolecon.2007.09.021

Westheimer, J., & Kahne, J. (2004). What kind of citizen? The politics of educating for democracy. *American Educational Research Journal*, *41*(2), 237–269. doi:10.3102/00028312041002237

KEY TERMS AND DEFINITIONS

Collaborative Problem Solving: Collaborative problem solving and the circular flow of ideas among peers, which is often the norm within this digital native generation.

Constructivism: Constructivist environments promote higher-order learning by engaging the individual in creating personally meaningful artifacts that can benefit and be shared with others within a community.

Multi-User Environments: Multi-user environments (MUEs), in terms of education, provide the opportunity for students to take on the role of producers, as opposed to consumers, of knowledge.

Participation: Participation in digital environments allows students to explore civic identities by participating in events and discourse related to civic issues in new and unique ways across global communities.

Participatory Citizen: A participatory citizen is actively engaged in his/her community and is working to solve social and environmental problems (e.g. pollution in the rivers).

ENDNOTES

[1] All names of schools, teachers and students are pseudonyms.

[2] Total score of the assessment was 40 points.

This work was previously published in Design and Implementation of Educational Games: Theoretical and Practical Perspectives, edited by Pavel Zemliansky and Diane Wilcox, pp. 189-205, copyright 2010 by Information Science Reference (an imprint of IGI Global).

Chapter 13
Learning Sociology in a Massively Multi-Student Online Learning Environment

Joel Foreman
George Mason University, USA

Thomasina Borkman
George Mason University, USA

ABSTRACT

Is it possible to enhance the learning of sociology students by staging simulated field studies in a MMOLE (massively multi-student online learning environment) modeled after successful massively multiplayer online games (MMOG) such as Eve and Lineage? Lacking such a test option, the authors adapted an existing MMOG—"The Sims Online"—and conducted student exercises in that virtual environment during two successive semesters. Guided by questions keyed to course objectives, the sociology students spent 10 hours observing online interactions in TSO and produced essays revealing different levels of analytical and interpretive ability. The students in an advanced course on deviance performed better than those in an introductory course, with the most detailed reports focusing on scamming, trashing, and tagging. Although there are no technical obstacles to the formation and deployment of a sociology MMOLE able to serve hundreds of thousands of students, such a venture would have to solve major financial and political problems.

INTRODUCTION

Sociology 101 is one of those ubiquitous general education courses taken annually by a million or more disinterested undergraduates who frequently cram and forget rather than form a deeply learned ability to see their lives through the lens of the sociological perspective. Part of the problem is the large lecture and the academic preference for paper based displays of learning, both of which enfeeble sociology's great potential for learning by doing.

DOI: 10.4018/978-1-4666-5071-8.ch013

As is the case with all college courses, the teaching of sociology is a loosely regulated cottage industry that lacks any national standards and is in the hands of personnel who rarely have had any formal training as instructors. It comes as no surprise that the quality of instruction is variable and inconsistent. The typical introduction to sociology is a conventional and familiar dosage of lecture, textbook reading, term paper, and written examinations—a mix that encourages short-term learning and rote repetition of the course content. The better versions feature small classes and teacher orchestrated discussions that encourage students to make connections between what they already know and what they are learning and thereby increase the likelihood of a more meaningful and enduring experience. Better yet are those classes that use simulations and other similarly engaging devices to ensure that students understand the material well enough to apply it analytically to real or fabricated social situations. Such classes are, unfortunately, in a small minority.

What to do about it? The success of online learning management systems (LMS) where students are able to "meet" and interact in cybernetic space suggests to some that we will one day see a convergence between such spaces and the much more sophisticated (from a functional and technological perspective) massively multiplayer online games (MMOG) like Lineage, Eve, and Guild Wars. With that possibility in mind, one can begin to imagine sociology courses that convene online in pedagogically designed spaces (a massively multi-student online learning environment or MMOLE) where students would spend much (if not all) of their time learning by doing.

WHY SOCIOLOGY?

For those who believe a college education should have demonstrable utilitarian benefits (rather than the vague "intellectual enrichment" of late adolescents), the study of sociology is a promising competitor for continuation in the general education requirements. Sociology studies how and why people behave as they do. It deconstructs naïve beliefs about the organization of human relations and replaces them with the ability to "see" the systematic ways that social systems distribute power and wealth and enable individual actions. Students endowed with such a vision and having to interact every day with other humans in small groups and complex bureaucracies are better able to make their social systems work for, rather than against, them. A student who is able to describe the relationship, say, between values, social status, and the reward system in a college fraternity, takes from a course on sociology benefits unavailable to a student who can define these abstract terms but not recognize them embodied in action.

Despite these formidable benefits and our high regard for them, we would not cede a permanent general education requirement to sociology. These valuable slots should be earned—through consistently excellent instruction. That is, the potential of the sociological perspective flows from what students have learned, retained, and are able to apply in their lives outside the classroom. And we have no reason today to believe that most (or even many) Sociology 101 students leave the course with its lessons secured in long-term memory.

As such, Sociology 101 is a perfect candidate for reformation as an MMOLE modeled after successful massively multiplayer online games. Immersed in such an MMOLE (one that predictably and consistently achieves a set of appropriate learning goals), students would develop their understanding of sociological principles as the result of their structured interactions within a set of simulated social scenarios. Rather than read in a textbook (or hear from a lecturer) about social mobility or the effect of gender on employment or the relation between caste and success, the student would experience, study, and have to negotiate controlled simulations of these social issues.

THE ONLINE SIMS (TSO) AND THE SOCIOLOGICAL FIELD STUDY

Following this line of reasoning, we considered how we might employ the extant online learning environments like WebCT and BlackBoard, and concluded that they are at this time relatively primitive systems informed more by the prevailing print technology of the past than the immersive audio/visual animations of the future. As we lacked a generous patron who could underwrite the $10 million plus cost to build and deploy a sociology MMOLE, we decided to figure out how to adapt an existing MMOG for pedagogical purposes.

The Online Sims seemed like a good candidate. It is the massively multiplayer online version of the extremely successful Sims gaming franchise. It is not violent, and thus appeals to female students; its "gameplay" is mostly about the kind of ordinary social interactions that characterize the housebound leisure of contemporary Americans. TSO allows geo-distributed players to build houses; to visit and interact with others through chat and a limited repertoire of physical behaviors (like kissing and dancing); to earn "Simoleans" and spend them on house furnishings and "games within the game;" and to develop skills that serve mainly to distinguish between the serious and casual players. Moreover, it is not difficult to learn the rules of engagement. About 20 hours of online exploration confirmed this judgment and convinced us that the environment could support an engaging and novel student exercise in the application of sociological principles.

After selecting an avatar body type and clothing, a newbie player sees a 30,000 foot view of the Sims world, selects a region, then a neighborhood, and then a specific building. This is rather like a visual descent that concludes with the player's avatar at the front door of a dwelling. As the dwellings in this view have no roofs, the player can see into them and then move about freely to take advantage of whatever amenities and activities are provided by the dwelling "owners." The TSO management system provides popularity lists that categorize the nominal intent of a property (which owners can define in terms such as "romance" and "skill building") and indicates which ones are open and accessible at a given time of the day. Since owners earn Simoleans just by attracting visitors, it is easy for newbies to "teleport" into numerous properties where they may observe or interact with diverse groups of other player avatars.

As such, TSO is an accessible (if limited) social system that can provide a constrained field study for sociology students at any time of their choosing. Although we would have preferred to microdesign the student exercise to assure predictable outcomes, to do so would have required far more time, online exploration, and imagination than we were able to devote to the project. We felt certain that the students would be engaged by and benefit from a simulated field study guided by a few questions keyed to course objectives, and we subsequently conducted two iterations. The first took place in fall 2003 with a single class of Sociology 101 students who gained access to The Sims Online with the free one-month subscription EA was offering at the time as a lure for new customers. The second iteration took place a year later (fall 2004) with participation from another group of Soc 101 students and a more advanced group in a Sociology of Deviance course. For this iteration, EA provided three months of free access.

SOCIOLOGY 101

The assignment instructions for the Sociology 101 exercise are as follows:

- Install the TSO software and spend approximately 10 hours exploring the online social system.
- Keep a journal in which you note the times you log on to the game, sites visited, what you did, friends you made, and skills you developed.

- Analyze your game experience in relation to the sociological ideas and concepts in the textbook and course lectures.
- Write a 5-page report that:
    - Describes the game as a constructed society. For example, what values of American society are exhibited explicitly or implicitly in the game? What aspects of social structure, social interaction, social networks, or group dynamics does TSO manifest? How are marriage and family handled? How is social class exhibited?
    - Suggest ways that the game might be used by students to apply sociological principles in future semesters.
    - Critique the game, given what you have learned about American society and sociology.

Our expectation, that 10 hours would be sufficient for students to learn how to navigate the game interface and still have time for substantial observer participation, proved to be correct. This was true for both iterations, though more so for the second because we included in-class demonstrations of the game experience. The real time demo, projected on a large screen, not only eased the student's entry and orientation experience, it motivated interest. Quite a few of the students had never seen an MMOG before and were fascinated to see the avatars moving independently through the game space and communicating with one another (via chat) even though the actual players were widely removed from one another in "real" space.

The student produced reports revealed several levels of analytical or interpretive ability. At the base level of achievement, common to all the participants, students matched distinct sociological terms with their instantiations in TSO. Most of the introductory students operated at this level and picked terms from a list of 15 American values that included progress, achievement and success, individualism, material comfort, democracy, humanitarianism, equality, and racism and group superiority. The terms identified most often were achievement, individualism, material comfort, and humanitarianism.

The students rarely dealt with the issue of racial/ethnic diversity or discrimination, which are major topics in sociology and to which the course instructor paid extensive attention. A few students remarked that they found all the Sims they encountered to be "white" even though colored skin tones were available as a choice of avatars, but that was the extent of their analysis. The tonal variation in the avatar skins is the game designers' only concession to the physical distinctions that are so important in stratifying real world social systems. Otherwise, the avatars are uniformly "attractive" in that they are of medium height; are in a 19-30 year age range (approximately); are slender, fat free, and well toned; and have no physical detractions, neither small ones like pimples nor more significant ones like deformed appendages. While one student (herself full bodied) observed that the TSO designers' decisions in this regard efface the discrimination in jobs, education, and leisure social relations faced by the overweight, many of the students claimed uncritically that TSO was "true to life."

TSO is, in fact, a simplified and idealized world lacking the diversity, complexity, and organizational elaboration of real life. The world contains no aging process, no old people, no fixed social classes, no unavoidable health problems, no adverse climatic conditions, no pain, and no irreversible death. It contains no government (other than the controls of the game monitors), no police, no military, and no corporations. Running out of money does not matter, and the rudimentary economy allows players to earn "Simoleans" mainly through entry level manufacturing jobs. Without any occupational diversification (other than manual labor and property ownership), the social system simply will not generate the job prestige hierarchies one finds in real world social systems. TSO is basically a classless meritocracy in which anyone willing to work hard enough can acquire the material accoutrements (large and well-furnished homes) of an upper class. Although the TSO management system publishes popularity lists and provides eight different lot sizes (thus differentiating between the sizes of dwellings), these distinctions benefit visitors as much as owners since attracting and entertaining the former is what increases the wealth of the latter. This structure—along with the general irrelevance of education, family history, income level, ethnicity, and occupational prestige—inhibits the formation of status hierarchies and the social complications wrought by such matters as income discrepancy and its systematic effects on homelessness, upper class status, and the like. Had TSO exhibited such social complications, we believe our introductory students would have had an easier time applying concepts learned in class.

A notable exception is the student who applied what he had learned about social networks, which the instructor discussed and illustrated in class and in an extra reading on the well known Milgram Small World Phenomena experiment. Social networks are integral to the game and may be used to advance a player's mobility in TSO (as this student pointed out) through the agency of "friendship webs." When one player becomes friends with another player, that friend is added to the player's friendship web, which is always available for any player, upon contact, to access and read. As in Milgram's small world experiment, players can see how connected they and their group of friends are to others in the game. The student observer in question reports that he spent his online time tracing the social network connections linking the friendship webs of the people he met. He visited the people listed in the friendship webs of his first encounters, looked at their webs for other contacts, and proceeded in similar fashion through the network. His application is the most sophisticated of any introductory student in that he understood how social networks applied in TSO and then used his knowledge to further his understanding of and action in the game. Such a learning experience, we believe, can be transferred to the real world social networks that successful people navigate so well.

SOCIOLOGY 310: SOCIOLOGY OF DEVIANCE

In 2004, our second iteration of the TSO simulated field study included the students in an advanced course—Sociology of Deviance. The course, which defined deviance as behavior that a social group regards as unacceptable and attempts to prohibit with negative sanctions, required students to enter at least three TSO properties and to observe to what degree and in what situations a controlling group defines and sanctions deviant behavior. The juniors and seniors in the course performed, as one would expect, on a much higher level than the students in Soc 101 and earned a higher percentage of A's (7/15) on their papers. They received these grades for several reasons. (1) They were able to distinguish between the rules of the game (i.e., "terms of service") and the deviance constructed by various actors in local circumstances within the game. (2) They described deviant behaviors and sanctions invoked at specific properties where they visited and observed. (3) They generalized appropriately (using the course definitions) about social control and transgressions.

In performing these sorts of observation and analysis, the students were acting within the long-standing tradition of social science and empirical research. What follows is an aggregate description of the more interesting instances of deviance observed and documented by the members of Soc 310.

SOME STUDENT FINDINGS

Most students found it easy to distinguish between the TSO instituted regulations and those maintained and enforced at specific properties developed by gamers themselves. The TSO institutional regulations (known as the "terms of service") are formally published on the Web site and must be agreed to as part of the enrollment process. The terms of service are diverse and include such matters as the illegal publication online of copyrighted material, the improper use of complaint submission buttons, and unwelcome harassment of other players. None of the students directly witnessed or produced violations so flagrant as to warrant the major means of enforcement available to TSO—banishment from the game. TSO is, after all, a virtual world. Since its virtual inhabitants are not subject to the kinds of physical damage or deprivations encountered in the real world, it is hard to imagine a form of deviant behavior that would count as more than a slight and passing emotional disturbance. Accordingly, there is no police force visible in the game and the monitoring and reporting of perceived infractions is generally up to the players themselves.

A case in point is the scamming, which often victimizes newly arrived players, who are sometimes referred to as "newbies." One such newbie reported being flattered and pleased when another Sim asked him to be his roommate, with the provision that the newbie buy some expensive appliances for the household—at a cost of 5,000 Simoleans, the amount of "money" with which each newbie starts the game. The next time the newbie logged on, he learned that the property was up for sale and that all the profits (including his 5,000 Simolean investment) would go to the owner.

This sort of random and individualized predation, which is more of a *caveat emptor* than a punishable transgression, depends on a discrepancy between the knowledge of the scammer and the victim. A group that refers to itself as the "F.U. Mafia" exploits such discrepancies on a much larger scale through their "scam houses." Scam houses are rather like a small Las Vegas casino where Sim visitors can play the "games of chance" permitted by TSO's terms of services. In one such game, a Sim pays 1,000 Simoleans to pick a card describing an activity he must perform in order to increase his stake. The problem is, the Sim does not know that the activity (in this case, eating 10 virtual snacks in 20 TSO minutes) cannot be performed successfully because of speed limitations built into the system. When the victim Sim predictably fails the challenge, the scam house owner pockets the 1,000 Simolean stake plus the cost incurred for the snacks. And the victim Sim often leaves without even realizing he or she has been scammed.

SIM SEX

TSO property owners can control deviant behavior within the confines of their domiciles by banishing perceived offenders and preventing them from returning. Our students observed property owners who invoked their banishment privilege for a variety of relatively minor infractions when visitors or roommates broke the house rules, disturbed the peace, quarreled with other members, used profane language, or engaged in deviant sexual behavior.

Simulated sexuality drew attention from a number of student observers, surely because of its inherent interest, but also because of its ambiguous status online. Although lacking the kinesthetic and consequences of real world sex, Sim sex remains an emotionally charged social interaction marked by various degrees of deviance and negative sanctions. The TSO designers have been quite careful to limit sexual activity to fully clothed hugging and kissing. But players have discovered, for example, that it is possible for two avatars (the player's online representatives) to get into the same bed together, an act that counts as a metaphor for intercourse. The more inquisitive players are likely to discover a hack produced by a rogue software company that allows a user to see other Sims in various states of nudity. Sims can also join each other in a hot tub where they can wash each other, kiss, and play. The latter activity, as witnessed by one student, entailed one Sim submerging while another moaned suggestively. It is not entirely clear whether this exploitation of game enabled behaviors to simulate sexual activity constituted a deviation from the game developer's intentions. However, a female Sim who was in the hot tub at the time was apparently offended (as one might be when witnessing a real world enactment in such a situation) and left the hot tub.

Since Sim coupling and Sim sex is not graphic or consequential, negative sanctioning, as in the case immediately above, usually takes the form of verbal admonishment or avoidance. One student observed an interaction in which a male Sim approached a female Sim and asked her to marry him. When the conversation revealed that the male was 14, the female (who was the house owner) wrote, "You're too young to be in this property." Nevertheless, she allowed the early adolescent to "marry" and "have sex with" another female Sim who was present. This is a clear deviation from what would be acceptable in the real world, but given the absence of real world physicality and consequences, the owner apparently felt that sanctions were unnecessary. What did strike her as an unacceptable transgression, however, was one couple's on screen publication of a sexually explicit conversation. This immediately elicited from her the threat of banishment from the house.

TRASHING AND TAGGING

One of the more activist student observers inadvertently discovered two artfully practiced forms of social deviation when he attempted to find out what would trigger sanctions at a particular property. In the first instance, he was ejected from a property after maliciously displacing some of the owner's virtual laptops. This act, he later learned, was a mild form of "trashing"—a deviant behavior practiced systematically by a local group calling itself the Irish Mafia. (The Irish Mafia and similar groups have formed spontaneously and are not game elements devised by TSO designers. More about this will follow.) In his second provocation, when his insults directed toward a female Sim failed to elicit a negative response, he "marked" her as his enemy. "Marking," more commonly known within TSO as "tagging," exploits the part of the game interface that displays a Sim's network of associates—the "friendship ring." Most Sims cultivate a completely positive "friendship ring" (one without any enemy "tags") because it affects their desirability as friends and roommates.

In tagging, the mafias can add a message, so when a user hovers over the enemy's name, it usually bears the stamp of the mafia, further spreading the group's infamy. The red links created by enemy tags are stigma symbols for innocent Sims and function as status symbols for the mafias.

The self-styled "Irish Mafia" reportedly uses both tagging and trashing to control neighborhoods and burnish its notoriety. A typical *modus operandi* entails the issuing of an extortion threat (pay us part of the revenue generated by your business or we will make trouble for you) when a new property owner enters an Irish Mafia neighborhood. If the new owner resists, the gang "tags" the property to deter potential visitors—thus depriving the owner of revenue generated by visitors. The red links created by the enemy tags and visible to others, simultaneously serve to stigmatize the innocent Sim while spreading word of the dubious achievements of the F.U. Mafia. Should this tactic fail to motivate a property owner, Mafia members known as hydras (derivation unknown) conceal their criminal associations, befriend the owner in the hope that he or she will eventually trust them enough to share "building privileges." Property owners' building privileges, as the term indicates, allow them to erect an edifice and furnish it with all of the items one would expect to find in a typical house plus a number of specialties that are unique to the TSO world. Building privileges also allow those who have them to delete any and all of the objects in and parts of the house, which is exactly what the Mafia "hydras" do (in effect, destroying the house) when a deceived and unsuspecting property owner is off-line. One particularly malicious variation of this Sim crime occurs when the Mafia "hydras" use a house's floor tiles to spell out profanities that are visible to any Sim exploring the terrain from a high-level neighborhood view. The Mafia then reports this infraction to the TSO management, resulting (supposedly) in the termination of the innocent property owner's account.

CONCLUSION

The student observations we have been describing persuade us that enough forethought, expertise, and money could produce a viable sociology MMOLE that would utterly transform the student learning experience. It would cost a lot to build (perhaps as much as $20 million) and to upgrade periodically (like Sim City or the Madden football game franchise) so that successive versions maintain a technological edge year after year. But because so many students take sociology every year, a potentially robust market exists to manage the costs.

What is the likelihood that this form of experiential learning would produce enduring conceptual skills for participating students? The answer is that the structure of the MMOLE would have to be designed to assure substantial, specific, and significant learning. It would do so by combining the content of sociological research (i.e. its empirical findings) with the relevant elements from videogames produced purely for leisure time activity and that have competed successfully for hundreds of hours of their users' attention. These elements include competition, sociality, graphic dynamism, and reinforcement systems—to mention but a few. Perhaps most important would be the organization of learning activities into a "level progression" whereby students would have to complete one level (requiring a demonstration of learning) before moving on to subsequent higher levels. As the student advances, the learning would accumulate and aggregate in a hierarchy leading to sociological competence. Winning the game, that is, ascending to and exiting from its highest level, would be the same as a very active demonstration of the desired sociological know-how. No external assessment would be required.

There are, of course, significant political obstacles to massive student participation in a sociology MMOLE. First, the professors and universities whose services could be displaced by a digital learning environment would surely object. The second political problem arises from the wide range of sociological studies. Sociology (along with anthropology) is probably the most inclusive social science as it covers governments and political behavior, economic behavior, history, as well as family relations, education, religion, small groups, bureaucratic organizations, social movements, social identity, race/ethnic relations, and social stratification. It ranges from the most macro level (global societies, individual societies, communities, neighborhoods, families, social networks) to the most micro (interactions between two people). Because of the breadth, depth, and inclusiveness of sociology, typical introductory courses vary widely in their content. As a consequence, a major political effort would have to be undertaken within the sociology profession itself to build agreements about the specific learning goals of an MMOLE catering to students in all parts of the nation. We believe that such an effort would be difficult but salutary.

Whether such an MMOLE would stand alone or serve as a component in a hybrid course is uncertain because sociology is a set of concepts that form a very specific perspective from which a researcher observes, collects data, and theorizes about social life. It is true that any successful member of a social order must have learned about the way it operates, but concepts and theories are needed in order to test that knowledge, to make it explicit, and to reproduce it uniformly for others to use. Under the best of circumstances, sociological concepts migrate into the public realm where we can see that terms like "siblings," "significant others," "reference group," and "sandwich generation" inform the way that the masses think about and act upon the raw data of their experience. Any game that would teach sociology would have to work in a similar manner. It would have to build conceptual structures or lenses through which the students would come to understand such matters as race relations, social stratification, and family relationships. The construction of the lenses (i.e., the delivery of conceptual information) could be done in a conventional fashion, externally through lectures or class discussions with a sociology professor. Or, the concept building activity could be built into the game as a help function or set of intellectual power-ups. For example, student players unable to solve a particular social problem might have recourse to in game functions that teach them how to design and use, say, a survey in order to get ahead in the game.

Assuming that these problems could be addressed, we can imagine an annual group of 200,000 student users whose semesterly subscription fees (say $100/student—a very reasonable figure) would generate $20 million every year. In a non-profit model, that should be enough to cover the cost of the initial construction of a sociology MMOLE, pay for annual maintenance, and an upgrade every few years. Far-fetched though this vision may now appear, we believe it will be realized one day.

KEY TERMS AND DEFINITIONS

Massively Multistudent Online Learning Environment (MMOLE): An e-learning system taking the form of a Multiuser Virtual Learning Environment (MUVLE) in which many users take part at the same time.

Massively Multiplayer Online Game (MMOG): An environment, usually based on a virtual world, where many users take part at the same time, each influencing that environment through their own actions within it.

Learning Management System (LMS): An e-learning system for managing educational programmes where they are many individuals participating for which separate statistics such as grand-points are needed to be understood in relation to that individual.

Massive Open Online Course (MOOC): A MOOC is a neologism and buzzword for a Massively Multistudent Online Learning Environment (MMOLE) that is based around an open-source distribution model.

Tagging: The terms for a phonemonon where the part of the game interface that displays a person's contacts or buddy-list for instance is exploited.

Trashing: A deviant behavior practiced systematically by a local group calling itself the Irish Mafia.

This work was previously published in Games and Simulations in Online Learning: Research and Development Frameworks, edited by David Gibson, Clark Aldrich, and Marc Prensky, pp. 49-58, copyright 2007 by Information Science Publishing (an imprint of IGI Global).

Chapter 14
The Applicability of Gaming Elements to Early Childhood Education

Holly Tootell
University of Wollongong, Australia

Alison Freeman
University of Wollongong, Australia

ABSTRACT

Many educators and technology developers advocate the use of gamification in educational environments. However, it is important to evaluate the applicability and value of gaming elements to the environments in which they are being implemented. Early Childhood Education (ECE) presents a unique educational context framed by national curricula and philosophical approaches that influence the adoption of technology, and therefore, gamification as an approach to enhancing learning through intrinsic motivation and engagement. This chapter evaluates the applicability and value of gaming elements to the use of technology in Early Childhood Education (ECE). Various definitions of gamification, particularly in the context of education, are considered. Six tenets of ECE and the concept of play are explored to inform an analysis of the appropriateness of gaming elements to ECE.

INTRODUCTION

This chapter evaluates the applicability and value of gaming elements to the use of technology in early childhood education (ECE). The chapter will firstly discuss gamification and consider the various definitions and understandings of the concept, particularly in the context of education. The chapter then provides an overview of ECE, including various national curricula and the two dominant philosophical approaches to ECE. Six tenets of ECE and the concept of play are explored.

DOI: 10.4018/978-1-4666-5071-8.ch014

Using Knewton's (2012) popularised Infographic 'The Gamification of Education', each of the elements of gaming are considered in relation to ECE, and the relevance of gamification in ECE environments is evaluated.

GAMIFICATION

Defining the Concept

Gamification has become a popular practice in many contexts, including enterprise, health, education, advertising and the military, with varied levels of acceptance and success (Deterding, Dixon, Khaled, & Nacke, 2011). Despite widespread discussion and application of gamification, and broad agreement on many key aspects of the concept, there is no single definition agreed by both practitioners and researchers (Erenli, 2012).

Even within a single context such as education, definitions vary (Muntean, 2011). The many definitions of games and gamification, and the embedded characteristics of some of these definitions, were considered in the context of learning by Erenli (2012). Based on a review of the literature, the following definition of gamification within the context of education was proposed (Deterding, Dixon, Khaled, & Nacke, 2011, p.10): "Gamification is the use of game elements in contexts that had originally no link to game related elements" (Erenli, 2012).

Kapp (2012) noted the importance of defining the basis of gamification (i.e. the 'game') in the context in which the game is 'played', and therefore the context in which the gamification is applied. The importance of context is supported by the work of Tootell et al. (2013) which considers the need to examine the use of any technology within its social context of use, linking this to the critical theory idea of 'lifeworld' (Habermas, 1984). Kapp defined a game in a learning context as "a system in which players engage in an abstract challenge, defined by rules, interactivity, and feedback, that results in a quantifiable outcome often eliciting an emotional reaction." (2012, p.7) and hence defined gamification as "using game-based mechanics, aesthetics and game thinking to engage people, motivate action, promote learning, and solve problems." (2012, p.10)

Gamification in Formal Education

The ability for integration of elements of gamification into learning experiences is enhanced with increased availability of natural user interface technologies such as iPads and interactive whiteboards, with over 1.5 million iPads already in use in educational programs worldwide (Tootell, et al., 2013). Despite this widespread availability of such technologies, the need for appropriate teacher training to maximize the usefulness of the devices has been highlighted (Beeland, 2002; Glover & Miller, 2001; Kaufman, 2009; Smith, Higgins, Wall, & Miller, 2005) and presents a challenge for both educators and educational institutions.

Given that the percentage of Internet users who engage in social gaming is continually increasing, the impact of games and gamification will be significant in the lives of today's youth. There was an estimated 118.5 million social gamers in the US and UK in 2011 (an increase of 71% based on the same report in 2010) (The Guardian, 2011), and it is estimated that over 70% of Global 2000 organisations will have at least one gamified application by 2014 (Gartner, 2011). A clear understanding of the concept of gamification is therefore essential for educators as they seek to connect more closely with learners and provide them with learning experiences that are aligned with future career opportunities.

This paper will consider whether this high level of engagement with games can be leveraged to enhance learning outcomes in early childhood education (0-6 years) by analysing the characteristics of gamification and comparing these to desired learning outcomes of ECE.

The use of games for learning is not new, with three types of games identified for reaching educational goals in a more engaging way: classic edu-tech or edutainment games, games developed by students themselves and gamified courses (Klopfer, Osterweil, & Salen, 2009; Muntean, 2011).

Edu-tech games are generally classified as 'serious games', which are games "for training… characterized by their specificity and applicability for particular work-related purposes" (Klopfer, et al., 2009, p.20). While some authors include the element of 'fun' in their definition of serious games, others imply simply that these are 'games' because they are not reality, and therefore they provide the benefits of a game environment (e.g. the ability to fail without real consequences). Serious games are popular in industries such as health and the military. Serious games are distinctly different from gamified courses and the gamification of learning, which applies elements of games to learning experiences. A further classification is that of learning games, which "target the acquisition of knowledge as its own end and foster habits of mind and understanding that are generally useful or useful within an academic context" (Klopfer, et al., 2009, p.21) and can be used in formal learning environments, informal learning environments (e.g. museums) and by self-learners. This research is concerned with only formal learning environments.

Klopfer et al. (2009) provides a list of ways games can be integrated into a learning environment, describing them as types of 'systems' that could be added to the environment: authoring systems; content systems; manipulating systems; trigger systems; gateway systems; reflective systems; point of view (POV) systems; code systems; documentary systems; ideological systems; research systems and assessment systems. Many of these can be adapted to games that are developed by students themselves. The method of description suggests that the authors view these 'games' as separate, non-integrated units within the learning environment. While Klopfer et al. (2009) present principles for learning game design, they do not specify the characteristics of gamification that justify its applicability to the classroom.

In contrast to the approach of game 'systems', Muntean (2011) advocated the extension of existing e-learning facilities to achieve gamification of an e-learning course. This study recommended the inclusion of the following gamification elements, in addition to existing course elements, for the gamification of an e-learning course: a customisable avatar, notifications, appointments, application of the cascading information principle for content delivery, points and levels to reflect achievement and hence learner status, a leaderboard and publication of top scores, bonuses, badges for non-academic achievement, the ability to convert points into virtual goods, feedback to indicate progression within the course, social elements, and the use of anticipation, all of which contribute to an overall flow of learning. It could be argued that such facilities could be incorporated into any digital learning environment; they are not only useful in dedicated e-learning systems.

The Positive Technological Development (PTD) framework, based on constructionism and positive youth development, was proposed by Bers (2012). Focussed on children, the PTD framework incorporates three components – individual assets, technology-mediated behaviours or activities, and applied practice – to inform the design of digital spaces that "help children gain the technological literacies of the 21st century while developing a sense of identity, values and purpose" (Bers, 2012, p.4). The individual assets (intrapersonal and interpersonal) component and the technology-mediated behaviours or activities component are both described in terms of a number of qualities. The qualities from these two components are mediated by the context in which the technologies are used. Links can be seen between many of these qualities and the characteristics of gamification in education, as described below under the heading 'Elements of gaming that may be applicable to early childhood education'. However, it should be noted that some qualities in the framework (for example, 'creativity') require thoughtful gamification design and do not clearly map to the characteristics of gamification in education discussed below.

Even assuming that gamification is applicable and valuable to some learning environments, it is important to critically analyse its usage in different learning contexts. It is likely that different game strategies and elements of gamification will be required to match the needs of varied learning environments and hence the specified learning outcomes (Kapp, 2012).

For the purposes of this research, it is necessary to identify the characteristics of gamification that are broadly accepted as being relevant to education. There is a lack of literature on agreed characteristics for gamification in the context of learning despite the significant level of discussion about the topic in practice and the media. One summary of gamification in education that has been widely circulated and discussed is Knewton's 'The Gamification of Education' Infographic (Knewton, 2012). This Infographic includes a list of elements of gaming [that we can] harness for educational purposes. Applying the definition that gamification involves "the use of game design elements in non-game contexts" this list of elements relevant to education has been selected as the *structure* for analysis of the applicability of gamification to ECE. A description of each of these elements is presented below.

The following section provides an overview of the educational philosophies on which ECE is based.

EARLY CHILDHOOD EDUCATION

National Frameworks and Technology

The adoption of technology in early childhood education (ECE) is of international interest. Technology in education is prompted by the United Nations Educational Scientific and Cultural Organisation (UNESCO) as a way of addressing "access, inclusion and quality" (UNESCO, 2011). The International Society for Technology in Education has produced a set of standards to assist teachers in their preparation of technology-supported learning opportunities for students (ISTE, 2008).

The integration of information and communication technology (ICT) is a key element in the new Australian National Curriculum (ACARA, 2012), the United States' National Association for the Education of Young Children 'Technology and Interactive Media as Tools in Early Childhood Programs' policy statement (NAEYC., 2012), and the United Kingdom's National Curriculum (UK Department for Education, 2012).

According to Turja et. al. (2009), guidelines for technology education in ECE curricula are mostly very general, or fragmented, or missing altogether. They suggest that technology be explicated in curriculum documents so appropriate focus can be given to its integration.

In the Australian context, the Early Years Learning Framework explicitly ties two learning outcomes to technology. Outcome 4: Children are confident and involved learners, states that children should have "access to technology". Outcome 5: Children are effective communicators, states that "technology should be child friendly" (DEEWR, 2009).

Te Whariki, which guides New Zealand ECE, identifies technology as a component of early literacy goals. The literacy outcome related to technology use states: "Children develop: experience with some of the technology and resources for mathematics, reading, and writing" (Blaiklock, 2011, p.63; Ministry of Education, 1996, p.78).

The UK Early Years Foundation Stages curriculum clearly documents the role of technology. In the learning objective of Understanding the World: Technology, "children recognise that a range of technology is used in places such as homes and schools. They select and use technology for particular purposes". In the learning objective of Expressive Arts and Design: Being Imaginative, "children use what they have learnt about media and materials in original ways, thinking about uses and purposes. They represent their own ideas, thoughts and feelings through design and technology, art, music, dance, role-play and stories" (Department for Education, 2012).

The use of technology within ECE in the United States is guided by a position statement by the National Association for the Education of Young Children and the Fred Rogers Center (2012). This position statement considers positive and negative influences of introducing technology into ECE curriculum. The position presented is that "Technology and interactive media are tools that can promote effective learning and development when they are used intentionally by early childhood educators, within the framework of developmentally appropriate practice (NAEYC., 2009), to support learning goals established for individual children" (NAEYC., 2012, p.5).

Learning Outcomes of Early Childhood Education

There are six basic tenets of early childhood curriculum models:

- Child-centred
- Active learning
- Plenty of time for children to pursue interests
- Studies or projects
- Reading with dialog, questions, discussions
- Creative, open-ended experiences

The Australian Early Years Learning Framework (EYLF), US National Association for the Education of Young Children (NAEYC), New Zealand Te Whariki and UK Early Years Foundation Stage (EYFS) all recognise these as features of a comprehensive approach to early childhood learning.

Child-Centred

Child-centred education experiences are essential to the healthy growth and development of children. It has been recognised in the UK Early Years Foundation Stages that "children learn by leading their own play, and by taking part in play which is guided by adults" (Department for Education, 2012). The child is free to investigate and experience things and "have a go". By allowing a child this freedom, each child is able to develop and learn using different methods and at different rates that are suitable for their individual needs. The New Zealand Te Whariki framework echoes these freedoms, and goes further to link this independent development to the empowerment of the child. "Children experience an environment where their play is valued as meaningful learning and the importance of spontaneous play is recognised" (Ministry of Education, 1996). In addition to empowerment, the Australian Early Years Learning Framework state that as children "participate in everyday life, they develop interests and construct their own identities and understandings of the world" (DEEWR, 2009). In these experiences children are able to develop their emerging autonomy, inter-dependence, resilience and a sense of agency.

Active Learning

Active learning refers to activities that engage and challenge children's thinking using real-life and imaginary situations. The UK EYFS identifies active learning as activities where children are concentrating and continue trying if they encounter difficulties, and enjoy achievements (Department for Education, 2012). Te Whariki extends this to include the learning experienced through responsive and reciprocal relationships with people, places and things. Active learning is a process in which they develop "working theories for making sense of the natural, social, physical and material worlds" (Ministry of Education,

1996). The Australian EYLF promotes active learning as the development of a range of skills and processes such as "problem solving, inquiry, experimentation, hypothesising, researching and investigating" (DEEWR, 2009)

Time for Children to Pursue Interests

In alignment with child-centred experiences, it is essential that children are allowed time to pursue interests. This time is separate from intentional learning activities. It is also known as free play. In this style of play children can learn to transfer and adapt what they have learned from one context to another. They can also use relationships with people, places, technologies and natural and processed materials to resource their own learning (DEEWR, 2009). Given the opportunity to pursue interests at their own pace, children experience an environment where they learn strategies for active exploration, thinking and reasoning (Ministry of Education, 1996).

Studies or Projects

In ECE, the opportunity to engage in common studies of projects gives children experiences of prolonged interest and staged learning. The studies or projects often arise from a child initiated topic which can then be used as a stimulus for the whole class. The EYLF notes that children can "use information and communication technologies to access information, investigate ideas and represent their thinking" (DEEWR, 2009).

Reading with Dialog, Questions, Discussions

The practice of reading with dialog, questions and discussions is a learning experience that is used more as children grow older. It is considered in the UK EYFS that "as their development allows, it is expected that the balance will gradually shift towards more activities led by adults, to help children prepare for more formal learning" (Department for Education, 2012). The Australian EYLF identifies three criteria that are relevant to reading: children engage with a range of texts and gain meaning from these texts; children express ideas and make meaning using a range of media; and children begin to understand how symbols and pattern systems work (DEEWR, 2009).

Creative, Open-Ended Experiences

The provision of creative, open-ended experiences allows children to be unrestricted by rules and expectations, which is in alignment with child-centred, active learning. The UK EYFS believes that by allowing children to create and think critically, they have and develop their own ideas, make links between ideas, and develop strategies for doing things.

Philosophies of Early Childhood Education

Reggio Emilia

Core to the Reggio Emilia philosophy is that the child is competent (Rinaldi, 1995). This moves the provision of early childhood education from a need to a right (Hewett, 2001; Soler & Miller, 2003), therefore establishing the need for competent, learned teachers (Bredekamp, 1993; Gandini, 1993). Within the Reggio philosophy, the teacher is considered a co-constructor with the children and documenter to the learning process. The relationship between the child, parent and teacher is central to the well-being of the child learner (Malaguzzi, 1993). All participants have an active role in informing the educative process.

The curriculum of Reggio Emilia is emergent (Malaguzzi, 1993; Rinaldi, 1995). General goals and hypotheses about projects and activities get made, and then adapted based on the children's interests. The documentation of these activities is central to the Reggio philosophy. The documentation is designed for several audiences: parents – to engage with their child's experiences; for teachers to understand the children and evaluate their own work. For the children, the documentation creates an awareness that their effort is valued (C. Edwards, Gandini, & Forman, 1995; Gandini, 1993).

The Reggio Emilia philosophy has been informed by Dewey, Vygotsky and Piaget. Malaguzzi states that the Reggio approach has gone beyond the Piagetian views of the "child constructing knowledge from within, almost in isolation". It is here that the social construction of knowledge becomes apparent with the emphasis on relationships with peers and adults (Rinaldi, 1995). The teacher as a guide and facilitator is consistent with Vygotsky's Zone of Proximal Development (ZPD), where the adult provides necessary scaffolding to assist children in their learning and development (Hewett, 2001).

Dewey has influenced Reggio Emilia philosophy. The following quote reflects the significance of interaction and participation in the learning process. "Play is not to be identified with anything which the child externally does. It rather designates his mental attitude in its entirety and in its unity. It is the free play, the interplay of all the child's powers, thoughts, and physical movements, in embodying, in a satisfying form, his own images and interests" (Dewey, 1956).

The Montessori Method

The Montessori Method was first implemented in Rome, Italy in 1907. It is the first method designed specifically for mass-marketed dissemination and replication. It does not draw on the works of other theorists like the Reggio Emilia approach. The method was first developed to work with "deficient" children with success in enabling "an inferior mentality" to grow and develop, and then considered that if applied to normal children, it would help "set free the personality" (Goffin & Wilson, 2001, p.39).

Montessori proposed a much prescribed set of learning environments and approaches that only she was fit to train. The "prepared environment" was critical to learning in the early childhood Montessori approach. The belief held was that children would engage with spontaneous activity, thus letting the educator know when they were ready for the next step in their formation. Early childhood education should take place in a specially created environment, the kindergarten, or the "child's garden" (Montessori & Gutek, 2004, p.11).

A criticism of Montessori is that the teacher or "director" is too prescriptive in the education of the child. However, proponents of the Montessori approach strongly argue that this is not true. The role of the director/directress in the original writings was for the adult to be responsible for placing the child in a learning situation where their own natural curiosity could be engaged. There is very limited role for teacher-directed learning in this model (Berliner, 1974).

PLAY

Play is considered to be a part of young children's life, which is recognised as a contributor to a child's social, personal, linguistic, physical, cognitive, moral, creative and artistic development (Synodi, 2010). Farné (2005) believes that although play continues throughout the human life cycle, it is during childhood that play has a "specific and deep educational role". There are three main components to understanding play as pedagogy: child-directed play, teacher-directed, and mutually-directed play. Historically, early childhood curriculum has associated play with child-centred pedagogy (S. Edwards & Cutter-Mackenzie, 2011), however in recent years there are a number of other perspectives that are being drawn in, including the importance of teacher interactions in play-based activities, and the significance of the nature of the dynamic relationship between children (learners) teachers and content (Ball & Forzani, 2007; Grieshaber, 2008) .

Play environments are critical to the success of play as learning. Play environments that include replicas of the adult world objects assist children in developing different modes of play: practice/functional play, make-believe/symbolic play and dramatic/role play (Turja, et al., 2009).

A firm definition of play in the context of early childhood education is difficult to find (Fleer, 2008; Johnson, Christie, & Wardle, 2005), however there are some descriptors commonly used:

- Active, exploratory
- Intrinsically motivated
- Carried out 'as if'
- More focused on process than on product, and
- Relatively free of external rules yet reflecting experiences and contexts (Stovers, 2011)

Fleer believes that the breadth of contributing theories to childhood play result in most childhood activities and behaviours being able to be described as 'play' (Fleer, 2008). A consequence of there being no firm definition is that 'play' can sometimes be caught up in political deliberations. For example, the OECD has avoided the term 'play' and instead has referred to "the child's agency and natural learning strategies" (OECD, 2006; Stovers, 2011). Vygotsky's 1966 theory of the role of play in the mental development of children provides directions for re-thinking how we have conceptualised play (Fleer, 2008).

Stovers (2011, p.66) believes that separating play from learning is as political as Dewey's efforts in the early 20th Century to recognise the learning that happens when children play. In the development of the Australian Early Years Learning Framework, Sumsion et. al. (2009) were concerned with finding the correct political significance of the role of play. They noted that a role of the EYLF was "to assist educators to recognise the opportunities that complex understandings of play present for helping children explore different ways of being, and to challenge injustice and bias" (2009, p.10).

Applicability of Gamification to Early Childhood Education

Knewton's (2012) 'The Gamification of Education' Infographic, which has initiated much discussion about the role and value of gamification in education, includes a list of 'elements of gaming [that we can] harness for educational purposes' (see Figure 1). Given that this summary of gamification in education has been widely circulated and discussed via educator groups on social media, and that this list of elements largely reflects the diverse characteristics of gamification discussed in the literature, this list of elements relevant to education has been selected as the *structure* for analysis of the applicability of gamification to ECE. The following analysis will consider whether the identified characteristics are appropriate for use in ECE.

Elements of Gaming That May be Applicable to Early Childhood Education

Progression: See Success Visualised Incrementally

The division of content into chunks, and the recording of progress based on these chunks, allows the learner to maintain an awareness of their progress. ECE embraces studies or projects on a topic – these studies encourage children to explore an area of interest in a prolonged manner, building both skills and knowledge through a range of interactions and experiences. Therefore, the organisation of material in ECE provides opportunities for delivering some or all elements of such studies using gaming elements.

ECE philosophies require that learners receive relevant, appropriate, timely, non-threatening feedback; this aligns closely with the feedback systems embedded in traditional gaming (i.e. participation in any type of game, which may or may not involve technology).

When designing game-based learning objects, it is important to develop systems that assess children's learning as they engage with the technology, to ensure they are meeting appropriate and expected progress (Nemeth & Simon, 2012). Learners' progress, and hence the evaluation of their learning, can be represented through levels, points, or even visual reward.

Knewton's Infographic identified two specific gaming elements that could be used to reflect progression in technology-based educational experiences.

Levels: Ramp Up and Unlock Content

As children demonstrate their increased knowledge and/or skills (through completing activities and acquiring points), they are rewarded by being promoted to higher levels. This increases the child's status within the game and is an indication of progression through content (Muntean, 2011). By gradually increasing the difficulty of the learning experience by delivering more detailed content or requiring the application of more highly developed skills, children's learning is scaffolded. This contributes to the development of competency within the child, as described in the Reggio Emilia philosophy.

Points: Increase the Running Numerical Value of Your Work

Learners are rewarded with points when they complete an activity or assessment. Learners can also be rewarded with points (or other items linked to a learner's status – for example, a badge) for positive non-academic contributions such as providing support to another learner or making a valuable contribution (Muntean, 2011). Points are usually visible to other learners. These points and badges serve as a continual motivator and status indicator of both academic achievement and behavioural contributions within a game. Collaboration provides the opportunity for the child to share knowledge, thereby increasing their awareness of relationships and social structures.

Investment: Feel Pride in Your Work and the Game

A personal profile (game terminology: avatar) gives each learner a unique online presence. The creation and customization of this profile (for example, by assigning it a picture, name and preferences) gives the learner an online presence that he/she can 'own'. This avatar concept is an essential element of gamification; in some cases, an avatar may simply be a username.

A learner's pride in their work is one of the main features of the 'individual assets' component of the PTD framework (Bers, 2012).

Knewton's Infographic identified five specific gaming elements that could be used to support this concept of investment in educational contexts.

Achievements: Earn Public Recognition for Completing Work

Activities attempted and completed by a learner are recorded on the avatar profile using points. Points information is typically public within the game, published on a leaderboard or through a list of top scores (game terminology: leaderboard, top scores). This encourages a focus on positive results (Muntean, 2011). Implementation of such features in ECE must be thoughtfully considered, given the importance of constructive feedback as opposed to rewards.

Publishing of points information also increases social interaction around the game because it

Figure 1. The Gamification of education (© 2013, Knewton. Used with permission).

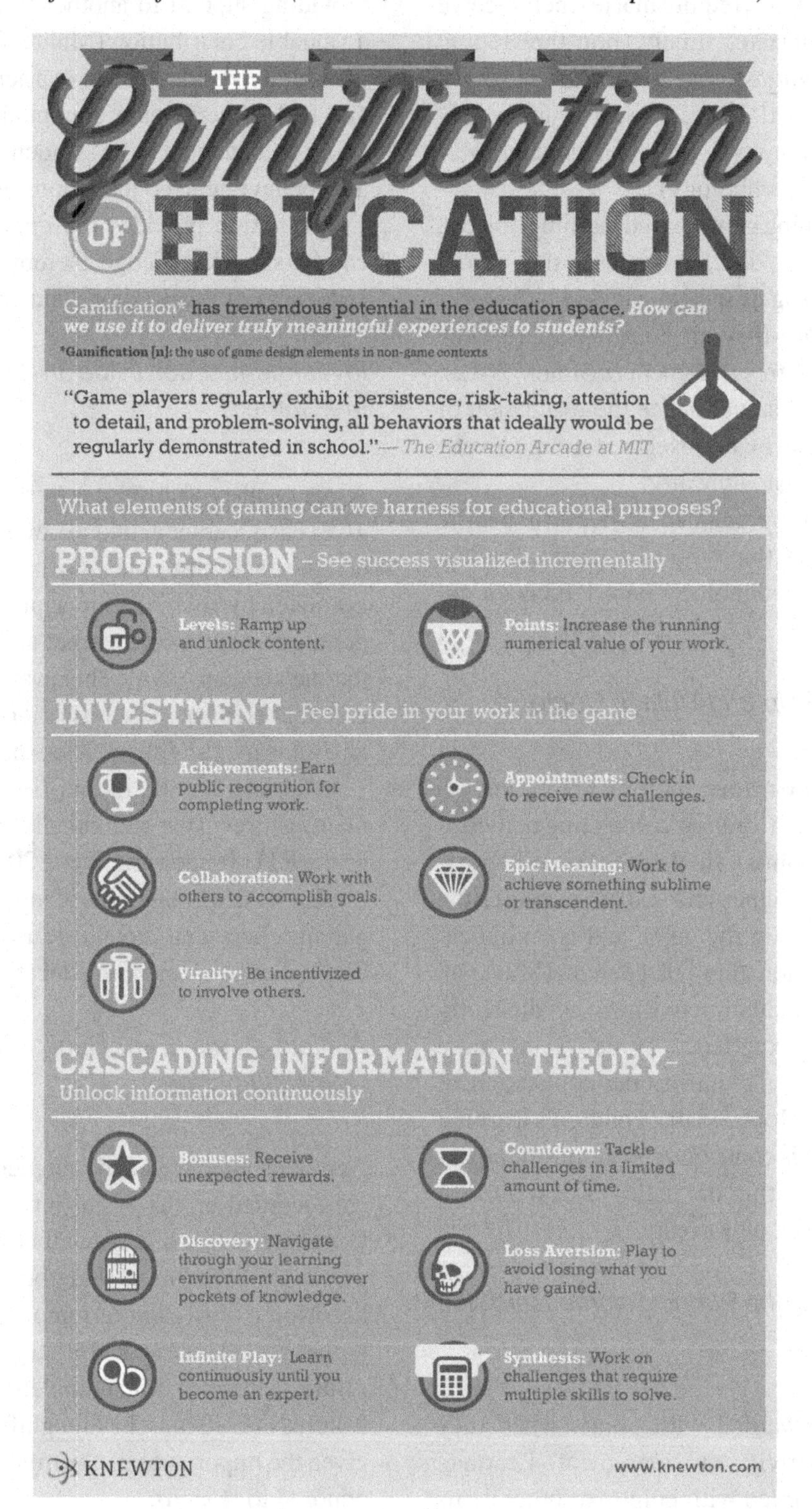

encourages learners to discuss their progress with others (Muntean, 2011), and may also motivate learners through peer comparison. Again, this information must be communicated in a developmentally appropriate way.

Given the importance of communication between educators, children and their carers, achievements recorded through gamified interactions could be used to share progress updates between members of each child's learning community.

Appointments: Check in to Receive New Challenges

Intentional and regular interaction with technology is now a key component of early childhood curriculums, and one aspect of designing developmentally appropriate classrooms (Simon & Nemeth, 2012). The use of deadlines or scheduled appointments, as is common in gaming, can be applied in educational contexts to encourage users to regularly engage with the game (Muntean, 2011). These appointments, in a ECE context, may be managed by educators. Learners may be required to complete a specified level or activity, or gain a specified number of points, within a set timeframe or by a set date. They may also be rewarded with bonuses based on their points at a specified time. This tool acts as a motivator for continual learning. Push notifications are used by some games to contact learners directly. These notifications act as reminders to engage with the game, increasing initiation of game engagement that is independent of the instructor. Such activities would require modification to be useful to ECE, however they may be useful in high-tech environments where children log in to technology systems regularly (for example, signing in upon arrival at the educational institution in the morning).

Collaboration: Work with Others to Accomplish Goals

A learner's avatar can belong to a group and have access to closed group information (Muntean, 2011) (for example, notifications, news and updates about other group members or shared interest information). Bers' (2012) PTD framework refers to the importance of children's ability to use technology to accomplish a goal ('competence'), to assist others with their use of technology ('caring') and to use technology for form and maintain positive relationships ('connection'). It also highlights the role technologies can play in learners "interchanging thoughts, opinions, or information" (Bers, 2012, p.12) ('communication'). There must be a balance between child-initiated technology experiences and other interpersonal experiences involving both small and large groups and offline collaborations (Simon & Nemeth, 2012), with the development of social skills essential for children.

Epic Meaning: Work to Achieve Something Sublime or Transcendent

Some learning environments offer learners the ability to convert their points or badges (game terminology: badge) into 'virtual goods' or be transferred into various types of financial compensation (Muntean, 2011). When these rewards appeal to the learners, they will act as high level motivators and enhance learner engagement. In traditional gaming, epic meaning almost always equates to a personal gain. This focus on praise or personal gain does not align well with ECE, where educators are concerned with providing constructive feedback (Nemeth & Simon, 2012).

The concepts associated with the gaming idea of 'epic meaning' can be applied to ECE if we consider that epic meaning can also equate to altruistic outcomes. Gaming elements can be used to teach children that small actions can have a significant impact in real world environments. This understanding of epic meaning aligns with the concepts underlying the increasingly popular cooperative games for social change. The PTD framework also highlights the importance of children understanding that technology can contribute to solving larger problems that benefit society ('contribution') (Bers, 2012).

Virality: be Incentivized to Involve Others

The publication of learner's scores (game terminology: leaderboard, top scores) encourages learners to discuss their progress with others (Muntean, 2011) and this information may be shared as a status symbol. Effective bonuses can incentivize participation. From a social perspective, (Bers, 2012) identifies the ability of technologies to enhance collaboration and caring, and to engage in community building ('collaboration' and 'community building').

In the ECE context, children must develop a range of basic skills before being offered opportunities to engage in information exchange and communication using technology. When ready, these exchanges could include interactions between the software, the child, classmates, the teacher and other members of the school community (Simon & Nemeth, 2012). Such exchanges at an early childhood level are likely to be concerned more with social interactions than issues of status.

Cascading Information Theory: Unlock Information Continuously

Activities, topics and courses can be divided into the smallest chunks of coherent content, based on cascading information theory (Muntean, 2011). Learners can absorb this content at a high level, or have the ability to navigate more deeply to discover more. The achievement of learning outcomes embedded in each chunk of content is demonstrated by a learner being awarded points (game terminology: points) for the learning tasks in that chunk of content. This design approach fits closely with the use of studies or projects, allowing educators to deliver small components of information that fit with the comprehension and attention span of children. The delivery of information in small chunks with increasing levels of challenge can be used to guide children through their personalized learning experience (Simon & Nemeth, 2012).

Knewton's Infographic identified six specific gaming elements that could be used to demonstrate this cascading information theory in education.

Bonuses: Receive Unexpected Rewards

Anticipation is a significant motivator (Muntean, 2011) which can be enhanced by the inclusion of surprises for learners (game terminology: bonus). Bonuses can be used as an additional reward when a learner completes a significant or difficult activity (Muntean, 2011). In mutually-directed play, children engage with their teacher in the play experience. This overarching structure of 'play', within which the teacher provides stimulus to dynamically redirect learners when necessary, creates a structure for unexpected interactions and positive experiences. These unexpected, positive and often seemingly insignificant experiences, are viewed by children as rewarding and motivating.

Countdown: Tackle Challenges in a Limited Amount of Time

Cambourne's (1988) Conditions of Learning specify that all learners need time and opportunity to use and practice new learning in realistic ways. This view is widely supported in educational literature, including for ECE. Imposing time limitations on learning experiences is one element of gaming that should not be applied to ECE. Child-directed play should be relatively free of externally imposed rules; it should be active and exploratory (Stovers, 2011). Time limitations are incompatible with this approach.

Discovery: Navigate Through Your Learning Environment and Uncover Pockets of Knowledge

Knewton's description of 'discovery' suggests that learning journeys are limited to pre-defined content which can be uncovered by learners. This description contradicts the usual view of a 'game' where users can explore freely and find content in an unstructured manner. While Knewton's description does not match this usual understanding of discovery in gaming, the idea of discovery is highly relevant to ECE. Children are encouraged to explore areas of personal interest and engage with new material as it is revealed to them. In ECE, it is important that discovery makes use of various devices, software, and apps that encourage creative thinking and offer multiple divergent learning paths (Simon & Nemeth, 2012). This use of technology to build creative, open-ended experiences is one area lacking development to date.

This idea of 'discovery' is distinct from the concept of navigation, which enables students to be aware of the next step so they know what to expect (Muntean, 2011). Navigation provides a structure on which learners can base their mental processing of the content and skills developed in the game. Such structures are essential for guiding children and promoting active learning.

Loss Aversion: Play to Avoid Losing what You have Gained

The concept of loss aversion is not relevant to ECE. Confident children must be able to learn through experience and be comfortable making mistakes. In a gaming context, mistakes equate to loss of points, rewards of status. ECE is concerned with positive reinforcement and constructive feedback, teaching children that unsuccessful experiences are still valuable learning opportunities.

Bers' (2012) PTD framework refers to the importance of the "opportunity of making choices about our behaviors, [ability to] explore "what if" situations, [and to] take action in the digital world, and experience its consequences" (Bers, 2012, p.12). This description ties more closely with the traditional view of gamification which purports the value of being able to 'fail' without real world consequences as well as the ability to 'explore' an idea as opposed to being tied in to formal structures. Cambourne (1988) acknowledges this as "approximation: where learners must be free to approximate desired study, as mistakes as essential for learning to occur".

Infinite Play: Learn Continuously Until you Become an Expert

The completion of learning activities allows the learner to build skills and/or knowledge. The completion of evaluation activities allows the learner to demonstrate their acquired skills and/or knowledge. Both learning activities and assessment activities can be used to assign the learner rewards (Muntean, 2011) (game terminology: points). While perseverance of learners is linked to confidence in the PTD framework (Bers, 2012), technology-based gaming needs to be moderated in ECE. The Early Childhood Environment Rating Scale – Revised (A+ Education Ltd, 2012) recommends that no more than 20 minutes per day should be spent sitting at a device to play educational games. On the other hand NAEYC (2012) do not prescribe a specific time limit of use, instead relying on the teacher to use their professional judgement to monitor engagement with the technology. This is supported by Simon and Nemeth (2012) who advocate the development of classroom-based systems to monitor children's use of technology, and hence ensure that they are spending appropriate amounts of time engaging in a range of choices.

Synthesis: Work on Challenges that Require Multiple Skills to Solve

The PTD framework's technology-mediated behaviours component (Bers, 2012) lists a range of technology-facilitated activities that children can undertake to build desired behaviours. Many of these activities (for example, 'content creation', 'creativity' and 'communication') can be combined to build challenges that require multiple skills to solve. When children complete independent technology-based activities in small groups (i.e. without adult facilitation), educators can use this as stimulus for child-based reflection about what they learnt or experienced (Simon & Nemeth, 2012), thereby developing a range of communication skills.

CONCLUSION

This chapter has provided an overview of the various national curricula and the dominant philosophical approaches that underpin early childhood education. It is these curricula and philosophies that influence the selection and use of technology for children in educational contexts. Therefore, these documents must be used as the basis for determining whether gamification provides appropriate and valuable learning opportunities for children.

The analysis of gaming elements in relation to these ECE documents has identified numerous gaming elements that can be used to build and reinforce desired ECE outcomes within accepted philosophical approaches, and hence facilitate appropriate gamification of learning experiences. However, the application of gamification should take the following issues into account:

- Design and development processes require guidance from ECE professionals to ensure suitable learning outcomes are embedded and interactions are philosophically appropriate.
- Some gaming elements contradict the outcomes and accepted approaches of ECE, and should be completely avoided.
- To date, one significant area of deficiency in gamification has been encouraging creativity. While the idea and language of gamification implies the experiences of 'fun' and 'play' associated with more traditional games, gamification has been largely absent from discussions of creativity within the educational domain specifically (Knewton, 2012). Given that 'creative, open-ended experiences' are one of the six tenets of early childhood curriculum models, developers and early childhood educators are faced with the challenge of designing gamified interactions that enhance creativity for young children.
- The selection of content and processes to gamify is a matter for early childhood educators to determine based on the needs of each educational institution. Some early childhood outcomes, such as the tenet of "Reading with dialog, questions, discussions", are not served by the elements of gaming directly; however, they provide sources of content and processes to which these gaming elements can be applied.

Most importantly in the ECE environment it is important to remember that "the use of technology should not outshine or replace any other experiences or opportunities" (Simon & Nemeth, 2012, p. 31). Technology is becoming an increasingly integrated component of the lives of young children. There are many elements of gamification that can be used to enhance learning, however, there is a need to ensure that a focus on gamification, or any singular approach to learning motivation and engagement, is not dominant or implemented in isolation, but instead used to complement other childhood activities such as creative play, real-life exploration, physical activity and social interactions.

REFERENCES

A+ Education Ltd. (2012). *The early childhood environment rating scale – Revised (ECERS-R)*. Retrieved 14 March, 2013, from http://www.ecersuk.org/4.html

ACARA. (2012). *The Australian curriculum.* Retrieved 20 February, 2013, from www.australiancurriculum.edu.au

Ball, D., & Forzani, F. (2007). What makes education research 'educational'? *Educational Researcher, 36*(9), 529–540. doi:10.3102/0013189X07312896

Beeland, W. (2002). Student engagement, visual learning and technology: Can interactive whiteboards help?. *Action Research Exchange, 1.*

Berliner, M. S. (1974). Montessori and social development. *The Educational Forum, 38*(3), 295–304. doi:10.1080/00131727409338116

Bers, M. (2012). *Designing digital experiences for positive youth development: From playpen to playground.* New York: Oxford University Press. doi:10.1093/acprof:oso/9780199757022.001.0001

Blaiklock, K. (2011). Curriculum guidelines for early literacy: A comparison of New Zealand and England. *Australian Journal of Early Childhood, 36*(1), 62–68.

Bredekamp, S. (1993). Reflections on Reggio Emilia. *Young Children, 49*(1), 13–17.

Cambourne, B. (1988). *The whole story natural learning and the acquisition of literacy in the classroom.* London: Ashton Scholastic.

DEEWR. (2009). *Belonging, being & becoming: The early years learning framework for Australia.* Retrieved 10 February, 2013, from http://deewr.gov.au/early-years-learning-framework

Department for Education. (2012). *Statutory framework for the early years foundation stage.* Cheshire, UK: Department for Education.

Deterding, S., Dixon, D., Khaled, R., & Nacke, L. (2011). *From game design elements to gamefulness: Defining gamification.* Paper presented at the MindTrek'11. Tampere, Finland.

Dewey, J. (1956). *The child and the curriculum, and the school and society.* Chicago: University of Chicago Press.

Edwards, C., Gandini, L., & Forman, G. (1995). Introduction. In C. Edwards, L. Gandini, & G. Forman (Eds.), *The hundred languages of children: The Reggio Emilia approach to early childhood education* (pp. 3–18). Norwood, NJ: Ablex Publishing Corporation.

Edwards, S., & Cutter-Mackenzie, A. (2011). Environmentalising early childhood education curriculum through pedagogies of play. *Australian Journal of Early Childhood, 36*(1), 51–59.

Erenli, K. (2012). *The impact of gamification: A recommendation of scenarios for education.* Paper presented at the 15th International Conference on Interactive Collaborative Learning. Villach, Austria.

Farné, R. (2005). Pedagogy of play. *Topoi, 24*(2), 169–181. doi:10.1007/s11245-005-5053-5

Fleer, M. (2008). A cultural-historical perspective on play: Play as a leading activity across cultural communities. In I. Pramling-Samuelsson (Ed.), *Play and learning in early childhood settings.* Dordrecht, The Netherlands: Springer. doi:10.1007/978-1-4020-8498-0_1

Gandini, L. (1993). Fundamentals of the Reggio Emilia approach to early childhood education. *Young Children, 49*(1), 4–8.

Gartner. (2011). *Gartner predicts over 70 percent of global 2000 organisations will have at least one gamified application by 2014.* Retrieved 12 February, 2012, from http://www.gartner.com/newsroom/id/1844115

Glover, D., & Miller, D. (2001). Running with technology: The pedagogic impact of the large-scale introduction of interactive whiteboards in one secondary school. *Journal of Information Technology for Teacher Education, 10*(3), 257–278. doi:10.1080/14759390100200115

Goffin, S., & Wilson, C. (2001). *Curriculum models and early childhood education: Appraising the relationship*. Upper Saddle River, NJ: Prentice-Hall Inc.

Grieshaber, S. (2008). Interrupting stereotypes: Teaching and the education of young children. *Early Childhood Education and Development*, *19*(3), 505–518. doi:10.1080/10409280802068670

Guardian. (2011). *PopCap games survey claims sharp rise for social gaming in US and UK*. Retrieved 10 February, 2013, from http://www.guardian.co.uk/technology/appsblog/2011/nov/14/popcap-social-games-survey

Habermas, J. (1984). *The theory of communicative action*. Boston: Beacon Press.

Hewett, V. (2001). Examining the Reggio Emilia approach to early childhood education. *Early Childhood Education Journal*, *29*(2), 95–100. doi:10.1023/A:1012520828095

ISTE. (2008). *National educational technology standards for teachers*. Retrieved 12 February, 2013, from http://www.iste.org/standards/nets-for-teachers/nets-for-teachers-2008

Johnson, J., Christie, H., & Wardle, F. (2005). *Play, development, and early education*. Boston: Pearson Education, Inc.

Kapp, K. (2012). *The gamification of learning and instruction: Game-based methods and strategies for training and education*. San Francisco, CA: John Wiley & Sons.

Kaufman, D. (2009). How does the use of interactive electronic whiteboards affect teaching and learning? *Distance Learning*, *6*(2), 23–33.

Klopfer, E., Osterweil, S., & Salen, K. (2009). *Moving learning games forward: The education arcade*. Cambridge, MA: Massachusetts Institute of Technology.

Knewton. (2012). *The gamification of education infographic*. Retrieved 12 February, 2012, from http://www.knewton.com/gamification-education/

Malaguzzi, L. (1993). For an education based on relationships. *Young Children*, *49*(1), 9–12.

Ministry of Education. (1996). *Te whāriki: He whāriki mātauranga mō ngā mokopuna o aotearoa early childhood curriculum*. Wellington, New Zealand: Learning Media Limited.

Montessori, M., & Gutek, G. L. (2004). The Montessori method: The origins of an educational innovation: Including an abridged and annotated Ed. of Maria Montessori's The Montessori Method. Oxford, UK: Rowman & Littlefield Publishers, Inc.

Muntean, C. (2011). *Raising engagement in e-learning through gamification*. Paper presented at the 6th International Conference on Virtual Learning ICVL 2011. Bucharest, Romania.

NAEYC. (2009). *Developmentally appropriate practice in early childhood programs serving children from birth through age 8*. Retrieved 14 March, 2013, from www.naeyc.org/files/naeyc/file/positions/position%20statement%20Web.pdf

NAEYC. (2012). *Technology and interactive media as tools in early childhood programs serving children from birth through age 8*. Retrieved 2 March, 2013, from http://www.naeyc.org/files/naeyc/file/positions/PS_technology_WEB2.pdf

Nemeth, K., & Simon, F. (2012). *Preschool curriculum and technology crosswalk*. Paper presented at the EETC Early Education and Technology for Children. Budapest, Hungary.

OECD. (2006). *Executive summary: Starting strong II: Early childhood education and care*. Retrieved from http://www.oecd.org/edu/school/startingstrongiiearlychildhoodeducationandcare.htm

Rinaldi, C. (1995). The emergent curriculum and social constructivism: An interview with Lella Gandini. In C. Edwards, L. Gandini, & G. Forman (Eds.), *The hundred languages of children: The Reggio Emilia approach to early childhood education* (pp. 101–111). Norwood, NJ: Ablex Publishing Corporation.

Simon, F., & Nemeth, K. (2012). *Digital decisions: Choosing the right technology tools for early childhood*. Lewisville, NC: Gryphon House.

Smith, H., Higgins, S., Wall, K., & Miller, J. (2005). Interactive whiteboards: Boon or bandwagon? A critical review of the literature. *Journal of Computer Assisted Learning*, *21*(2), 91–101. doi:10.1111/j.1365-2729.2005.00117.x

Soler, J., & Miller, L. (2003). The struggle for early childhood curricula: A comparison of the English foundation stage curriculum, Te Wha¨riki and Reggio Emilia. *International Journal of Early Years Education*, *11*(1).

Stovers, S. (2011). *Play's progress? Location play in the educationalisation of early childhood in Aotearoa New Zealand. (Doctor of Philosophy)*. Auckland, New Zealand: Auckland University of Techology.

Sumsion, J., Barnes, S., Cheeseman, S., Harrison, L., Kennedy, A., & Stonehouse, A. (2009). Insider perspectives on developing: Belonging, being & becoming: The early years learning framework for Australia. *Australian Journal of Early Childhood*, *34*(4), 4–13.

Synodi, E. (2010). Play in the kindergarten: The case of Norway, Sweden, New Zealand and Japan. *International Journal of Early Years Education*, *18*(3), 185–200. doi:10.1080/09669760.2010.521299

Tootell, H., Plumb, M., Hadfield, C., & Dawson, L. (2013). *Gestural interface technology in early childhood education: A framework for fully-engaged communication*. Paper presented at the Hawaii International Conference on System Sciences (HICSS). Maui, HI.

Turja, L., Endepohls-ulpe, M., & Chatoney, M. (2009). A conceptual framework for developing the curriculum and delivery of technology education in early childhood. *International Journal of Technology and Design Education*, *19*(4), 353–365. doi:10.1007/s10798-009-9093-9

UK Department for Education. (2012). *Statutory framework for the early years foundation stage*. Retrieved 10 February, 2013, from https://www.education.gov.uk/publications/eOrderingDownload/00267-2008BKT-EN.pdf

UNESCO. (2011). *ICT in education*. Retrieved 23 April, 2012, from http://www.unesco.org/new/en/unesco/themes/icts/

KEY TERMS AND DEFINITIONS

Computers and Children: The exploration of how computers and children interact in early childhood education.

Early Childhood Education: Education experiences encountered before the age of formal schooling, usually 0-5 years, but can be 0-8 years.

Educational Technology: The application of technology to educational experiences.

Game-Based Technologies: The application of technologies to the gaming experience, such as through gamification.

Play: A contributor to a child's social, personal, linguistic, physical, cognitive, moral, creative and artistic development.

Positive Technological Development: The application of a framework focused on children that draws the individual, technology and practice together.

Chapter 15
From Chaos Towards Sense:
A Learner-Centric Narrative Virtual Learning Space

Torsten Reiners
Curtin University, Australia

Lincoln C. Wood
Auckland University of Technology, New Zealand & Curtin University, Australia

Jon Dron
Athabasca University, Canada

ABSTRACT

Throughout educational settings there are a range of open-focused learning activities along with those that are much more closed and structured. The plethora of opportunities creates a confusing melee of opportunities for teachers as they attempt to create activities that will engage and motivate learners. In this chapter, the authors demonstrate a learner-centric narrative virtual learning space, where the unrestricted exploration is combined with mechanisms to monitor the student and provide indirect guidance through elements in the learning space. The instructional designer defines the scope of the story in which the teacher and learner create narratives (a sequence of actions and milestones to complete a given task), which can be compared, assessed, and awarded with badges and scores. The model is described using an example from logistics, where incoming orders have to be fulfilled by finding the good and delivering it to a given location in a warehouse. Preliminary studies showed that the model is able to engage the learner and create an intrinsic motivation and therewith curiosity to drive the self-paced learning.

INTRODUCTION

Stories are one of the oldest means of passing on information and experiences to others. Storytellers combine words with gestures and expressions, creating illusions, using intonation to build up suspense to finally reach full immersion in the narrative. Storytelling is art; the canvas being the mind and the words the crayons to draw the picture. Storytelling is connective; it requires an audience with whom we can share. Storytelling is creative; we hear words and sounds, see

DOI: 10.4018/978-1-4666-5071-8.ch015

gestures and expressions, but we also combine these shared impressions with our personal experience, understanding, and knowledge to our very individual story. Storytelling is an effective mean to convey "information in a compelling and memorable way" (Neal, 2001) and the "original form of teaching" (Pederson, 1995). "[] It's our desire to still employ the mood and storytelling tools inherited from film and theatre" (Björke, 2003). Similar to the film industry, instructional designers have to adopt and use the technology in the way it is designed; not enforcing old beliefs and thoughts and methods on it. The narrative has to be sculpted and designed specifically to express the narrative in its environment.

Yet we have to ask ourselves if story telling is teaching? Are teachers story tellers? With all due respect to the numerous teachers worldwide and their never-ending effort to transform the classroom into a learning space full of stories and adventures, we can see that they are often not. The classroom is just a space, the story "provides relevance and meaning to the experience. It provides context." (Kapp, 2012; p.41). Instead, the system that these teachers work within is seemingly more concerned with the continuous equalisation of courses world-wide; predefining years ahead what has to be taught, which text book is to be used, and how the learners have to demonstrate the *successful* transfer to their heads; being assessed in uniform tests at times most convenient for the institutions; at least if we assume to be trapped on the lower levels of Blooms taxonomy (Bloom, 1956). While stories are still told by engaged teachers, the systems that they work within have forgotten to include the audience of the stories. We expect that all canvases show the same picture, not guiding the audience through the story but dictating what is important and how to interpret it. With no intention for discussion; we shall emphasise the governmental and administrative drag towards programs like "No Child Left Behind" or strict uniformity and comparability in undergraduate and master programs to simplify the transfer process between educational institutions and smoothen the transition to the working place (Noddings, 2007). Teaching and learning is not about the laziest way, but the best way to engage the learner in understanding and critical thinking (Friedman, 2005); one of the primary concerns for educators to achieve (Boyle-Baise & Goodman, 2009).

The endemic passivity within classrooms is disturbed by giving the listener the power of influencing the storyline by being asked to make decisions at key points. An example is the role playing game 'Dungeons & Dragons', in which a group of characters (each controlled by one player) undergo an adventure in a fantasy context. A storyteller (Dungeon Master) is responsible to pursue the story, play different roles in the story, and challenge the players with tasks like fighting, entering dungeons, or seeking treasures. The dungeon master is capable of controlling the story in any direction; being both the master of the scope and the given objectives. The lecturer can do the same in a classroom; can allow learners to explore the learning space without restrictions, yet having selected activities to provide a scope to keep learners on track. The supervisor or manager in an industrial context can monitor employees' activities to achieve the objectives, and just gently (or with a brusque attitude!) provides them with guidance to ensure that activities are finished on time; so that employees' efforts are not wasted.

The storytelling becomes more complex if we extract the storyteller; the component with the most direct influence on the learner (Bauman & Briggs, 1990). Learners may not attend a classroom session but engage in a self-paced learning process; e.g., learning in a distance educational environment (Gregory et al., in press; Moore & Anderson, 2012). The basic but often used model is to merely provide (or 'dump') all materials within a learning management system with some general instructions to proceed and succeed in assessments and examinations. The environment, in this case a rather unattractive and limited one, becomes the story teller, with the learner doing the journeying as an influencing participant in the story (Danilicheva et al., 2009). A different kind of environment can be found in games, where the focus is on storytelling and the learning process is woven into the story itself. Compared to the first generation of computer games, modern games embed massive stories to feed the player with background information for an extended immersion (Kapp, 2012). An interesting example is the open world that the company RockStar created with the critically acclaimed GTA-series (Grand Theft Auto[1]). The environment is designed to be fully explored by the player, while the main storyline is blended into the normal *life* within the city; e.g., characters walk on the street unsuspiciously interacting with other non-player characters (NPC). The player is reminded in different ways, however, that there is a main task to accomplish. This can be as subtle as a reminder message from a NPC up to a more immediate 'drag' towards the original story and main objectives. In other games, such impetus can be provided through the inclusion of timers, or changes within the landscape and environment that progressively restrict the range of actions that the player is able to take, forcing them by default to complete the required tasks as there is nothing else to do!

What can educators and instructional designers learn from these game-based examples? Simply this: it is not enough to merely create a narrative; the scope of the narrative must be suitably wide for learners to engage a sense of curiosity and develop intrinsic motivation for learning, while being limited to enable the instructor to ensure completion of learning objectives and course outcomes. It is relevant to track the learner and match its path against the expected narrative as designed by the instructional designer in the role of a game master (Broussard, 2012). It is not required to match the given narrative exactly but 1) the objectives has to be fulfilled; 2) extra actions need to have a (learning) benefit (e.g. sitting on a chair vs. studying the manual of a vehicle to be used); 3) key actions have to performed (in the right order) (e.g. entering data in the system to keep a protocol); and 4) a score has to be calculated based on the actions done and expected.

In this chapter, we reflect on how the concept of stories and learner-centric narratives can be used to design a learning space in the domain of operations and supply chain management. In the following section, we discuss the terminology regarding narratives and stories as used in the literature and investigate current trends regarding how to increase engagement and motivation of learners within a defined learning space. We continue with a brief description of the processes in a warehouse to fulfil orders. The example is used to explain how an instructional designer can use the story to create potential narratives to guide the learner towards the learning objectives, how the learning process can be analysed, and how gamification elements can trigger the curiosity of the learner. The focus in all chapters is in logistics and supply chain. The chapter is concluded with an outlook on how the learning space will evolve and contribute new ways of immersing the learner in the future.

BACKGROUND

This section depicts the terminology used to describe the stories and narratives; terms that are commonly used in the literature whereas we slightly adjust them with respect to virtual environments and gamification. Figure 1 further visualises the relations between the terms to support a wider understanding. The term ***story*** as used in the beginning of the chapter sets the overall scope and constraints of what to cover and what to exclude during the story telling (also called a bounded learning (purpose) or action (interactive) space). The story is the ***setting*** in which the ***actors*** will *live* their very unique narrative(s); including all required properties and elements. In the above example of the Dungeon & Dragons, the story is

Figure 1. Visualisation of story and narrative

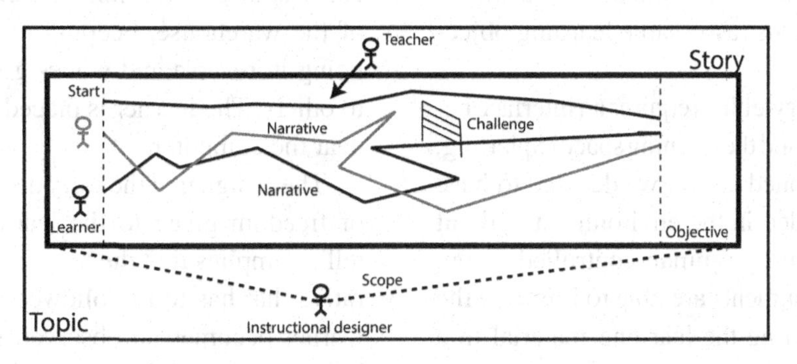

the box with all parts to play the game; i.e., the objectives and charts to describe the behaviour of possible enemies on the yet untold/unformed/unmodelled journey. The story is told by ***narratives***; unique paths through the story which also enliven the story and "unfold in space" (Nitsche & Thomas, 2003). The story itself is designed by someone in the role of an instructional designer; with narratives being created by the teacher (expert-knowledge-based design of model answer as well as suggestions how to traverse the story; defining ***milestones*** as a sequence of actions in the scenario, with a high level of continuity between one portion of the scenario and another) and the learner (guided by the teacher's narrative but forming its own perception and awareness around it). This provides structure for a teacher to meet assessable ***learning objectives***, but also the flexibility which encourages motivation and volition in learning. From a technical perspective, we consider both narratives (teacher and learner) to be the same; whereas one objective in the learning process might be the alignment of both narratives (see also the final discussion in Section "Story, Scenario, and the Gamified Nudge." Narratives support the process of understanding and building cognitive structures (Riedle & Young, 2003; Bruner, 1990). Interactions in a 3D virtual world already provide an engaging environment to have stories and narratives resulting from the activities and interactions of avatars in this space. Narratives are either pre-scripted (ready to reveal their sequences of milestones and activities over and over again), or use exploration and goal-oriented triggers to multiply the possible narratives that learners can indirectly choose from; e.g., GTA which was mentioned above as an example.

Danilicheva et al. (2009) distinguish plot-based storytelling (narrative being created by the teacher to be followed by the learner) and character-based storytelling (narrative is dynamically created by the interaction of the learner with the environment and intelligent computer-controlled avatars). Nitsche and Thomas (2003) use the term *Story Map*: the learner explores the virtual environment and maps the space and the story as part of this process. The story is tied to the navigation in the space (Murray, 1997) rather than predefined and orchestrated by the teacher. Recorded actions during the learning process can be seen as dynamic narratives being developed while moving in the (learner-centric) bounded learning space. Having a restricted virtual space using avatars has the benefit of having comprehensive recordings of all actions and situations. These memories can be static and consist of different sequences of actions, images, or statistics; or dynamic by recorded movies; so called Machinima. And all recorded memories are also memories for the learner; remembering and reflecting the past. Kapp (2012) lists four key elements for a vicarious experience, which are implemented in the described environment below. That is, characters (actors; i.e. learners), plot (story and narratives), tension (milestones with feedback on achievement), and solutions (assessable learning objectives).

The virtual storytelling requires an interface between the learner and the learning space (Spierling, 2002). As mentioned above, we decided to have the story embedded in the environment without a real-life storyteller or human-controlled avatar. 3D virtual environments are able to immerse the learner by projecting the learning material in a learning space that is as close to the real world as possible, as all interactions are mapped "as natural as personal contact" (Danilicheva et al., 2009) with the environment; minus the real-world risks and safety concerns. The current deficit lies in the control through keyboard and mouse and is a major concern for the realism; yet, current and future technology is changing the human-computer-interaction dramatically; see for example Oculus Rift[2], Google Glass[3], and the upcoming Kinect[4] or Leap Motion[5]. Another technology is so called bots or virtual (intelligent actors): computer-controlled avatars with different stages of artificial intelligence regarding interactivity or capability to become a completely autonomous storyteller. Bots can be used to increase the authenticity, the reality; providing interactive elements in the story and guiding the learner through dialogues or hints. The challenge is to make the learner believe that the virtual actor is real and not just an animated script. That is, providing an illusion of realism (Perlin, 2003). As the scope of the chapter is on narratives, we refer to Reiners et al. (2013) and Wood & Reiners (2013) for further discussions and examples.

Figure 2 shows a story. The setting is a small warehouse with just three shelves to stock different goods on pallets or in barrels. Through the door, the learner can see a straddle carrier moving a container to a container bridge; implying that the warehouse is located on a container terminal. Right outside is a ute vehicle. In addition, there are three forklifts that look very similar and also an improvised table with a notebook. The audio consists of general noises; e.g., engines outside of the warehouse, beeping to indicate vehicles going in reverse and voices giving some orders to others. The learner is placed in the warehouse near the computer.

The designer of the story decides on the degree of freedom given to the learner. Passive story telling implies that the teacher is defining a narrative that has to be followed step by step. The learner becomes an observer, not being able to influence the path, speed, stop, or detours to other areas of interest. The other extreme is an open world, with no limitation how to continue the narrative; allowing the learner to get lost in irrelevant parts of the story. In the example shown in Figure 1, the passive story telling implies that each step to do to fulfil the orders is pre-orchestrated and just presented to the learner. This learning approach has the disadvantage that it is not about understanding but following instructions step by step. The other extreme is that the learner is just getting the orders in a random order and has to fulfil them without any further help. This requires the students to understand the story and how elements can be used to achieve the learning outcome. There is no support besides a final (summative) feedback when the outcome is achieved. An intermediate approach is about the learner starting its own narrative to progress through the story, but being tight to a *narrative rubber-band* to define a maximum range for explorative action (e.g., time or distance) before we nudge the learner back to an intended goal-oriented narrative. Suitable feedback mechanisms create learning opportunities through reflection. Motivation of the student is enhanced with 'open' elements that are constrained/bounded by the story design. Completing a narrative with required skill means completion to a required level of mastery (e.g., with a suitably low level of mistakes), on time (e.g., within a suitable length of time). Many academics can recall particular PhD students who suffered from 'globalitis' or the desire to address a large range of materials in their

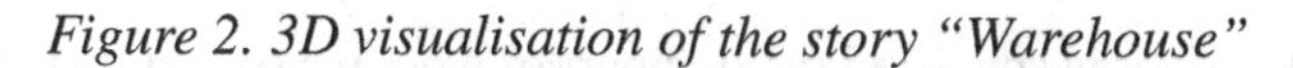

Figure 2. 3D visualisation of the story "Warehouse"

dissertation. However, most supervisors at this point try to gently guide the students to restrict their scope. This is to ensure that the project is workable, can be completed within a defined time period, and is largely achievable given the student's capabilities and resources. In much the same way, any student or actor in a system can have their activities gently guided back on track, ensuring that their progress is ultimately moving them in one direction, towards what it is that they define as success. Here, we should focus on the soft and gentle hints to motivate them. This can be completed with fewer hints near the conclusion (on higher levels); these hints can also be adaptive and used to drag them back on track.

Learning should not be about memorising; it should be about engaging, being immersed, in the learning process. The first question a teacher must answer is how to ignite the intrinsic motivation, build up tension and make the learners curious to explore the learning space; yet to provide an environment with a safety net and support to handle every possible situation. Detaching the real learner from the environment is a very first step as all risk is projected to the avatar. The next step is about embedding elements to invoke fun and passion for the learner (Reiners & Wood, 2013) and to trigger curiosity in exploring the learning space in a self-paced learning process. In our research, we focus on *curiosity*, *gamification,* and *guidance*; among other objectives like authenticity and technology (Reiners et al., 2012).

Curiosity is driven by freedom to move around. The "virtual door" to move to the next scenario opens only when the learner has achieved the learning objectives for the present scenario (i.e., completed the sequence of actions in order). Learners should be provided with scaffolding for learning but also with the opportunity to engage in other, related, activities as long as these support the attainment of the predefined learning objectives. (Note: there are ways of modelling this and determining whether a learner's activities are within scope of the narrative or outside the scope; however, this is not the focus of the chapter.)

Many of the outcomes of games are desired in a gamified system and this is driven by gamification being "the use of game design elements in non-game contexts" (Deterding, 2011, p. 10) with the explicit intention of creating a fun-filled, playful environment to encourage passionate engagement by users. This is done by taking a page out of the game-designer's handbook and adopting "the *motivational properties* of games and layering them on top of other learning activities, integrating the human desire to communicate and share accomplishment with goal-setting to direct the attention of learners and motivate them to action" (Landers & Callan, 2011, p. 421, emphasis added). This clarifies that an activity is not turned into a game; it retains its core essence as a learning activity. Instead, other motivating game-based elements are 'layered' or scaffolded around the activity. A careful configuration of varying combinations of *building blocks* creates these layers of gamification onto regular processes. These concepts lead to a range of interchangeable terms that are used synonymously, primarily "*behavioral games*, *funware*, *applied gaming*, *productivity games*, the *game layer* of a process, or *playful design*" (Deterding, 2011). At present, gamification is the term that appears to be gaining the most traction in the media and publications. Why is gamification and gaming so appealing to learners? In a nutshell: Humans love puzzles and finding solutions to challenges, want to pass obstacles and feel the joy of winning (Hokkanen, 2009). And people may gamify the world every day by making their daily drive to work more enjoyable; imagining the story of other travellers, counting green lights in a row, doing a countdown for the traffic light turning green again, guessing if colleagues are already in, etc.

Gamification includes, for example, multiple recordings, rewinds, point scoring. It differs from the 'regular' gamified system which is extremely linear or 'fixed'; it provides an open yet bounded space for learning. The importance is that a user can deviate somewhat in this open space before the designed boundaries 'nudge' them back on track for completion of learning objectives. As long as the outcome is achieved, the specifics of 'how' this is achieved are less relevant, although in this case it is specified (as it is part of a course or lecture) but the students don't need to do this religiously as though they were actually employed in the job at present. Scores can be completion times (which can be improved in multiple attempts); getting killed or making fatal mistakes removes points or increases the completion time. It is important to emphasise that the combination of virtual environments and gamification provides a further advantage for the learner: failure. In games, failing is encouraged by the designer as the continuous replay of certain parts trains to become better. Players are motivated to repeat situations over and over just to figure out the right solution and try to improve in the score (McGonigal, 2011). Fujimoto (2012) reports, that the failure rate in some games is about 80%; by which the player is even more engaged to invest more energy to master the current task.

Most gamification solutions tend towards a points-, badges-, and leaderboard-based approach; this can easily be added to existing applications. In theory, it can promote a competitive atmosphere, full of rivalry, as users compete to outdo one another. In practice, such an approach can produce stunningly negative dynamics, as unintended consequences spring forth from the thoughtless application. Consider the situation where you recently joined the learning space. You see that there are some well-recognised users, with very competitive scores and a collection of badges that would make a boy scout green with envy. While this inspires some users, it can prove to be a complete turn-off for others. In this chapter, we limit the gamification of the learning space to badges, scores, and a leaderboard; yet elaborate further mechanics in the conclusion as suggested by Reiners and Wood (2013).

Ultimately, the overall storyline or sequence of activities takes a participant from the start to the conclusion (objective) of an activity and this sequence can be altered somewhat; however, due to the often sequential or hierarchical nature of activities required to generate an outcome, there is frequently a limit to the amount of flexibility and how these activities are undertaken. Therefore, there must be some observer, or system controller, who is able to monitor and observe the user activities, compare these to what should be undertaken, and then gently nudge so that desired outcomes are attained. The observer does not have to be a human being and does not have to be visible to the players.

STORY: A DAY IN THE WAREHOUSE

A simple business process, fundamental to effective supply chain and business management, is the order fulfilment process. The flow of materials through the supply chain is connected by this process, which shakes materials from one firm and gets them moving to another. It can be completed by a single person with reference to others, or through a computer system providing access to required data. In a given small warehouse, a store person may be required to receive an order, check inventory level, physically collect and assemble constituent parts or materials, and package for despatch. Figure 3 provides an overview of the action-based, complete sequence of actions in the proper order.

1. **Receive Order:** The skills, the specific actions required in order, are: check computer (to see if there is a new order or in response to an audio notification of an order receipt). The main focus here is being aware that orders come in and how to obtain the order. The specific action is to move to the computer and access the 'order receipt' screen.
2. **Check Credit Rating of the Customer:** If their credit rating is not adequate, inform the customer that you cannot fulfil the order. If their credit rating is good continue to the next step.
3. Read the next line of the order and begin to process the order.
4. Check stock level, ensuring there is suitable stock for the order. If there is inadequate stock, inform that customer that you cannot fulfil the order. If there is adequate stock, continue to process the order.
5. **Identify Stock Location:** This may be done from memory after working through the scenario a couple of times or it may be done by visually sighting the stock location. If the stock is located on the ground, it will be necessary to locate and use the pallet mover. If the stock is not located on the ground (i.e., it is higher up on a rack) then a forklift must be located and used. When the moving equipment is in place, continue process.
6. **Move Stock to the Despatch Area:** When this has been done, check whether the order is complete. If it is not, return to step 3 and check the next line on the order. If it is complete, continue to finalise the order.

Each of these actions must be completed in sequence. A user is not able to complete the work by going to a random location and taking items to the despatch area. The correct items must be located and brought to the correct area (the despatch area). Even with this basic sequence of activities it becomes immediately apparent that there are improvements that can be made. There may be five different items on the order and while the first four are in stock, the final item may be out of stock. In this case, if the user has completed this sequence of actions they will have completed a number of steps before this becomes apparent. A learning lesson here for the student is to consider their activities. In this case, it may be wise to see whether all of the items are available in stock, before wasting time moving around the warehouse for the first time (NB: some warehouses are large!).

Thus, there are limits and boundaries included by the virtual environment place which restrict the learners' movements and actions, while within these structures the learner has freedom to work with certain actions, even while these may be in the wrong order to complete the scenario immediately.

Figure 3. The process of order fulfilment

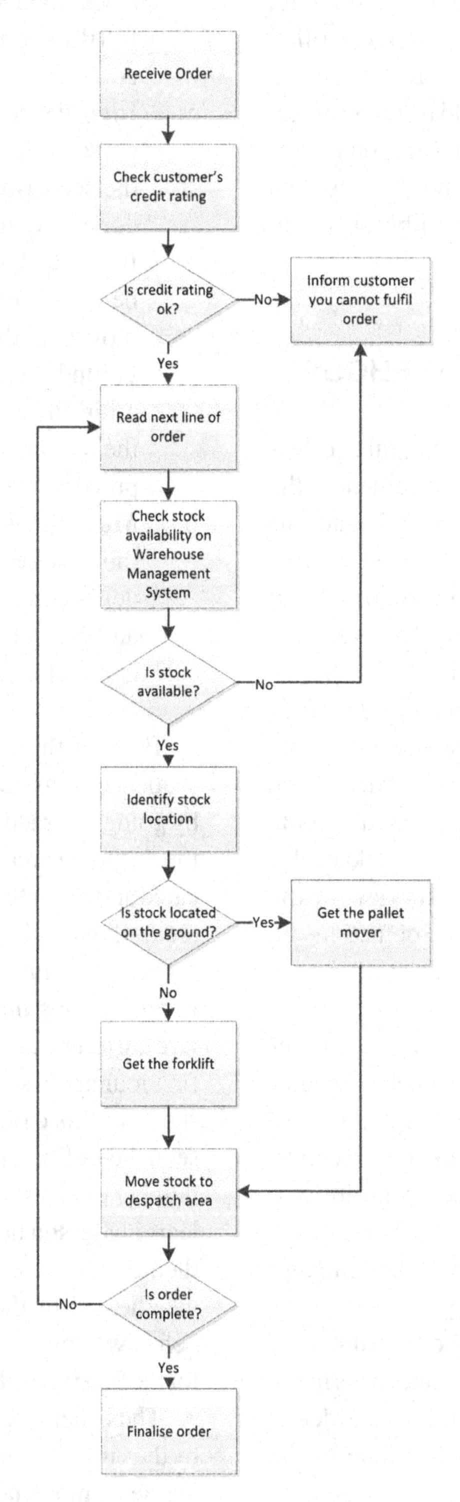

STORY, SCENARIO, AND THE GAMIFIED NUDGE

The learning space with its boundaries and elements is used in the exploration by the learner. The instructional designer maps the anticipated learning outcomes to real-world objects and their attributes. Regarding the previous example, the boundaries are given by the surrounding warehouse and the assumed constraints of not being able to move objects outside for a certain distance; including the learner. The objects are communication devices (computers), shelves, equipment to move objects, products, and some representation for a delivery region. In the example shown in Figure 2, we use the ute vehicle as the location to deliver the goods to; other options would be a packaging machine, despatch area, or new location in the shelves.

A sequence of scenarios can be created; each scenario having multiple narratives representing an expert solution as specified by the teacher. The first option would be a highly scaffolded approach, where the narratives are constrained and allow for only little flexibility for the learner to achieve the objectives. This would focus on understanding basic tasks for completing the learning objectives. Later narratives are less regulated and consist only of the learning objective. This provides the learner with the opportunity to explore new methods of addressing the problems and exercise their creativity in how to structure some actions to achieve desired outcomes more quickly. In the following example, we use an approach with a maximum flexibility for the learner in the beginning; allowing the exploration of the learning space and experimentation with the given elements. Implemented triggers observe the activities at certain key areas and provide the necessary nudge to ensure that the student stays focused on the learning objectives.

Depending on the granularity of the scenario, the teacher is building a possible sequence of tasks and actions to achieve the learning objectives (narrative of the scenario). Tasks describe an activity on a higher conceptual level than actions and allow the learner to perform in their own chosen way. One possible task is the transport of a pallet of products, which can be done by using a forklift or by carrying each item by itself. Both means of transport come to the same final constellation in the warehouse (pallet is moved), but one is more time consuming and does not take advantage of previous investment in technology like the forklift. Figure 4 shows a possible narrative for the learning objective "fulfil the next order". The narrative contains of 4 milestones (1: read, interpret, and understand order on computer; 2: choose the best matching forklift or pallet mover; 3: by go to the correct shelf and load the right good; and 4: load the good on the ute). The milestones are linked by tasks as an expert would perform them. In addition, triggers are defined to evaluate the correct behaviour (T0: general trigger to check on security equipment and fulfilling the constraints, e.g. driving speed; T1: check on the correct selection of

Figure 4. Story and the model narrative of the teacher

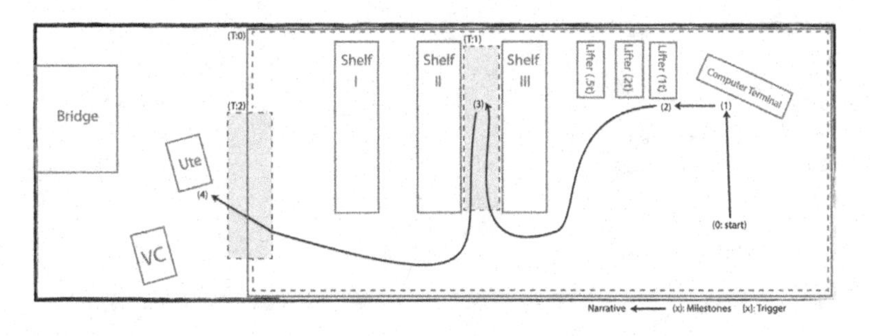

forklift as one of them is too weak and one is too large; T2: check on the right manoeuvre to pass the gate, e.g., slowing down and looking for cross traffic before leaving the warehouse. This narrative is used to evaluate the activities being observed from the learners during the learning process.

Note that scaffolded narratives can include further activities or actions (being a short (atomic) activity). For example, instead of just enforcing the usage of a specific forklift, the learner has to do a safety check before operating it. Tasks and actions are associated with competencies and skills; yet we focus in this chapter on the milestones and derived assumptions about what happened in-between; see also Fardinpour et al. (under review) for discussion how to record, analyse, and assess actions using action taxonomies.

Before starting the narrative, the learner is given a general briefing about the story; yet the details of how to proceed may stay unrevealed and need to be explored. The scenario includes objects for which the learner either has the corresponding skills (accessing a computer and receiving the latest order) or can achieve it within the setting of the story (calling the boss to ask about access to the computer). The exploration of the learning space is intended to be unrestricted in the following example; as illustrated Figure 5.

Figure 5 visualises the narrative of a learner in the learning space. There are several observations that distinguish this narrative from the narrative done by the expert. Note, that the following description includes the reactions from the system based the area and passed trigger:

1. Without any hint about what has to be done, the learner explores the space; i.e., looking at the vehicle and talking to a co-worker. Here, the learner did not ask questions and therefore did not receive any hints. By leaving the warehouse, the learner activated trigger T2; providing the first nudge back towards the learning objective. That is, the driver of the ute is asking about his order and when it is getting ready (the need for the driver to leave adds so-far not known time restrictions to the task). If the learner continues to walk away from the warehouse, the driver can start shouting to underline the importance of getting the order. Having interactive bots in the scenario can engage curiosity; causing the learner to ask questions about the order; which then results in an answer like "Sorry, I am just the driver, but my boss emailed you today". The answer does not present the final solution but pushes the learner towards the computer in the warehouse. Repetitively questioning could result in the driver calling the boss to get further information about the order; which is then displayed to the learner.
2. Knowing about the order, the learner walks back to the computer to find out more details. Even though the order is displayed, the system is not capable of verifying if all

Figure 5. One possible narrative of a learner in the learning space

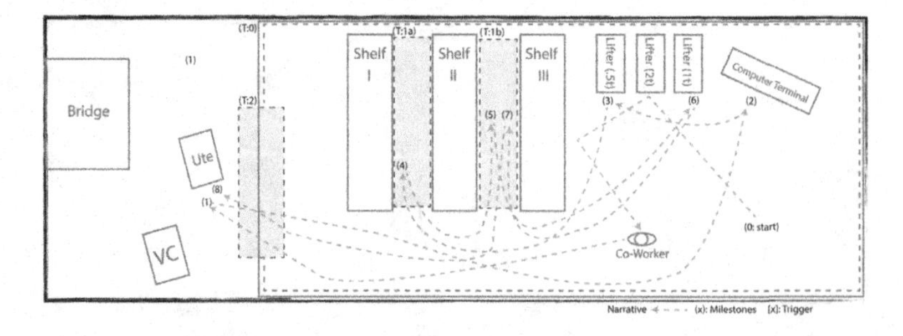

information is read and understood by the learner.

3. The learner picks a forklift; thus seem to have understood that the order cannot be retrieved otherwise. The forklifts are marked with their maximum weight, indicating that the selection has to be based on some criteria.
4. The learner activates trigger T1a; which verifies if the current order is in the associated ail. In this example, this is not the case and the system displays the order marking the position in the warehouse, thus showing the learner important features of an order as well as guiding the learner to the correct shelf.
5. The next activated trigger is T1b; verifying the right shelf but the wrong forklift as the order is weighs more than the allowed 0.5t carrying load of the forklift. Again, the system displays the order, indicating the information about the weight. The learner has to drive back and chose another forklift with the correct attributes. Note that there are no triggers for the forklifts as the selection itself is not *wrong* as long as it is not used for a specific task.
6. The next activated trigger is T1b; this time the shelf and the forklift match the order and the order can be loaded on the forklift. Note that this process depends on how the scenario is designed and if the exact operation of the forklift is part of the learning process. Otherwise, the learner is offered an option to automate this part.
7. The final activity is the transport of the order to the ute; successfully finishing this narrative.

Successful is relative and represented by different gamification components; i.e. scores and badges. The score could be time based, thus a small score would indicate a better (more efficient) performance; the badges are received for doing something special like selecting the right vehicle after reading the order or by staying below a certain distance to fulfil the order. Even though this summative feedback is generally motivating in games causing players to repeat the level until the lowest known score is beaten, we target for formative feedback to support the process of understanding the scenario. To do this, we analyse the recorded narrative (milestones) of the users and compare it to the narrative of the expert.

In Figure 6, both narratives are aligned by their milestones. That is start (0), reading the order at the computer (1), choosing the correct forklift (2), find the shelf that contains the order (3), and delivery to the ute (4). From a technical perspective, milestones are recorded when a trigger is activated (leaving the warehouse), the state of an object changes (communicating with the ute driver, using a forklift), or a specific area is reached. For each milestone, the state of the learning space is recorded; i.e., the position of all avatars and objects to calculate a delta and making assumption of activities; e.g., the learner moved objects, experienced the driving of the vehicles, or talked to other avatars. To reduce the amount of data, we do not keep close records for

Figure 6. Alignment and matching of two narratives based on their milestones. Note that the distance between the milestones does not reflect the time needed to achieve them.

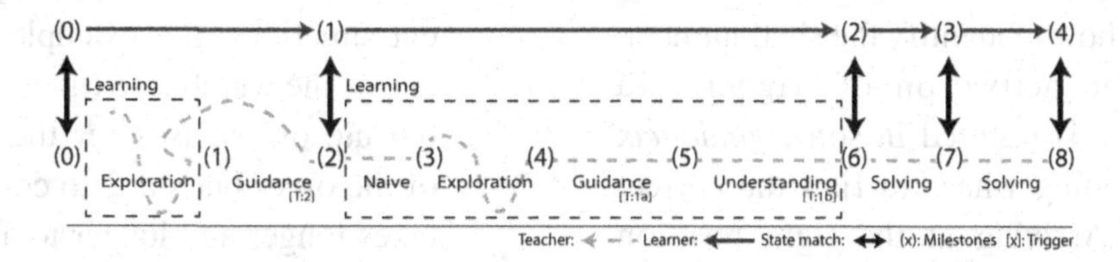

all objects but of the learners themselves (positions and executes actions). Unfortunately, it is impossible to observe activities in the real world (looking up information in a book) or learn more about the decision-making process rather than analysing the resulting activity. In this example, the system is able to interpret a learner's narrative in the following way:

1. Activity between milestone 0 and 1 [0-1]: The only observed change in the scenario was the position of the learner as well as the walked distance. Furthermore, the activation of trigger T2 conflicts with the teacher's milestone 1, indicating that the learner is not on the expected learning path. The trigger T2 is associated with the ute driver, thus starting the communication including some hints on how to proceed. This activity can be associated with *exploration* and *learning* about the scenario.
2. [1-2]: The next milestones matches the expectations; thus the hints by the driver worked and the learner was guided back to the expected narrative.
3. [2-3]: The next milestone is created when the learner selects a forklift. Based on the expected selection according to the teacher narrative (using the 1t forklift rather than the 0.5t version), it is feasible to assume that the learner either did not read/understood the order or did not verify the specification of the forklifts (naïve selection).
4. [3-4]: The activation of trigger T1a and the driven distance indicates another *exploration* of the learning space; trying to find the location of the order. It is feasible to assume that the learner did not read/understand the position information on the order or does not know how to identify the shelf number.
5. [4-5]: The activation of trigger T1a (Milestone 4) resulted in some *guidance*; e.g. explaining where to find the correct position. Arriving at the right position (Milestone 5) activates trigger T1b, which verifies the correct position but the wrong forklift; as the 0.5t version would flip over if loaded with the heavier order.
6. [5-6]: As the immediately following milestone matches the expected milestone on the teacher's narrative (after some guidance in the previous activity), including the selection of a new vehicle; we can assume that the learner understood what was wrong and what should be the next step to achieve the learning objectives. Note that the learning objectives were never directly stated but only hinted at the driver or visible by the incoming order on the computer.
7. [6-7][7-8]: The match between the milestones on both narratives indicates that the learner understood the learning objectives and is in the process of solving the learning activities.

The desired outcomes of the scenario are those elements that must be accomplished for a 'successful' outcome for the learner. This is often tied to 'learning outcomes' in education. Within this remit there are certain things a student must undertake, while other activities are optional, or the space between doing one activity and another is elastic, so that one activity must not necessarily follow from the other. The described story telling approach and the usage of comparative narratives promotes an authentic learning process.

The described understanding of how narratives can be embedded in a story has several advantages:

- The story is an ideal framework for gamification and can be embedded in multiple places without being immediately visible to the learner. The guidance can be done via virtual actors which respond to different situations. For example, if the learner leaves the warehouse (noticed by triggers), the ute driver asks for the current status of the order but starts to complain when it takes longer and longer to actually deliver the good. Another time-based event could be virtual phone calls by a supervisor asking how the work is coming along (and reacting accordingly to the answer of the learner), announcements through speakers, or colleagues explaining why and how the job has to be done. If all works well, the learner can receive badges (summative feedback) or advices on how to improve the next time (formative feedback). Time itself represents a score which the learner has to maximise to achieve leadership on a score board; e.g. a counter ticking down to represent the time when the ute driver has to leave.

- Milestones represent the states of the learning space. In this model, we are not directly observing every single activity of the learner but allow for a maximum of freedom to make a decision. Preliminary studies showed that the delta of two states is sufficient to deduce an assumption of the major process that was done. The learning space is focused on the achievement of learning goals, not necessarily the tasks themselves. As mentioned before, we consider failure as a critical component of the learning process. The learner is encouraged to find the best solution (most likely the narrative of the teacher) as this will be the one with the highest score as it is using the resources efficiently. Note that the score can also punish usage of extra material (fuel, electricity), damage (forklift is used with too heavy goods), or not finishing milestones (e.g., randomly picking a good and delivering it to the ute; without finding out what the order was).
- There is an opportunity to provide formative assessment according to the milestones; being fulfilled, missed, or not in the expected sequence. Placed triggers in the learning space provide sufficient insight into the learning process; i.e., if the milestones on the learner narrative can be compared to the narrative given by the teacher. Rather than feeding the knowledge to learners, the learner is guided from one milestone to the next; having a free choice of how to gain the knowledge. Most important in such learning spaces are methods to identify the need for help; e.g., based on time or "distance" to an expected outcome. Distance could be measured in the number of object states that do not match the expectation; e.g.m too many pallets were moved to fulfil the order.

CONCLUSION

Teachers need to focus on the provision of skills and knowledge, yet this requires significant design capability. We present the learner-centric bounded learning space, structuring the approach to create an interactive and layered series of narratives within a story. While the tools are forthcoming to achieve this in a simplified manner, here we present an overview of how these tools work and can be used to create environments that are designed to motivate through developing openness in a way that encourages curiosity and, through gamification, rewards attempts, efforts, and success in achieving the learning objectives. This contributes to literature in the 'authentic learning' space as it allows suitably complex scenarios to be developed, and it contributes to literature in the 'motivational learning' space as the provision of a bounded space enables experimentation and flexibility in activities undertaken towards attaining outcomes. Throughout the chapter, we used teacher as the neutral form of someone passing on knowledge to another person called learner. Even though the described learning space is developed with a university setting in mind, the underlying concept of storytelling and applicability of the learning space is transferable to other educational settings.

Future research is required to define and refine methods to automate the assessment of 'completion' of a sequence of tasks and ensuring that there is some way to measure how closely related learners' actions are to the outcomes that are specified in the learning objectives. This is required to suitably create the 'gamified nudge' to bring the learner 'back on track' in the sequence of tasks and actions. First experiments showed an overall satisfaction of being able to explore rather than follow instruction. Learners reported that they felt immersed in the environment and could identify themselves with the avatar; yet we are interested to increase the level of immersion as well as authenticity. With Second Life, we noticed various problems with respect to the physics (forklift jumps on shelves and boxes can be "pushed" over long distances). Therefore, we look at different game engines like Unity; which include realistic visualisation and behaviour of objects and vehicles.

ACKNOWLEDGMENT

Support for the production of this publication has been provided by the Australian Government Office for Learning and Teaching (Development of an authentic training environment to support skill acquisition in logistics & supply chain management, ID: ID12-2498). The views expressed in this publication do not necessarily reflect the views of the Australian Government Office for Learning and Teaching.

REFERENCES

Bauman, R., & Briggs, C. (1990). Poetics and performance as critical perspectives on language and social life. *Annual Review of Anthropology*, *19*, 59–88. doi:10.1146/annurev.an.19.100190.000423

Björke, K. (2003). Seizing power: Shaders and storytellers. In *Proceedings of ICVS 2003* (Vol. 2897, pp. 3–11). Berlin, Germany: Springer.

Bloom, B. S., Engelhart, M. D., Furst, E. J., Hill, W. H., & Krathwohl, D. R. (1956). *Taxonomy of educational objectives: The cognitive domain*. New York: Longman.

Boyle-Baise, L., & Goodman, J. (2009). The influence of Harold O. Rugg: Conceptual and pedagogical considerations. *Social Studies*, *1*, 31–40. doi:10.3200/TSSS.100.1.31-40

Broussard, J. (2012). Making the MMOst of your online class. In *Proceedings of 28th Annual Conference on Distance Education & Learning* (pp. 1–5). IEEE.

Bruner, J. (1990). *Acts of meaning*. Cambridge, MA: Harvard University Press.

Danilicheva, P., Klimenko, S., Baturin, Y., & Serebrov, A. (2009). Education in virtual worlds: Virtual storytelling. In *Proceedings of International Conference on CyberWorlds* (pp. 333–338). IEEE.

Deterding, S., Dixon, D., Khaled, R., & Nacke, L. (2011). From game design elements to gamefulness: Defining gamification. In *Proceedings of the 15th International Academic MindTrek Conference: Envisioning Future Media Environments* (pp. 9–15). Tampere, Finland: ACM. doi:10.1145/2181037.2181040

Fardinpour, A., Reiners, T., & Dreher, H. (2013). Action-based learning assessment: Evaluating goal-oriented actions in virtual training environments. In *Proceedings of ASCILITE 2013*. ASCILITE.

Friedman, T. (2005). *The world is flat*. New York: Farrar, Straus and Giroux.

Fujimoto, R. (2012). *Games and failure*. Retrieved from http://shoyulearning.wordpress.com/2012/05/20/games-and-failure/

Gregory, S., Lee, M. J. W., Dalgarno, B., & Tynan, B. (2013). *Virtual worlds in online and distance education*. Athabasca, Canada: Athabasca University Press.

Gregory, S., Reiners, T., & Tynan, B. (2010). Alternative realities: Immersive learning for and with students. In H. Song (Ed.), *Distance learning technology, current instruction, and the future of education: Applications of today, practices of tomorrow* (pp. 245–272). Hershey, PA: IGI Global.

Hokkanen, L., Kataja, E., Kaukomies, O., Linturi, A., Pylkkö, I., & Salohalla, L. (2011). The future of gaming: technology, behaviour & gamification. In A. G. Paterson (Ed.), *They are everywhere* (pp. 46–82). Retrieved from http://booki.cc/they-are-everywhere/

Kapp, K. M. (2012). *The gamification of learning and instruction: Game-based methods and strategies for training and education*. New York: Pfeiffer.

Landers, R. N., & Callan, R. C. (2011). Casual social games as serious games: The psychology of gamification in undergraduate education and employee training. In *Serious Games and Edutainment Applications* (pp. 399–423). London: Springer. Retrieved from. doi:10.1007/978-1-4471-2161-9_20

McGonigal, J. (2011). *Reality is broken, why games make us better and how they can change the world*. London: Random House.

Moore, M. G., & Anderson, W. G. (2012). *Handbook of distance education*. London: Routledge.

Murray, J. H. (1997). *Hamlet on the Holodeck: The future of narrative in cyberspace*. Cambridge, MA: MIT Press.

Neal, L. (2001). Storytelling at a distance. *elearn Magazine*. Retrieved from http://elearnmag.acm.org/featured.cfm?aid=566979

Nitsche, M., & Thomas, M. (2003). Stories in space: The concept of the story map. In *Proceedings of ICVS 2003* (Vol. 2897, pp. 85–93). Berlin, Germany: Springer.

Noddings, N. (2007). *When school reform goes wrong*. New York: Teachers College Press.

Pederson, E. M. (1995). Storytelling and the art of teaching. *Forum*, *33*(1).

Perlin, K. (2003). Building virtual actors who can really act. In *Proceedings of ICVS 2003* (Vol. 2897, pp. 127–134). Berlin, Germany: Springer.

Reiners, T., Dreher, C., & Dreher, H. (2012). Transforming ideas to innovations: A methodology for 3D systems development. In A. Hebbel-Segger, T. Reiners, & D. Schäfer (Eds.), *Alternate realities: Emerging Technologies in education and economics*. Berlin: Springer.

Reiners, T., Gregory, S., & Knox, V. (2012). Virtual bots, their influence on learning environments and how they increase immersion. In S. Gregory, M. J. W. Lee, B. Dalgarno, & B. Tynan (Eds.), *Virtual worlds in open and distance education*. Athabasca, Canada: Athabasca University Press.

Reiners, T., & Wood, L. C. (2013). Immersive virtual environments to facilitate authentic education in logistics and supply chain management. In Y. Kats (Ed.), *Learning management systems and instructional design: Metrics, standards, and applications* (pp. 323–343). Hershey, PA: IGI Global. doi:10.4018/978-1-4666-3930-0.ch017

Riedl, M. O., & Young, M. (2003). Character-focused narrative generation for execution in virtual worlds. In *Virtual storytelling: Using virtual reality technologies for storytelling* (Vol. 2897, pp. 47–56). Berlin: Springer. doi:10.1007/978-3-540-40014-1_6

Spierling, U. (2002). Digital storytelling. *Computers & Graphics*, *26*(1), 1–66. doi:10.1016/S0097-8493(01)00172-8

Wood, L. C., Teräs, H., Reiners, T., & Gregory, S. (2013). *The role of gamification and game-based learning in authentic assessment within virtual environments*. Paper presented at the HERDSA 2013. New York, NY.

Wriedt, S., Reiners, T., & Ebeling, M. (2008). How to teach and demonstrate topics of supply chain management in virtual worlds. In *Proceedings of ED-MEDIA 2008: World Conference on Educational Multimedia, Hypermedia & Telecommunications* (pp. 5501–5508). Chesapeake, VA: AACE.

ADDITIONAL READING

De Freitas, S. (2008). *Serious Virtual Worlds: A scoping study (Prepared for the JISC e-Learning Programme)*. Serious Games Institute.

Frazel, M. (2010). *Digital Storytelling Guide for Educators*. International Society for Technology in Education.

Ohler, J. (2007). *Digital Storytelling in the classroom: New Media Pathways to Literacy, Learning, and Creativity*. Corwin Pr Inc.

Reiners, Torsten, & Wood, L. C. (Eds.). (2014). *Gamification in Education and Business*. New York: Springer.

Reiners, T., Gregory, S., & Dreher, H. (2011). Educational assessment in virtual world environments. In J. D. Yorke (Ed.), *Meeting the Challenges. Proceedings of the ATN Assessment Conference 2011, 20-21 October* (pp. 132–142). Perth, Western Australia: Curtin University.

Schroeder, R. (2002). *The Social Life of Avatars*. New York: Springer. doi:10.1007/978-1-4471-0277-9

Wood, L. C., & Reiners, T. (2012). Gamification in logistics and supply chain education: Extending active learning. In P. Kommers, T. Issa, & P. Isaías (Eds.), IADIS International Conference on Internet Technologies & Society 2012 (pp. 101–108). Presented at the IADIS 2012, Perth, Australia.

KEY TERMS AND DEFINITIONS

Gamification: The use of game-based mechanics and game-based design elements in non-game settings to engage users and encourage achievement of desired outcomes through motivation of users.

Immersion: The state of consciousness where someone perceives an artificial or virtual space as real; often described as a degree of suspension of disbelief which correlates with the perception of the virtuality being real.

Narrative: Unique paths through the story which also enliven the story and "unfold in space" and support the process of understanding and building cognitive structures. Narratives are either pre-scripted (ready to reveal their sequences of milestones and activities over and over again), or use exploration and goal-oriented triggers to multiply the possible narratives that learners can indirectly choose from.

Story: The story sets the overall scope and constraints of what to cover and what to exclude during the story telling (also called a bounded learning (purpose) or action (interactive) space). The story is the setting in which the actors will *live* their very unique narrative(s); including all required properties and elements

Storytelling: Storytelling is an effective mean to convey information in a compelling and memorable way.

Virtual Agent: So-called bots are non-player characters, alternatively known as "animated pedagogical agents" that are guided by the compute but behaving, as we would expect, as other humans

Virtual World: It is a computer-based, immersive, 3D multi-user virtual environment that imitates real (or imaginary) life, experienced through a graphical representation of the user.

ENDNOTES

1. http://www.rockstargames.com/grandtheft-auto
2. http://www.oculusvr.com
3. http://www.google.com/glass/start
4. http://www.xbox.com/kinect
5. https://www.leapmotion.com

Chapter 16
Background Music in Educational Games:
Motivational Appeal and Cognitive Impact

Stephanie B. Linek
Leibniz Centre of Economics, Germany

Birgit Marte
University of Graz, Austria

Dietrich Albert
University of Graz, Austria

ABSTRACT

Most game-designers likely stick to the assumption that background music is a design feature for fostering fun and game play. From a psychological point of view, these (intuitive) aspects act upon the intrinsic motivation and the flow experience of players. However, from a pure cognitive perspective on instructional design, background music could also be considered to be redundant information, which distracts from learning. The presented study investigated the influence of background music (present vs. not present) within an educational adventure game on motivational (intrinsic motivation, experienced flow) and cognitive variables (cognitive load, learning success). The results suggest a high motivational potential of background music. However, neither positive nor negative effects on learning were detected. Thus, background music can be considered as a motivating design element of educational games without negative side-effects on learning.

INTRODUCTION

Background music is an important design feature in games and is mainly implemented for enhancing immersion and game play. Since educational games aim at establishing a playful, enjoyable form of learning, background music can be conceptualized as a source of motivation to play and to learn. However, from a theoretical point of view background music can have different effects on learning which will be explained in more detail in the following subchapters. On the one hand

DOI: 10.4018/978-1-4666-5071-8.ch016

it might foster learning via motivation, on the other hand it might either distract from learning or be neutral with respect to cognitive variables. So far, systematic research on background music in educational games, which considers motivational as well as cognitive variables is rather spare (Richards, Fassbender, Bilgin, & Thomson, 2008; for an overview see Zehnder & Lipscomb, 2006). The aim of the presented study is to give first evidence on the motivational and cognitive effects of background music with respect to game-based learning.

THEORETICAL BACKGROUND

The presented study concentrates on two main aspects of background music: On the one hand the motivational appeal of background music is investigated and on the other hand the cognitive impact of background music is disputed. Accordingly, the theoretical section is twofold: First, the relevant theories and findings on the motivational appeal of background music are explored. Second, the cognitive impact of background music is disputed from a theoretical perspective. Based on the presented theoretical considerations, the section closes with the derived research questions of the presented investigation.

Motivational Appeal of Background Music

Background music as a design feature of (educational) video games is often mainly implemented for fostering fun and game play. These aspects can be theoretically conceptualized as intrinsic motivation and the flow experience of players. In this section we provide first a psychological theory on motivation, and second outline the theoretical framework of the so-called Flow Theory.

Among the most prominent theories concerning motivation is the Self-Determination Theory (Deci & Ryan, 1985) that focuses on the degree to which human behavior is volitional or self-determined. Within this approach different types of motivation can be distinguished, whereas the most essential distinction is made between intrinsic motivation and extrinsic motivation. Intrinsic motivation refers to the fact that individuals do something because it is inherently interesting or enjoyable. This type of motivation has emerged to be essentially important for education. On the other hand, extrinsic motivation refers to doing something because it leads to a separable outcome (Ryan & Deci, 2000a). With respect to this differentiation, the implementation of background music mainly refers to the concept of intrinsic motivation.

A psychological concept that is closely related to motivation refers to the so-called flow experience, which is said to be intrinsically rewarding and can be seen as a kind of hedonic state. Principally, the Flow Theory (Csikszentmihalyi, 1990; Csikszentmihalyi, Abuhamdeh, & Nakamura, 2006) is a unified framework on subjective experiences in the course of mastering everyday challenges. According to this approach the quality of subjective experiences is determined by the balance between the perceived challenges of the task on the one hand, and the perceived necessary skills for mastering the task on the other hand. If challenges and skills are both at a low level, the person experiences apathy. If the perceived challenges are lower than the perceived skills, the person is bored. If the perceived challenges overrun the perceived skills, the person feels anxiety. If the perceived challenges as well as the perceived skills are simultaneously high, the person experiences flow or flow in consciousness, respectively, and this represents the highest quality of subjective experience. The experience of flow or being immersed in an activity, respectively, is intrinsically rewarding and can be seen as a hedonic state. This notion implies the following assumptions: If the person is bored (skills > challenges) then he/she should seek new challenges. On the other hand, if the person is overloaded and anxious (skills

< challenges) he/she will try to enhance his/her skills and knowledge, or in other words he/she will engage in learning activities. The assumption about the balance between challenges and skills being crucial for the intrinsic motivation and the enjoyment of the activity received empirical support by Moneta and Csikszentmihalyi (1996), as well as by other authors (Asakawa, 2004; Schweinle, Meyer, & Turner, 2006) even though there are also some inconsistent findings.

The concept of Flow Theory is similar to other ancient approaches (cf. Moneta & Csikszentmihalyi, 1996), namely the Yerkes-Dodson-Law (Hebb, 1955; Yerkes & Dodson, 1908) according to which pleasure as well as learning is supposed to be optimized at a medium level (of arousal, challenge, complexity). This level guarantees that it neither bores nor overwhelms the individual. The absolute level (i.e. "medium") depends on the person's individual abilities and characteristics, respectively.

The empirical research on the impact of music within video games on the intrinsic motivation and the flow experience of the players is still in its infancy and the findings in this context are rather scarce (Zehnder & Lipscomb, 2006). Accordingly, the study presented by Zehnder, Igoe, and Lipscomb (2003; cited after Zehnder & Lipscomb, 2006) was thought of as a first pilot study within this field. Although no main effect of the presence or absence of background music was found, data analyses on a more fine-grained level revealed that the background music influenced the verbal ratings of the different game segments. These results can be interpreted as first evidence that the impact of background music in games might be rather complex and should be studied in a more detailed way with respect to various motivational and cognitive variables.

To conclude, the potential of music as motivating factor in (educational) games is indeed accepted, however, systematic research findings are rather spare (Richards et al., 2008). Another aspect is that by explicitly stressing the balance between challenges and skills, not only the motivating flow experience, but also the relationships to cognitive aspects and learning have to be addressed. The latter issue will be discussed in more detail in the subsequent section.

Cognitive Impact of Background Music

In general, different effects of background music on cognitive aspects are conceivable. From a pure cognitive perspective, background music is considered to be redundant information, which distracts from learning due to the limited capacity of the working memory. A theoretical approach that focuses on this limitation is the Cognitive Load Theory (Sweller, Van Merriënboer, & Paas, 1998). According to this approach the overall cognitive load is constituted by three different types of cognitive load: Intrinsic cognitive load (ICL), germane cognitive load (GCL), and extraneous cognitive load (ECL). Intrinsic load refers to the cognitive load that is caused by the learning task itself. The germane load corresponds to cognitive activities that are apt to foster learning, e.g., schema construction. Finally, extraneous load refers to cognitive activities that are irrelevant to learning and thus, should be avoided in order to have more free capacities for other learning relevant cognitive processes. From this differentiation it can be derived that, in order to maximize learning, an optimal instructional design should foster germane load and lower extraneous load. The intrinsic load cannot be changed by the instructional design itself.

Similar assumptions on the limited capacity of the working memory (with special focus on multimedia learning) can be found in the Cognitive Theory of Multimedia Learning devised by Mayer (2000) according to which background music is considered as redundant material or seductive detail (Harp & Mayer, 1998; Moreno & Mayer, 2000), that distracts from learning and thus, has a detrimental effect on the learning outcome.

Contrariwise, the so-called Mozart effect provided evidence in favor of background music. The Mozart effect described the phenomenon of better performance in special abilities while listening to Mozart's music (Rauscher, Shaw, & Ky, 1993). However, empirical research on the Mozart effect is quite inconsistent. It is not yet clear to which kind of music this effect refers (i.e., to Mozart only or also other background music; see also Richards et al., 2008) and if the effect is due to neurophysiological priming or rather to changes in mood and arousal (Asby, Isben, & Turkey, 1999; Cassity, Henley, & Marley, 2007; Schellenberg & Hallam, 2005; Steele, 2000; Thompson, Schellenberg, & Husain, 2001). Additionally, there are also contradictory findings and thus, the question arises if this effect really exists or might be due to statistical artifacts (Carstens, Huskins, & Hounshell, 1995; McKelvie & Low, 2002; Steele, Bass, & Crook, 1999; Stough, Kerkin, Bates, & Mangan, 1994).

Another differentiated view of the connection between background music, arousal, cognitive load and learning was presented by the study of Huk, Bieger, Ohrmann, and Weigel (2004). The respective findings indicate that the implementation of background music is advantageously for learners with high prior knowledge only and has slightly detrimental effects for inexperienced learners. The authors explained these results by a trade-off between beneficial effects in case of experienced learners as a result of mood and arousal on the one hand, and detrimental effects for inexperienced learners due to extraneous cognitive overload on the other hand. Another similar explanation of the so-called expertise-reversal effect was provided by McNamara, Kintsch, Songer, and Kintsch (1996; for an overview see Kalyuga, Ayres, Chandler, & Sweller, 2003).

To sum up, findings on the cognitive effects of background music in general and within video games in particular (e.g., North & Hargreaves, 1999) are rather inconsistent. Especially with respect to video games the empirical base on the cognitive effects of background music is rather spare (e.g., North & Hargreaves, 1999; Richards et al., 2008).

Summary and Derivation of Research Questions

As pointed out above, the existing base of research provides no clear evidence regarding the motivational and cognitive effects of background music within educational games. In the following we present a study that concentrates on these two possible effects of background music whereby divers motivational (intrinsic motivation and flow experience) and cognitive (cognitive load and learning success) variables are considered.

The main research question is twofold and regards to the motivational and cognitive impact of background music within educational adventure games:

1. Which impact has the implemented background music of an educational adventure game on the intrinsic motivation and the flow experience?
2. Which impact has the implemented background music of an educational adventure game on cognitive load and learning?

METHODOLOGY

The following section addresses the methodology of the conducted study including the description of the experimental environment in the form of an educational adventure game.

Participants

The sample of the study comprised 59 school children (38 males & 21 females) at the average age of M = 13.6 years (SD = .89). The pupils were from three different schools in Paris and participated voluntarily in the course of their regular school time.

Experimental Design

A two-group design was used to investigate the motivational and cognitive impact of background music with the presence (experimental group) or absence (control group) of background music as independent variable. The experimental group comprised n = 27 individuals (22 males & 5 females). The control group included n = 32 pupils (16 males & 16 females).

Experimental Environment and Procedure

As experimental environment a first chapter of an educational adventure game regarding optics (properties of light) was designed. The learning contents about the properties of light were directly integrated in the story of the game. The pupils slipped into the role of George (i.e., the avatar), who was designed from a fist-person/ego-perspective. This means the avatar itself was not graphically represented, only its hand was partly visible (e.g., when opening a door). The story starts with George arriving at the villa of the Natural Science Entertainment Park director in order to apply for an internship. In the course of the game he gets to know Lisa, who is the niece of the director (Uncle Leo). From Lisa George learns that her uncle Leo was kidnapped by the evil "Black Galileans" and that she herself is trapped in the basement of the villa after she succeeded in escaping the Galileans. On his way to help and free Lisa as well as to find Uncle Leo, George has to mange several tasks that require physical knowledge on optics to be solved. All along he is assisted by Lisa via a headset he finds inside the villa as well as by Galileo who appears as a ghost. The main task within the first chapter was to open a door with a light sensor in order to free Lisa. For the correct manipulation of the light sensor the players had to apply knowledge about the properties of light and the functionality of blinds. This knowledge could be acquired trough experimentation with blinds lying on a table, a screen, and a light source whereby the player was tutored by the ghost Galileo. The game was presented either with background music (experimental group) or without (control group). At average the pupils played 24 minutes.

All pupils were tested in groups. Prior to playing the game participants were given a pre-questionnaire that included several control variables reported in the next section. The game was then provided on laptops available in the classroom. The dependent variables as well as additional control variables were assessed immediately after the game by presenting a post-questionnaire to both groups. Both, the pre- and post-questionnaire were realized as paper-pencil tests.

Dependent Variables and Measurements

To assess the extent of intrinsic motivation we used two subscales of the short version of the multidimensional Intrinsic Motivation Inventory (IMI, Deci, & Ryan, 2004; Ryan, 1982). The IMI has its theoretical foundation in the Self-Determination Theory (Ryan & Deci, 2000a, 2000b) and was already successfully deployed in previous studies (e.g. Scheiter, Gerjets, & Catrambone, 2006). The original inventory comprises the six subscales Interest/Enjoyment, Perceived Competence, Effort/Importance, Pressure/Tension, Perceived Choice, and Value/Usefulness. However, for the purpose of the present study we used only the subscales for Interest/Enjoyment and Effort/Importance. Each subscale was represented by three statements (e.g. Interest/Enjoyment subscale: "I thought this was a boring game"), which had to be judged on a 7-point scale that ranged from 1 ("not at all true") to 7 ("very true"). The wording of the items was adapted with respect to the presented game. For each subscale the mean score was calculated.

The degree of flow an individual experienced during playing was assessed by means of the translated version of the short flow-scale devised by Rheinberg, Vollmeyer, and Engeser (2003; Rheinberg, 2004). This measurement comprises 10 statements that assess the overall flow experience (e.g., "I did not recognize how the time went by.") with the two subscales "smoothness" (6 items) and feeling "absorbed" (4 items). Another additional 3 items refer to feelings of "worry" (e.g., "I wasn't allowed to make mistakes."). Each statement had to be rated on a 7-point scale ranging from 1 ("not at all true") to 7 ("very true"). For evaluating the flow experience different sub-scores were calculated. For the flow-scale (flow items only) as well as the worry-scale (worry items only) the average value was computed. Furthermore, within the flow-scale two different mean values were conceived, one for the sub-scale smoothness and one for the feeling to be absorbed.

For the character-evaluation, evaluation of the story and the avatar's activities several questions were presented. After playing with the demonstrator, each pupil had to indicate on a 5-point scale ranging from 1 ("not at all") to 5 ("very much") how much they like the different game-characters (The NPCs Galileo and Lisa as well as the avatar George).

Additionally, the pupils were asked to rate the story of the game in general, and the activity of George in particular concerning a 5-point bipolar rating-scale with the attributes "boring" vs. "interesting".

A commonly used instrument to measure the extent of an individual's subjective perception of cognitive load required for performing certain tasks, and hence, the workload, is the NASA-TLX (Hart & Staveland, 1988) that is based on the Cognitive Load Theory (Sweller et al., 1998). In its original version this measurement collects information about intrinsic cognitive load, germane cognitive load, and extraneous cognitive load as well as the subjective expectation of success, and stress. For the purpose of the reported study the items of a modified version used by Gerjets, Scheiter, and Catrambone (2004) were adapted with respect to the presented game. For example, to assess how much stress contributed to the required cognitive demand, the individuals were asked "How stressed (insecure, discouraged, irritated, annoyed) did you feel during the game?". Each item had to be rated on a low-to-high continuum from 0 ("easy") to 10 ("demanding") that indicated how much the respective element contributed to the overall cognitive workload.

To assess the learning outcome, that is, to which degree the learning objectives taught in the game were actually acquired by the individuals four open-ended questions (free recall) about the embedded physics topics were posed to each learner (e.g., "What do you know about the properties of light?"; "What do you know about blinds?"). The answers to these open questions were rated according to three categories, "correct", "incorrect", and "irrelevant", for each of which a sum score was calculated.

In order to control whether other factors than the presence or absence of background sound influence the dependent measures, prior to the evaluation study various user characteristics and preferences were collected from both, the experimental and the control group. Among the assessed user characteristics were an ID (to assort data), gender, class, school, and age of the players. Moreover, several questions about game playing experiences (e.g. "How often do you play?") and preferences (e.g. "Which one do you prefer? Educational games or video games?"), about the attitude towards physics in general ("Do you like to learn physics?") as well as about music style attitudes (e.g. "How many hours a day do you listen to music?") and preferences (e.g. "What kind of music do you like to listen to?") were presented to the learners. These aspects were assessed either through open-ended questions, rating scales or multiple choice items. Further important collected variables that were assumed to have an impact were the school grades (especially in math and physics). Additionally, the players assigned to the experimental group with background music were asked about the "usefulness", the "liking" as well as the "appropriateness" of the background sound. Each question had to be answered on a 5-point scale that ranged from 1 ("yes, a lot") to 5 ("not at all"). The participants of the control group had to rate whether there should be music in the game on the same type of scale. This group was also asked what kind of music they would add to this game by means of an open-ended question. Finally, both groups had to indicate whether they would prefer to have the game with or without music.

RESULTS

In this section the influence of background music is explored by a comparison between the two experimental groups by means of t-test analyses. The means and standard deviations for the dependent variables can be found in Table 1.

Intrinsic Motivation

The conducted t-test reveals a significant positive effect of background music on interest ($t(51) = 2.62$, $p = 0.01$) as well as on the invested effort ($t(52) = 2.13$, $p = 0.04$). In the presence of background music pupils reported a higher amount of interest in the game and put more effort in the game compared to pupils who played the game without background music.

Flow Experience

The presence of background music had a significant positive influence on the overall flow-index ($t(49) = 2.28$, $p = .03$) as well as on both flow-subscales for 'smoothness' ($t(50) = 2.00$, $p = .05$) and 'being absorbed '($t(53) = 2.75$, $p = .01$). However, the subscale 'worry' was not affected by background music ($t(51) = -.13$, $p = .90$).

Character Evaluation, Evaluation of the Story and Evaluation of the Avatar's Activities

There were no significant effects of background music on the evaluation of the game characters. Also for the evaluation of the story of the game and the activities within the game there was no significant effect of background music found.

Cognitive Load, Stress, and Expectation of Success

For the data on cognitive load there was a significant effect for the reported GCL ($t(49) = 2.32$, $p = .03$). The presence of background music resulted in a lower reported GCL, that is, in the presence of background music the learners reported less cognitive activities which are considered to be supportive for learning compared to the game-version without music. However, the presence of background music had no significant effect on the reported ICL and ECL. There were no significant effects found for the reported stress or the expectation of success.

Learning Success

For the learning success there were no significant effects for the presence of background music found.

Control Variables

There were no significant differences between the experimental group and the control group regarding school grades, physics attitude, general music usage, and gaming-experience. The liking of the music and the other assessed control variables had no impact on the results on the reported t-test analyses.

Table 1. Means and standard deviations (in brackets) for dependent variables

		Without background music	With background music
Intrinsic motivation			
	Interest	3.46 (1.37)	4.52 (1.55)
	Effort	3.81 (1.00)	4.39 (0.96)
Flow			
	Flow (overall)	3.30 (1.25)	4.10 (1.17)
	Smoothness	3.13 (1.14)	3.78 (1.12)
	Absorbedness	3.58 (1.66)	4.77 (1.50)
	Worry	3.21 (1.40)	3.16 (1.57)
Evaluation Characters			
	Galileo	3.00 (1.41)	3.25 (1.49)
	Lisa	3.21 (1.31)	3.55 (1.44)
	George (avatar)	2.63 (1.20)	3.55 (1.29)
Evaluation Story of the Game		3.03 (1.30)	3.04 (1.20)
Evaluation Game-Activities		3.35 (1.14)	3.24 (1.30)
Cognitive Load			
	ICL	6.59 (2.42)	6.00 (2.13)
	GCL	6.33 (2.15)	4.75 (2.72)
	ECL	5.50 (2.56)	5.50 (3.42)
Stress		3.77 (2.99)	2.84 (3.01)
Expectation of success		5.73 (2.78)	5.45 (3.29)
Learning success			
	Correct answers	1.09 (1.12)	1.04 (1.74)
	Incorrect answers	0.13 (0.34)	0.15 (0.53)
	Irrelevant answers	0.75 (1.05)	0.37 (0.79)

DISCUSSION

Overall, the presence of background music has a significant positive influence on intrinsic motivation and the experienced flow within the game – which is of special importance for a voluntary spare-time activity.

Interestingly, background music had no effect on the reported ECL. This contradicts the Cognitive Load Theory and the seductive detail hypothesis according to which background music is thought as a distracting element that should cause ECL. However, there was a significant effect of background music on GCL in the form that players reported about less GCL in the presence of background music. That means, in the presence of background music, learners invested less cognitive activities that are considered as supportive for deeper understanding. According to the Cognitive Load Theory this should result in a lower level of learning success compared to the version without music. However, there was no negative effect of background music on the level of learning success found.

On the other hand, with respect to the positive influence of background music on the intrinsic motivation and the flow experience, one might argue that this in turn should have been resulted in a positive effect on learning. However, analogous, positive effects on learning could not be proved.

One possible explanation regarding the missing effect of background music on learning may lie in the data structure of this study that means the rather low level of learning. Accordingly, the data might include a floor effect and thus, are only of restricted predictability. However, besides the possibility of floor effects, there are other interpretations thinkable.

First, our findings conflict with studies, which found that intrinsic motivation is associated with better learning and performance (e.g., Benware & Deci, 1984; Deci, Schwartz, Sheinman, & Ryan, 1981; Grolnick & Ryan, 1987; Valas & Sovik, 1993). In this context overviews (e.g. Ryan & Deci, 2000b) indicate that intrinsically motivated students are more curious, and engage more in deep level learning, which is assumed to hold true for students of all age groups (e.g. Wolters & Pintrich, 1998). However, the findings of Martens, Guliker, and Bastiaens (2004) indicate that a higher intrinsic motivation does not foster more activities; rather students tend to different things in dependence of their level of intrinsic motivation and thus, higher intrinsic motivation does not automatically produce better learning results.

Another possibly codetermining factor that was not considered in the reported study is an individual's degree of cognitive arousal. Beneficial effects of an optimal level of arousal on learning are declared, for example, in the Yerkes-Dodson law (Hebb, 1955; Yerkes & Dodson, 1908). Even though, the reported amount of stress provided no evidence for different arousal levels due to music, for future investigations it may be insightful to include assessment instruments for the physiological arousal of individuals and to relate this measure to the degree of motivation and flow as well as the learning outcome and cognitive load.

To sum up, the different explanations discussed above underlie the complexity of the effect of background music. With respect to game-based learning the situation becomes even more complicated since the core idea of this new approach is the melting of fun and learning and the existing theories and research cannot be applied directly. Accordingly, the presented study can be seen as a practice-oriented design study which provides first insights and demonstrates which other possible factors might be taken into account.

Overall, the reported findings provide first evidence for the motivational potential of background music without detrimental (distracting) effects on learning. Thus, background music can be considered as a source of motivation within educational games without the danger of impairing learning success.

REFERENCES

Asakawa, K. (2004). Flow experience and autotelic personality in Japanese college students: How do they experience challenges in daily life? *Journal of Happiness Studies*, *5*(2), 123–154. doi:10.1023/B:JOHS.0000035915.97836.89

Asby, F., Isen, A., & Turken, A. (1999). A neuropsychological theory of positive affects and its influence on cognition. *Psychological Review*, *106*, 529–550. doi:10.1037/0033-295X.106.3.529 PMID:10467897

Benware, C., & Deci, E. L. (1984). Quality of learning with an active versus passive motivational set. *American Educational Research Journal*, *21*, 755–765. doi:10.3102/00028312021004755

Carstens, C. B., Huskins, E., & Hounshell, G. W. (1995). Listening to Mozart may not enhance performance on the revised Minnesota paper form board test. *Psychological Reports*, *77*, 111–114. doi:10.2466/pr0.1995.77.1.111 PMID:7501747

Cassity, H. D., Henley, T. B., & Markley, R. P. (2007). The Mozart effect: Musical phenomenon or musical preference? A more ecologically valid reconsideration. *Journal of Instructional Psychology*, *34*(1), 13–17.

Csikszentmihalyi, M. (1975). *Beyond boredom and anxiety*. San Francisco, CA: Jossey-Bass.

Csikszentmihalyi, M. (1990). *Flow: The psychology of optimal experience*. New York, NY: Harper Collins.

Csikszentmihalyi, M., Abuhamdeh, S., & Nakamura, J. (2005). Flow. In C. S. Dweck, & A. J. Elliot (Eds.), *Handbook of competence and motivation* (pp. 598–608). New York, NY: Guilford Publications.

Deci, E. L., & Ryan, R. M. (1985). *Intrinsic motivation and self-determination in human behavior*. New York, NY: Plenum. doi:10.1007/978-1-4899-2271-7

Deci, E. L., & Ryan, R. M. (2004). *Intrinsic motivation inventory*. Retrieved from http://www.psych.rochester.edu/SDT/ measures/IMI_description.php

Deci, E. L., Schwartz, A. J., Sheinman, L., & Ryan, R. M. (1981). An instrument to assess adults' orientations toward control versus autonomy with children: Reflections on intrinsic motivation and perceived competence. *Journal of Educational Psychology*, *73*, 642–650. doi:10.1037/0022-0663.73.5.642

Gerjets, P., Scheiter, K., & Catrambone, R. (2004). Designing instructional examples to reduce intrinsic cognitive load: Molar versus modular presentation of solution procedures. *Instructional Science*, *32*, 33–58. doi:10.1023/B:TRUC.0000021809.10236.71

Grolnick, W. S., & Ryan, R. M. (1987). Autonomy in children's learning: An experimental and individual difference investigation. *Journal of Personality and Social Psychology*, *52*, 890–898. doi:10.1037/0022-3514.52.5.890 PMID:3585701

Harp, S. F., & Mayer, R. E. (1998). How seductive details do their damage: A theory of cognitive interest in science learning. *Journal of Educational Psychology*, *90*(3), 414–434. doi:10.1037/0022-0663.90.3.414

Hart, S. G., & Staveland, L. E. (1988). Development of NASA-TLX (Task Load Index): Results of experimental and theoretical research. In P. A. Hancock, & N. Meshkati (Eds.), *Human mental workload* (pp. 139–183). Amsterdam, The Netherlands: North Holland. doi:10.1016/S0166-4115(08)62386-9

Hebb, D. O. (1955). Drives and the C.N.S. (conceptual nervous system). *Psychological Review*, *62*, 243–254. doi:10.1037/h0041823 PMID:14395368

Huk, T., Bieger, S., Ohrmann, S., & Weigel, B. (2004). Computer animations in science education: Is background music beneficial or detrimental? In *Proceedings of the World Conference on Educational Multimedia, Hypermedia and Telecommunications* (pp. 4227-4234).

Kalyuga, S., Ayres, P., Chandler, P., & Sweller, J. (2003). The expertise reversal effect. *Educational Psychologist, 38*, 23–31. doi:10.1207/S15326985EP3801_4

Martens, R. L., Gulikers, J., & Bastiaens, T. (2004). The impact of intrinsic motivation on e-learning in authentic computer tasks. *Journal of Computer Assisted Learning, 20*, 368–376. doi:10.1111/j.1365-2729.2004.00096.x

Mayer, R. E. (2001). *Multimedia learning*. Cambridge, UK: Cambridge University Press. doi:10.1017/CBO9781139164603

McKelvie, P., & Low, J. (2002). Listening to Mozart does not improve children's spatial ability: Final curtains for the Mozart effect. *The British Journal of Developmental Psychology, 20*(2), 241–258. doi:10.1348/026151002166433

McNamara, D., Kintsch, E., Songer, N. B., & Kintsch, W. (1996). Are good texts always better? Interactions of text coherence, background knowledge, and levels of understanding in learning from text. *Cognition and Instruction, 14*, 1–43. doi:10.1207/s1532690xci1401_1

Moneta, G. B., & Csikszentmihalyi, M. (1996). The effect of perceived challenges and skills on the quality of subjective experience. *Journal of Personality, 64*, 275–310. doi:10.1111/j.1467-6494.1996.tb00512.x PMID:8656320

Moreno, R., & Mayer, R. E. (2000). A coherency effect in multimedia learning: The case for minimizing irrelevant sounds in the design of multimedia instructional messages. *Journal of Educational Psychology, 92*, 117–125. doi:10.1037/0022-0663.92.1.117

North, A. C., & Hargreaves, D. J. (1999). Music and driving game performance. *Scandinavian Journal of Psychology, 40*, 285–292. doi:10.1111/1467-9450.404128

Prensky, M. (2005). Computer games and learning: Digital game-based learning. In J. Raessens, & J. Goldstein (Eds.), *Handbook of computer game studies* (pp. 97–122). Cambridge, MA: MIT Press.

Rauscher, F. H., Shaw, G. L., & Ky, K. N. (1993). Music and spatial task performance. *Nature, 365*, 611. doi:10.1038/365611a0 PMID:8413624

Rheinberg, F. (2004, July 5-8). *Motivational competence and flow-experience*. Paper presented at the 2nd European Conference on Positive Psychology, Verbania, Italy.

Rheinberg, F., Vollmeyer, R., & Engeser, S. (2003). Die Erfassung des Flow-Erlebens. In J. Stiensmeier-Pelser, & F. Rheinberg (Eds.), *Diagnostic von Selbstkonzept, Lernmotivation und Selbstregulation* (pp. 261–279). Göttingen, Germany: Hogrefe.

Richards, D., Fassbender, E., Bilgin, A., & Thompson, W. F. (2008). An investigation of the role of background music in IVWs for learning. *ALT-J Research in Learning Technology, 16*(3), 231–244. doi:10.1080/09687760802526715

Ryan, R. M. (1982). Control and information in the intrapersonal sphere: An extension of cognitive evaluation theory. *Journal of Personality and Social Psychology, 43*, 450–461. doi:10.1037/0022-3514.43.3.450

Ryan, R. M., & Deci, E. L. (2000a). Intrinsic and extrinsic motivations: Classic definitions and new directions. *Contemporary Educational Psychology, 25*, 54–67. doi:10.1006/ceps.1999.1020 PMID:10620381

Ryan, R. M., & Deci, E. L. (2000b). Self-determination theory and the facilitation of intrinsic motivation, social development, and well being. *The American Psychologist, 55*, 68–78. doi:10.1037/0003-066X.55.1.68 PMID:11392867

Schellenberg, E. G., & Hallam, S. (2005). Music and cognitive abilities in 10- and 11-year-olds: The blur effect. *Annals of the New York Academy of Sciences, 1060*, 202–209. doi:10.1196/annals.1360.013 PMID:16597767

Schweinle, A., Meyer, D. K., & Turner, J. C. (2006). Striking the right balance: Students' motivation and affect in elementary mathematics. *The Journal of Educational Research, 99*(5), 271–293. doi:10.3200/JOER.99.5.271-294

Steele, K. M. (2000). Arousal and mood factors in the "Mozart Effect". *Perceptual and Motor Skills, 91*, 188–190. doi:10.2466/pms.2000.91.1.188 PMID:11011888

Steele, K. M., Bass, K., & Crook, M. (1999). The mystery of the Mozart effect: Failure to replicate. *Psychological Science, 10*, 366–369. doi:10.1111/1467-9280.00169

Stough, C., Kerkin, B., Bates, T., & Mangan, G. (1994). Music and spatial IQ. *Personality and Individual Differences, 17*, 695. doi:10.1016/0191-8869(94)90145-7

Sweller, J., van Merriënboer, J. J. G., & Paas, F. G. W. C. (1998). Cognitive architecture and instructional design. *Educational Psychology Review, 10*, 251–296. doi:10.1023/A:1022193728205

Tompson, W. F., Schellenberg, E. G., & Husain, G. (2001). Arousal, mood, and the Mozart effect. *Psychological Science, 12*, 248–251. doi:10.1111/1467-9280.00345 PMID:11437309

Valas, H., & Sovik, N. (1993). Variables affecting students' intrinsic motivation for school mathematics: Two empirical studies based on Deci and Ryan's theory of motivation. *Learning and Instruction, 3*, 281–298. doi:10.1016/0959-4752(93)90020-Z

Wolters, C. A., & Pintrich, P. R. (1998). Contextual differences in student motivation and self-regulated learning in mathematics, English, and social studies classrooms. *Instructional Science, 26*, 27–47. doi:10.1023/A:1003035929216

Yerkes, R. M., & Dodson, J. D. (1908). The relation of strength of stimulus to rapidity of habit-formation. *The Journal of Comparative Neurology and Psychology, 18*, 459–482. doi:10.1002/cne.920180503

Zehnder, S., Igoe, L., & Lipscomb, S. D. (2003, June). *Immersion factor – sound: A study of influence of sound on the perceptual salience of interactive games.* Paper presented at the Conference of the Society for Music Perception & Cognition, Las Vegas, NV.

Zehnder, S., & Lipscomb, S. (2006). The role of music in video games. In P. Vorderer, & J. Bryant (Eds.), *Playing video games: Motives, responses, and consequences* (pp. 241–258). Mahwah, NJ: Lawrence Erlbaum.

KEY TERMS AND DEFINITIONS

Background Music: Background music is an important design feature in games and is mainly implemented for enhancing immersion and game play. Since educational games aim at establishing a playful, enjoyable form of learning, background music can be conceptualized as a source of motivation to play and to learn.

Cognitive Load Theory: Cognitive Load Theory suggests that the overall cognitive load of a person is constituted by three different types of cognitive load. Namely; Intrinsic cognitive load (ICL), germane cognitive load (GCL), and extraneous cognitive load (ECL).

Cognitive Theory of Multimedia Learning: Cognitive Theory of Multimedia Learning states that background music is considered as redundant material or seductive detail, that distracts from learning and thus, has a detrimental effect on the learning outcome.

Flow Theory: Flow Theory is a unified framework on subjective experiences in the course of mastering everyday challenges.

Learning Outcome: Aspects taught in a gamified learning system that are acquired by the individuals in terms of the post-education knowledge (e.g., "What do you know about the properties of light?"; "What do you know about blinds?").

This work was previously published in Developments in Current Game-Based Learning Design and Deployment, edited by Patrick Felicia, pp. 219-230, copyright 2013 by Information Science Reference (an imprint of IGI Global) and in the International Journal of Game-Based Learning, Volume 1, Issue 3, edited by Patrick Felicia, pp. 53-64, copyright 2011 by IGI Publishing (an imprint of IGI Global).

Chapter 17
Games and Simulations: A New Approach in Education?

Göknur Kaplan Akilli
Middle East Technical University (METU), Turkey

ABSTRACT

Computer games and simulations are considered powerful tools for learning with an untapped potential for formal educational use. However, the lack of available well-designed research studies about their integration into teaching and learning leaves unanswered questions, despite their more than 30 years of existence in the instructional design movement. Beginning with these issues, this chapter aims to shed light on the definition of games and simulations, their educational use, and some of their effects on learning. Criticisms and new trends in the field of instructional design/development in relation to educational use of games and simulations are briefly reviewed. The chapter intends to provide a brief theoretical framework and a fresh starting point for practitioners in the field who are interested in educational use of games and simulations and their integration into learning environments.

INTRODUCTION

It is unanimously acknowledged that we are living in the information age, taking part in the information society (Bates, 2000; Reigeluth, 1996). What makes these two emerging concepts possible is technology, or rather, the rate of progress that has been achieved in technology over the past 50 or so years (Molenda & Sullivan, 2003). Throughout this period, technology has been both the generator and the transmitter of information with an increasingly faster speed and wider audience each and every day. It now dominates most facets of our lives, penetrating into the conduct of normal daily life.

DOI: 10.4018/978-1-4666-5071-8.ch017

The field of education is not an exception in the permeation of technology. On the contrary, education has always been considered as potentially one of the most productive breeding-grounds for technology, where it would perhaps find its finest resonances and lead to revolutionary effects. Yet, high expectations regarding the revolutionary impacts of technology on education have hardly been realized so far. More specifically, instructional technology, or the use of technology in educational environments, has not contributed significantly to the realization of these expectations (Molenda & Sullivan, 2003; Russell, 2003). It may be argued that the relative ineffectiveness of instructional technology thus far has been caused by the application of the same old methods in new educational media—"New wine was poured, but only into old bottles" (Cohen & Ball, 1990, p. 334). The inconclusiveness of the research is illustrated by the Clark and Kozma debate, started by Clark's 1983 statement that media do not influence students' learning (Clark, 1983). Kozma (1991) counter-argued that learning and media are complementary and that interrelationships of media, method, and external environment have influence on learning. Both of them rationalized their arguments by calling on Russell's (2003) study on, so called, "no-significant-difference" research. Clark (1983, 1994a, 1994b) uses this phenomenon as evidence for his argument, whereas Kozma (1994) uses this phenomenon as indicative of insufficient evidence for his debate.

Current models and methods of instructional technology are insufficient to meet the consequences of the paradigm shift from industrial age to information age (Bates, 2000; Reigeluth, 1996, 1999). Consequently, instructional designers are faced with the challenge of forcing learning situations to fit an instructional design/development model rather than selecting an appropriate model to fit the needs of varying learning situations (Gustafson & Branch, 1997).

One of the possible novelties in instructional methods is the use of games. Indeed, it may possibly be wrong to call games a novelty in education, since young children, by nature, begin to learn through games and playing from their earliest years (Rieber, 1996). However, as they grow up, their play and games are being replaced by formal education, the transition of which does not always—especially nowadays—seem to be a sharp one to the extent that games are being used also in some educational environments, yet their success is questionable or at least not rigorously established. In another sense the use of games in education is not so much a novelty, because its history may be traced back well over a thousand years (Dempsey, Lucassen, Haynes, & Casey, 1998). It is now known that even in times before history, games and dramatic performances as representations of real life were effective as teaching tools. In our modern day, with the new technological advancements, I strongly believe that traditional games have been replaced by electronic games, and, in a similar manner, dramatic representations of old have been transformed into role-playing in simulation environments. Hence, electronic games and simulations have begun to enter contemporary formal education. In addition, the "already-present" new generation of learners have grown up with ever-present games. Prensky (2001) refers to them as the digital natives of the "game generation" (p. 65). He states that this new generation is different from the "digital immigrants" (people born before games were digital and ubiquitous) resulting from their different life experiences with games as a part of the "new media socialization" (Calvert & Jordan, 2001; Prensky, 2001, p. 65). Digital natives who play a lot of games are provided with skills, such as dealing with large amounts of information quickly even at the early ages, using alternative ways to get information, and finding solutions to their own problems through new communication paths. The new "game generation" prefers doing many things simultaneously by using various paths toward the same goal, rather than doing one thing at a time following linear steps. They are less likely to get stuck with frustration when facing a new situation; on the contrary they push themselves into a new situation without knowing anything about it and prefer being active, learning by trial and error, and figuring things out by themselves rather than by reading or listening. Lastly, they want to be treated as "creators and doers" rather than "receptacles to be filled with the content." Hence, the game generation is also referred to as the "intellectual-problem-solving-oriented generation" (Prensky, 2001, p. 76).

When the above issues are considered, it leads to three main bodies of questions, which shape the main focus and scope of this chapter:

1. What are games and simulations? What makes something a game or simulation? What are their educational uses? Do they really have an effect on learning?
2. What is happening in the instructional design/development (IDD) field? Is there a place for games and simulations in both the theory and the practices of IDD?
3. If games and simulations are useful educational tools, how can they be used in education? How can instructional designers take them into account, while designing learning environments? Are there any instructional design/development models (IDDMs) that would light up an instructional designer's path, guiding their journey to integrate games and simulations into their designs?

GAMES AND SIMULATIONS: WHAT ARE THEY?

Games and simulations are often referred to as experiential exercises (Gredler, 1996), in which there is "learning how to learn" that provides something more than "plain thinking:" beyond thinking (Turkle, 1984). Prensky (2001) defines games as "organized play" (p. 119). Heinich, Molenda, Russell, and Smaldino (2002) define a game as "an activity, in which participants follow prescribed rules that differ from those of real life [while] striving to attain a challenging goal" (p. 10). Dempsey, Rasmussen, and Lucassen (1996) define gaming in a basic sense as "any overt instructional or learning format that involves competition and is rule-guided" (p. 4). In my opinion, (except for Prensky's [2001] later and incessant emphasis throughout his book) these definitions are lacking two vital elements: fun and creativity. So my own definition of "game" becomes "a competitive activity that is creative and enjoyable in its essence, which is bounded by certain rules and requires certain skills."

As put forth by many researchers, several game genres can be distinguished, such as action, puzzle, educational, fighting/combat, sports, racing, role play/adventure, flight, shoot'em, platform games, business, board, word, general entertainment, fantasy violence, human violence, non-violent sports, sports violence, and simulation games (Alessi & Trollip, 2001; Funk, Hagan, & Schimming, 1999; Media Analysis Laboratory, 1999; Prensky, 2001; Yelland & Lloyd, 2001). Many researchers also assert that games have some characteristics such as "one or more players (decision makers), rules of play, one or more goals that the players are trying to reach, conditions introduced by chance, a spirit of competition, a strategy or pattern of action-choices to be taken by the players, a feedback system for revealing the state of the game, and a winning player or team" (Price, 1990, p. 52), "turn-taking, fantasy, equipment, and some combination of skill versus luck" (Alessi & Trollip, 2001, p. 271). Furthermore, Price (1990) categorizes "educational" games as academic games, which aim to teach and provide practice, while motivating the learners, and life simulation games, which are context simulation games including strict rules in real-life contexts, or open-ended life simulation games including flexible rules and goals in social science contexts.

A simulation is defined as an interactive abstraction or simplification of some real life

(Baudrillard, 1983; Heinich et al., 2002), or any attempt to imitate a real or imaginary environment or system (Alessi & Trollip, 2001; Reigeluth & Schwartz, 1989; Thurman, 1993). It is "a simulated real life scenario displayed on the computer, which the student has to act upon" (Tessmer, Jonassen, & Caverly, 1989, p. 89).

Although both games and simulations are terms that refer to different concepts, they have common characteristics, too. On the surface, both contain a model of some kind of system, and in both of them learners can observe the consequences of their actions, such as changes occurred in variable, values, or specific actions (Gredler, 1996; Jacobs & Dempsey, 1993). Jacobs and Dempsey (1993) state that the distinction between simulation and games is often blurred, and that many recent articles in this area refer to a single "simulation game" entity. One of them is Prensky (2001), who argues that "depending on what it is doing, a simulation can be a story, it can be a game, [and] it can be a toy" (p. 128).

Gredler (1996) identifies three important differences between the deep structure of games and simulations. Instead of attempting to win the objective of games, participants in a simulation are executing serious responsibilities with privileges that result in associated consequences.

Secondly, the event sequence of a game is typically linear, whereas, according to Gredler (1996), a simulation sequence is non-linear. The player or a team in many games respond to a content-related question and either advance or do not advance depending on the answer, which is repeated for each player or team at each turn. However, in a simulation, participants are confronted with different problems, issues, or events caused mainly by their prior decisions made at each decision point.

The third difference is the mechanisms that determine the consequences to be conveyed for different actions taken by the players. Games consist of rules that describe allowable moves, constraints, privileges, and penalties for illegal (non-permissible) actions. The rules may be totally imaginative, unrelated to real world or events. In contrast, a simulation is based on dynamic set(s) of relationships among several variables that change over time and reflect authentic causal processes. That is, the processes should possess, embody, and result in verifiable relationships.

According to Prensky (2001) simulations and games differ in that, "simulations are not, in and of themselves games. In order to become games, they need additional structural elements—fun, play, rules, a goal, winning, competition, etc." (p. 212). Depending on these definitions and characteristics, as an attempt to derive a general term, I will use *game-like learning environments*, which will be defined as "authentic or simulated places, where learning is fostered and supported especially by seamless integration of motivating game elements, such as challenge, curiosity, and fantasy."

EFFECTS OF GAMES AND SIMULATIONS ON LEARNING

Although the literature on games and simulations is accumulating day by day, the issue of whether games influence students' learning in a positive way is still vague. For instance, Molenda and Sullivan (2003) state that among problem solving and integrated learning systems, games and simulations are among the least used technology applications in education. However, there are some studies that describe the effects of games and simulations on discovery learning strategies; problem solving skills and computer using skills; and effects on students' intellectual, visual, motor skills and indicate how games and simulations impact student engagement and interactivity, which are important for learning environments.

Cole (1996) has shown that long-term game playing has a positive effect on students' learning (cited in Subrahmanyam, Greenfield, Kraut, & Gross, 2001, p. 16). Gredler states that intellectual skills and "cognitive strategies" are acquired during academic games (1996, p. 525). However, she also states that certain games require only simple skills such as recall of verbal or visual elements rather than higher-order skills and as a result, provide environments for winning by guessing (Gredler, 1994). Similarly, Prensky (2001) admits that especially with the non-stop speedy games, the opportunity to stop and think critically about the experience is lessened (Prensky, 2001; Provenzo, 1992). Csikszentmihalyi (1990) also supports the belief that during an enjoyable activity, insufficient amount of time is devoted for thinking and reflection.

Games are claimed to have cognitive development effects on visual skills including "spatial representation," "iconic skills," and "visual attention" (Greenfield, 1984, cited in Prensky, 2001, p. 45; Subrahmanyam et al., 2001, p. 13). Greenfield, deWinstanley, Kilpatrick, and Kaye (1994) claim that as players become more skilled in games, their visual attention becomes proportionally better.

Critical thinking and problem-solving skills (Rieber, 1996), drawing meaningful conclusions (Price, 1990), some inductive discovery skills like observation, trial, and error and hypothesis testing (Gorriz & Medina, 2000; Greenfield, 1984, cited in Prensky, 2001; Price, 1990), and several other strategies of exploration (Prensky, 2001; Provenzo, 1992) were other positive effects of games on learning.

Subrahmanyam et al. (2001) articulate that playing computer games can provide training opportunities for gaining computer literacy, which is consistent with Prensky's (2001) statement that games can be used in order to help people gain some familiarity with the computer hardware.

Games motivate learners to take responsibility for their own learning, which leads to intrinsic motivation contained by the method itself (Rieber, 1996). Malone (1980) and Malone and Lepper (1987) define four characteristics of games that contribute to increases in motivation and eagerness for learning. These are challenge, fantasy, curiosity, and control. Challenges in a game tend to fight students' boredom and keep them engaged with the activity by means of adjusted levels of difficulty. Fantasy in a game increases enthusiasm by providing an appealing imaginary context, whereas curiosity offers interesting, surprising, and novel contexts that stimulate students' needs to explore the unknown. Finally, the control characteristic gives learners the feeling of self-determination.

According to Rieber (1996), gaming elements have a relationship with enjoyable activities that enable the "flow" stage, a term coined by Csikszentmihalyi (1990). Thus, gaming activities have the potential to engross the learner into a state of flow and consequently cause better learning through focus and pleasant rewards (Prensky, 2001), while increasing their motivation and attainment (Rosas, Nussbaum, Cumsille, Marianov, Correa, Flores, et al., 2003).

Other characteristics that ensure the effectiveness of game-based learning are their engagement and interactivity, and active participation (Gredler, 1996; Prensky, 2001; Price, 1990; Provenzo, 1992). Games provide a great deal of highly interactive feedback, which is crucial to learning (Gredler, 1994; Malone, 1980; Prensky, 2001; Rieber, 1996). "Practice and feedback, learning by doing, learning from mistakes, goal oriented learning, discovery learning, task-based learning, question-based learning, situated learning, role playing, coaching, constructivist learning, multi-sensory learning" are applicable interactive learning techniques, when learning through games (Prensky, 2001, p. 157).

EDUCATIONAL USE OF GAMES AND SIMULATIONS

There is evidence that the use of games as instructional tools dates back to 3000 B.C. in China (Dempsey, Lucassen, Haynes, & Casey, 1998). Nevertheless, games and simulations did not become a part of the formal field of instructional design until the early 1970s, despite their entrance into the educational scene in the late 1950s (Gredler, 1996). Seels and Richie (1994) report that in those times audio-visual specialists saw the potential of games and simulations but not of video or electronic games.

Although computer games can be considered powerful tools for increasing learning (Dempsey, Lucassen, et al., 1998; Dempsey, Rasmussen, & Lucassen, 1996), there are two major problems that instructional designers encounter. One is that there are no available comprehensive design paradigms and the other is the lack of well-designed research studies (Gredler, 1996). Since the first problem will be handled in the following sections, at this point, it is proper to proceed with a discussion on the second problem.

While the literature on games and simulations is growing, a majority of the research studies report on perceived student reactions preceded by vague descriptions of games and simulations or on comparisons of simulations versus regular classroom instruction (Gredler, 1996). The more important questions that need further research remain unanswered (Dede, 1996; Dempsey, Lucassen, et al., 1998): How to incorporate games into learning environments? How do students learn best through games and simulations? What are the significant impacts of games and simulations on learning that differentiate them from other forms of online teaching?

Rieber (1996) argues that technological innovations provide new opportunities for interactive learning environments that can be integrated with and validated by theories of learning. Prensky (2001) underscores the need for change in instructional design by claiming that much of the instruction currently provided through computer assisted instruction and Web-based technologies does not contribute to learning, rather it subtracts. People do not want to be included in such learning "opportunities" offered via "new wine into old bottles" innovative technologies, unless they have to, since these learning "opportunities" possess still the same boring content and same old fashioned strategy as traditional education (pp. 92-93). Prensky (2001) puts forth that learning can best take place when there is high engagement, and he proposes "digital-game-based learning," which has potential for achievement of the necessary "high learning" through "high engagement" (p. 149). He states that high engagement, interactive learning process, and the way the two are put together will guarantee the sound working of digital game-based learning (Prensky, 2001).

Rieber (1996) states that, "Research from education, psychology, and anthropology suggests that play is a powerful mediator for learning throughout a person's life" (p. 43). In line with this statement, Prensky (2001) further claims that, "Play has a deep biological, evolutionarily important, function, which has to do specifically with learning" (p. 112). However, despite some important psychological and cultural relationships to games, the education profession has long been hesitant about the value of games as an instructional tool or strategy (Rieber, 1996). For instance, as the prevailing philosophy in education has changed over time, the attitude toward play changed accordingly, too. "In one era, play can be viewed as a productive and natural means of engaging children in problem-solving and knowledge construction, but in another era it can be viewed as wasteful diversion from a child's studies" (Rieber, 1996, p. 44).

The *seamless* integration of beneficial elements of games and simulations into learning, in an endeavor to create "game-like learning environments" seems promising and worth trying. Before discussing the instructional designer's concerns and reviewing instructional design/development models, I will first provide a brief look into the "instructional design/development" field to catch a glimpse of what is going on there.

INSTRUCTIONAL (SYSTEMS) DESIGN/DEVELOPMENT (IDD)

The need for the development of a linking science and the need for a "middleman" between learning theory and educational practice was first asserted by John Dewey in 1900 (as cited in Reigeluth, 1983), yet, when the origins of instructional design procedures are traced, it is seen that the first research efforts date back only to the time of World War II (Dick, 1987). Moreover, the need for a "middleman" was also put forth by Glaser (1971), who stated that an instructional designer must perform the interplay between theory, research, and application.

As the title seems to imply (i.e., is it "design" or "development," and is it "instruction" or an "instructional system"?), there is no consensus about the name and the definition of, what I choose to call "instructional design/development (IDD)." Basically, my concern here is "instructional design" as an activity rather than the most accurate name that refers to this activity. However, the term IDD is used here as a term of convenience, since it encompasses the width and the depth of these activities in a fairly acceptable manner. The literature shows an interchangeable use of instructional design, instructional systems design (ISD), instructional development (ID), and even instructional technology (IT) (Gustafson & Branch, 1997; Reigeluth, 1983; Schrock, 1995; Seels & Richie, 1994). Even though several attempts have been made to derive standardized definitions and terms (Gustafson & Branch, 1997; Schiffman, 1995; Seels & Richie, 1994), the results have not been widely adopted and used in the literature.

Reigeluth (1983) characterizes his views on instructional design as "concerned with understanding, improving and applying methods of instruction" (p. 7), contrasted with instructional development as being "concerned with understanding, improving and applying methods of *creating* [italics added] instruction" (p. 8). Furthermore, he states that instructional design produces knowledge of optimal blueprints about methods of instruction, whereas instructional development optimizes the process of developing the instruction and encompasses design, implementation, and formative evaluation activities. He also emphasizes that design theories are different from descriptive theories due to their prescriptive nature, in the sense that they offer guidelines, without attempting to spell out every detail and allow no variation (Reigeluth, 1983, 1997, 1999). On the other hand, Gustafson and Branch (1997) accept the Seels and Richie (1994) definition, which is "an organized procedure that includes steps of analyzing, designing, developing, implementing, and evaluating instruction" (p. 31). However, they declare that Seels and Richie (1994) have coined this definition for ISD, instead of instructional development. Shrock (1995) has also made a definition similar to that of Seels and Richie's (1994), yet for instructional development. Gustafson and Branch (1997) further characterize instructional development as "a complex, yet purposeful process that promotes creativity, interactivity and cyberneticity [communication and control processes]" (p. 18).

WHAT IS AN INSTRUCTIONAL DESIGN/DEVELOPMENT MODEL (IDDM)?

Gustafson and Branch (1997) define model as "simple representation of more complex forms, processes, and functions of physical phenomena or ideas" (p. 17). It provides a visual representation of an abstract concept (Schindelka, 2003), helps people to "conceptualize representations of reality" (Gustafson & Branch, 1997, p. 17), and "explains ways of doing" (Gustafson & Branch, 1998, p. 3).

In line with Reigeluth's (1983) opinions about instructional development, Gustafson and Branch (1997) have gone a step further and stated that instructional development models have at least four components, which are "analysis of the setting and learner needs; design of a set of specifications for an effective, efficient and relevant learner environment; development of all learner and management materials; and evaluation of the results of the development both formatively and summatively" (p. 12). They have also added that a fifth activity could be the distribution and monitoring of the learning environment across various settings, over an extended period of time. These components help instructional development models serve as "conceptual and communications tool" (p. 13). Gros, Elen, Kerres, Merriënboer, and Spector (1997, p. 48) state that, "instructional design models have the ambition to provide a link between learning theories and the practice of building instructional systems."

The origins of instructional design procedures can be traced to the first research efforts dating back to World War II (Dick, 1987). Gustafson and Branch (1997) state that instructional development models first appeared in 1960s and since then an increasing number of models have been published in the literature. Seels and Richie (1994) highlight the simplicity of the first instructional design models, which had only to master a few techniques and a fundamentally linear theory, since instructional science was an infant and many of the tools and theories of today were not conceivable. Since then, a variety of developments and trends have impacted instructional design practices (Reiser, 2001). However, the introduction of microcomputers in the 1980s has exerted the most significant effect on instructional design practices. With the advent of desktop digital media and the subsequent arrival of worldwide Internet access, discussions began for the need to develop new models of instructional design to accommodate the capability and interactivity of this technology (Merrill, Li, & Jones, 1990). Wide variations have emerged in models in terms of their purposes, amount of detail provided, degree of linearity in which they are applied, and quantity, quality, and relevance of the accompanying operational tools (Gustafson & Branch, 1997). This paradigmatic change has contributed to the instability of the terminology and shows that the field of IDD is not static; it has evolved in time and is still evolving. This is good, since a field that becomes static and uncreative is likely to become less prominent (Seels & Richie, 1994).

Since the 1990s, six factors have had significant impact on instructional design practices (Reiser, 2001). These are performance technology movement, constructivism, Electronic Performance Support Systems (EPSSs), rapid prototyping, increasing use of Internet for distance education/ distance learning, and knowledge management endeavors. However, to provide an account of these factors is out of the scope of this chapter, and Reiser's work should be consulted for a comprehensive discussion.

CRITICISMS ABOUT THE CURRENT STATE OF IDD AND IDDMS

Gustafson and Branch (1997) assert that there has been a cumulative increase in the number of published instructional development models since the 1960s. However, there seems to be little uniqueness in the structure of these models, although they are abundant in number. In other words, as time passes, models are enhanced in quantity, but not in quality (Gustafson & Branch, 1997, 1998).

Some writers have argued that the traditional instructional design models are resistant against substantial changes (Rowland, 1992) and are only fit to narrow, well-defined, and static scenarios, because they are process-oriented rather than people-oriented, and use clumsy, bureaucratic, and linear approaches (Gordon & Zemke, 2000; Jonassen, 1990; McCombs, 1986; Tripp & Bichelmeyer, 1990; You, 1993; Zemke & Rossett, 2002). Contrasting with these criticisms, others contend that over time, problems become apparent in the traditional ISD model and important and permanent modifications and additions are performed (Clark, 2002; Schiffman, 1995; Shrock, 1995).

The procedural stratifications and time-consuming practices of traditional ISD models have drawn much of the criticism. As an alternative, thinking of instructional development as a set of concurrent, overlapping procedures might help both to speed up the process and to overcome many limitations of the traditional instructional design models. One of the most well known examples is "prototyping" or "rapid prototyping," which is a design approach borrowed from the discipline of software engineering (Tripp & Bichelmeyer, 1990).

Both Prensky (2001) and Rowland, Parra, and Basnet (1994) assert that often instructional design is done by the book or by using an overly rationalistic view, which in turn produces "boring cookie-cutter outcomes" (Prensky, 2001, p. 83). These writers emphasize that a move toward more creative methodologies is necessary, in order to lead to flexible, creative solutions to unique situations.

Since the existing design theories have not reached perfection, there is need for new theories and models that will guide instructional designers in the use of ideas about learning founded in human development and cognitive science, and in taking advantage of new information technologies as tools for feedback and assessment or for instruction in general (Reigeluth & Frick, 1999).

Apart from technological changes, Reigeluth (1999) discusses a paradigm shift in education and training, a major shift from Industrial Age to Information Age thinking, which implies shifts in various attributes for instruction (see Table 1).

The change in paradigms, according to Reigeluth (1996), requires a shift from standardization to customization. New models of IDD need to make possible a unique learning experience for each learner, rather than trying to produce a single, clearly defined outcome for all learners. The need for customization is also consistent with Winn's (1997) and Jonassen Hennon, Ondrusek, Samouilova, Spaulding, Yueh et al.'s (1997) criticisms about the positivist basis of ID models. Both disapproved the way that a linear design process assumes the predictability of human behavior, the closed and isolated nature of learning situations, the responsibility for learning belonging to the instructor rather than the learner. New IDD models need to reflect the dynamic, complex, and non-linear nature of the design process, the changing contexts of learning in digital game-based environments, and the many and varied cognitive, emotional, and social differences in abilities among learners.

NEW TRENDS IN IDD AND IDDMS

This section explores a number of new alternative approaches that have been suggested for the improvement of the IDD process. Jonassen et al. (1997) suggest adapting new scientific models, such as hermeneutics, fuzzy logic, and chaos theory. Reigeluth (1996, 1999) suggests customized, learner-centered and social-contextual design conducted by user-designers, which is also articulated by Winn's (1997) matched timing of design and use of instructional material and Winn's (1996) statement of necessity to get help from the Human Computer Interaction discipline. Lastly, Hoffman (1997) offers the ideas of plasticity and modularity as a result of linking Reigeluth's (1983) Elaboration Theory (ET) and hypermedia. There are further suggestions, such as Gros et al.'s (1997) multimedia-facilitated IDD models that depend on multi-perspectival presentation of knowledge or Wilson, Teslow, and Osman-Jouchoux' (1995), and Wilson's (1997) adaptation of postmodernism to IDD field, which need to be further explored.

Hermeneutics emphasizes the importance of the socio-historical context in mediating the meanings of individuals creating and decoding texts; this implies that IDD must strive to introduce gaps of understanding, which allow the learner to create his/her own meanings (Jonassen et al., 1997). Other chapters in this book introduce the idea that new massively multiplayer online

Table 1. Key alterations with the shift from Industrial Age to Information Age

Industrial Age	Information Age
Industrial Society (Bates, 2000)	Information Society
Bureaucratic organization	Team-based organization
Centralized control	Autonomy with accountability
Adversarial relationships	Cooperative relationships
Autocratic decision making	Shared decision making
Compliance	Initiative
Conformity	Diversity
One-way communications	Networking
Compartmentalization	Holism
Parts-oriented	Process-oriented
Planned obsolescence	Total quality
CEO or boss "King"	Customer (learner) as "King"

learning environments entail new social processes that align well with social constructivist, hermeneutic philosophy, and methods.

Chaos theory finds order in the chaos of natural structures through looking for self-similarity and self-organization, patterns that are repeated at different levels of complexity through a structure, for example, a fractal. It can offer two alternatives to IDD: first complex, dynamic IDDMs that adjust to learners on the fly, and secondly due to its sensitiveness to initial conditions, consideration of learners' emotions, and related self-awareness, besides cognitive skills and self awareness (Cagiltay, 2001; Jonassen et al., 1997).

The last alternative that Jonassen et al. (1997) suggest is fuzzy logic. Fuzzy logic is based on the idea that reality can rarely be represented accurately in a bivalent manner. Rather, it is multivalent, having many in-between values, which do not have to belong to mutually exclusive sets. It is a departure from classical two-valued sets and logic, that uses "soft" linguistic (e.g., large, hot, tall) system variables and a continuous range of truth values in the closed interval [0, 1], rather than strict binary (True or False) decisions and assignments. Since the sequence of events within a project depends on human decisions, which is based on approximate reasoning of human beings, fuzzy logic can be well applied to IDD process.

The fuzzy logic perspective implies for IDD that behavior can be better understood probabilistically, using continua, rather than binary measures. Instead of having strictly bounded and sequenced phases, having intertwined phases, which have flexible and fuzzy boundaries, would be more advantageous in that it would allow designers to move freely in between phases throughout the entire IDD process. Jonassen et al. (1997) state that the more one moves away from deterministic approaches to thinking and designing toward more probabilistic ways of thinking, the more useful it becomes in providing methods for assessing "real-life" issues, where things are not black-and-white, but rather any number of different shades of color across the spectrum. Jonassen et al. (1997) further state that it is impossible to predict, let alone describe, what will happen in learning situations due to the elusive and complex nature of human consciousness, which is also consistent with Winn's (1996) opinion that although instructional designers would like them to do otherwise, people think "irrationally," and reason "implausibly." Both of these statements support the main definition of fuzzy logic. However, both researchers' studies lack more specific facets of fuzzy logic. More specifically, the set-theoretic facet of fuzzy logic implies the non-linear, dynamic IDDM phases, which have "fuzzy" rather than strict boundaries. This provides freedom for instructional designers to move back and forth throughout the design process and even conduct more than one activity at a time.

Depending on the previously mentioned shift to Information Age, Reigeluth (1999) also suggests an alternative to the linear stages of the ID process. The entire process cannot be known in advance, so designers are required to do "just-in-time analysis" (p. 15), synthesis, evaluation, and change at every stage in the ID process. However, this is not a newcomer to the field, since learner-centeredness and parallel process have been articulated by Heinich (1973) a long time ago (cited in Winn, 1996). Reigeluth (1999) further states that to be capable to meet the demands of the Information Age, the instructional designer should become more aware of the broader social context, within which the instruction takes place, and a point that is also made by various researchers as well (Dede, 1996; Jonassen et al., 1997; Kember & Murphy, 1995; Richey, 1995; Tessmer & Richey, 1997). For example, the instructional designer might consult more broadly with stakeholder groups to reach a common vision of the final instruction and the means to develop it. The social context can expand to include the learners, consistent with Kember and Murphy's (1995) suggestion that linking the learners to designers supports iterative improvement.

Lastly, Hoffman (1997) offered plasticity and modularity as a result of linking Reigeluth's (1983) Elaboration Theory (ET) and hypermedia. He states that the Web-like linking of ideas that characterizes hypermedia is more alike to the functioning of human cognition than is the traditional linear structure found in much educational programming. He further asserted that this kind of model for IDD could lead to the possibility of modularity and plasticity, which would bring along the ease to make changes in response to learner needs without changing the overall structure of the product and rapid development. It could also allow the customization from the user end to allow a more feasible learner control in like manner to that of a Web structure.

To sum up the whole discussion, IDD and IDDM should find alternative ways to catch up with the changing world of education due to changes in the world itself. The previously mentioned alternatives are thought to be useful and helpful to renew and strengthen the IDD field against the criticisms. It also reveals the fact that like the other disciplines, IDD also begins to evolve into a multidisciplinary discipline. Indeed, Jonassen et al.'s (1997) statement summarizes the main idea:

Like the chiropractor who realigns your spine, we might become healthier from a realignment of our theories. If we admit to and attempt to accommodate some of the uncertainty, indeterminism, and unpredictability that pervade our complex world, we will develop stronger theories and practices that will have more powerful (if not predictable) effects on human learning. (p. 33)

DESIGN MODELS FOR EDUCATIONAL USE OF GAMES AND SIMULATIONS

Theories that inspire game design include "Flow Theory of Optimal Experience" developed by Mihaly Csikszentmihalyi (1990) and "Activity Theory" developed by Alexey Leontiev, a student of Lev Vygotsky (Kaptelinin & Nardi, 1997). Moreover, there are some myths and principles to be taken into consideration during preproduction and production stages of game design proposed by Cerny and John (2002). Yet, there seem to be hardly any design models except for the instructional design/development model tailored for the creation of game-like learning environments, which is called the FIDGE model (Akilli & Cagiltay, 2006). Hence it is clear that there is a need for IDD models that will help and guide educators to design game-like learning environments, "which requires the ability to step outside of a traditional, linear approach to content creation—a process that is counter-intuitive to many teachers" (Morrison & Aldrich, 2003).

This section offers a brief review of different design principles and lessons learned from game design processes before briefly reviewing the FIDGE model. For instance, Amory, Naicker, Vincent, and Adams (1999) identified game elements that students found interesting or useful within different game types, which were the most suitable for their teaching environment and presented a model that links pedagogical issues with these identified game elements.

Prensky (2001) presents various principles for good computer game design and other important digital game design elements. For instance, he claims that good game design is balanced in terms of challenge, creative in terms of originality, focused in terms of fun, and has character in terms of richness and depth that make you remember it, tension that keeps the player playing, and energy that keeps you up all night (pp. 133-134). In addition to these elements, he further asserts that a game should have a clear overall vision with highly adaptive, easy to learn but hard to master structure offered via a very user-friendly interface. It should have a constant focus on the player experience that keeps the player within the flow state providing exploration, discovery, and frequent rewards, not penalties. It should provide mutual assistance, which means achieving one thing in the game helps to solve another, and the ability to save this progress (pp. 134-136). Lastly, as for digital game-based learning, he provides five questions to be asked during the process of designing, again with his emphasis for fun followed by learning. These five questions can be summarized as the appeal of games in terms of fun for other people too, who are not targeted as audience; the self-perceptions of users as "players" not as "students" or "trainees;" the level of addiction and prominence of the game among the players; the level and rate of improvement at player's skills; and the level of encouragement and enactment for players' reflection on their learning (p. 179).

The "Games-to-Teach" project carried by Massachusetts Institute of Technology proposes design principles for successful games design (MIT, 2003). These are designing educational action games by turning simulations into simulation games; moving from parameters to "power-ups [adjustments made on some traits of the character in the game, such as shifts in player speed, height, and so forth to enhance their attributes];" designing game contexts by identifying contested spaces, identifying opportunities for transgressive play [that enables players to experience new roles via "temporarily letting go of social/cultural rules and mores"]; using information to solve complex problems in simulated environments; providing choices and consequences in simulated worlds; and differentiating roles and distributing expertise in multiplayer games.

The most recent study on the subject with a promising design/development model is the "FIDGE model" (Akilli & Cagiltay, 2006). The model consists of dynamic phases with fuzzy boundaries, through which instructional designers move in a non-linear manner. The model's foundation in the fuzzy logic concept leads to a visualization of the model that is unlike traditional "boxes-and-arrows" representations (see Figure 1). There are two other sets of principles that underlie the model, which are related to socio-organizational issues for the design team and to the instructional design/development process itself. Table 2 summarizes the model in its essence.

All of these studies deserve appreciation, since educational games are mostly classified as "boring" by students. Moreover, they also show that endeavors are being suffered for and steps are being taken toward what Kirriemuir (2002) emphasized: "Computer games provide a medium that engages people for long periods of time, and gamers usually return to the same game many times over. There are obvious lessons here for the developers of digitally-based educational, learning and training materials."

Figure 1. The overall appearance of the FIDGE model (Source: Akilli & Cagiltay, 2006, p. 112; reprinted with permission from IOS Press; reprinted with permission from IOS Press)

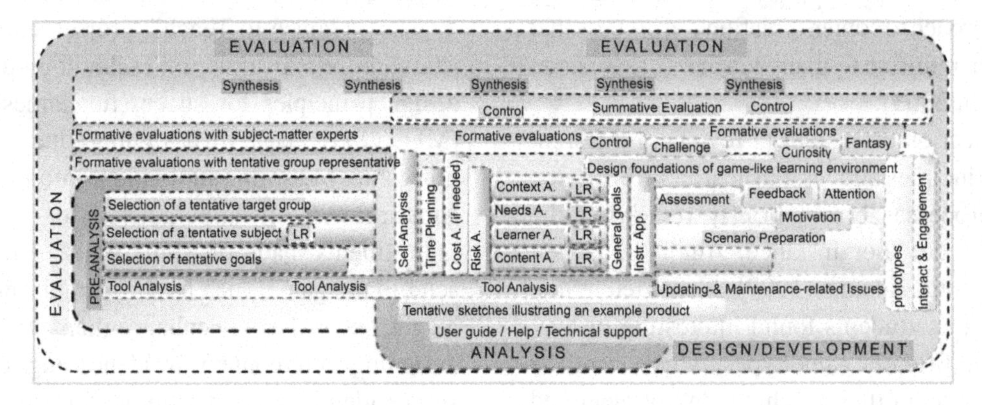

CONCLUSION

This chapter has provided a brief theoretical framework for the educational use of games and simulations and their effect on learning. It reviewed and addressed some of the main criticisms and new trends in the IDD and IDDM fields.

The characteristics of the "game generation," the importance of games for education, and criticisms about IDDMs' failure to meet these changing needs lead to the conclusion that instructional designers should strive to seamlessly integrate game elements into their designs and to create game-like learning environments, so that they can armor students for the future and build powerful learning into their designs. However, there seems to be a little number of design guidelines, and only one IDD model exists in the literature, to guide instructional designers through this painstaking process, which at the same time provides an already existent but newly discovered playground for the practitioners in the field.

New IDD models are needed to help designers create game-like learning environments that can armor students for the future and build powerful learning into their designs.

Table 2. Summary of the FIDGE model (Source: Akilli & Cagiltay, 2006, p. 110; reprinted with permission from IOS Press)

Issue	Its Property
Participants	All of actively participating learners and experts
Team	Multidisciplinary, multi-skilled, game-player experience
Environment	Socio-organizational, cultural
Process	Dynamic, non-linear, fuzzy, creative, enriched by games' and simulations' elements (fantasy, challenge, etc.)
Change	Continuous, evaluation-based
Evaluation	Continuous, iterative, formative, and summative, fused into each phase
Management	Need for a leader in the team and a well-planned and scheduled time management
Technology	Suitable, compatible
Use	By (novice/expert) instructional designers and educational game designers for game-like learning environments and educational games

REFERENCES

Akilli, G. K., & Cagiltay, K. (2006). An instructional design/development model for game-like learning environments: The FIDGE model. In M. Pivec (Ed.), *Affective and emotional aspects of human-computer interaction game-based and innovative learning approaches* (Vol. 1, pp. 93–112). Amsterdam, The Netherlands: IOS Press.

Alessi, S. M., & Trollip, S. R. (2001). *Multimedia for learning: Methods and development* (3rd ed.). Boston: Allyn and Bacon Publication.

Amory, A., Naicker, K., Vincent, J., & Adams, C. (1999). The use of computer games as an educational tool: Identification of appropriate game types and game elements. *British Journal of Educational Technology*, *30*(4), 311–321. doi:10.1111/1467-8535.00121

Bates, A. W. (2000). *Managing technological change. Strategies for college and university leaders*. San Francisco: Jossey-Bass.

Baudrillard, J. (1983). *Simulations*. New York: Semiotext. [e]

Cagiltay, K. (2001). A design/development model for building electronic performance support systems. In M. Simonson & C. Lamboy (Eds.), *Annual Proceedings of Selected Research and Development [and] Practice Papers presented at the 24th National Convention of the Association for Educational Communications and Technology*, Atlanta, GA (pp. 433-440). Bloomington, IN: Association for Educational Communications and Technology. (ERIC Document Reproduction Service No: ED 470175).

Calvert, S. L., & Jordan, A. B. (2001). Children in the digital age. *Applied Developmental Psychology*, *22*(1), 3–5. doi:10.1016/S0193-3973(00)00062-9

Cerny, M., & John, M. (2002, June). Game development. Myth vs. method. *Game Developer Magazine*, 32-36.

Clark, R. C. (2002). The new ISD: Applying cognitive strategies to instructional design. *Performance Improvement Journal*, *41*(7), 8–14.

Clark, R. E. (1983). Reconsidering research on learning from media. *Review of Educational Research*, *53*(4), 445–459. doi:10.3102/00346543053004445

Clark, R. E. (1994a). Media will never influence learning. *Educational Technology Research and Development*, *42*(2), 21–29. doi:10.1007/BF02299088

Clark, R. E. (1994b). Media and method. *Educational Technology Research and Development*, *42*(3), 7–10. doi:10.1007/BF02298090

Cohen, D. K., & Ball, D. L. (1990). Relations between policy and practice: A commentary. *Educational Evaluation and Policy Analysis*, *12*(3), 331–338.

Csikszentmihalyi, M. (1990). *Flow: The psychology of optimal experience*. New York: Harper Perennial.

Dede, C. (1996). The evolution of constructivist learning environments: Immersion in distributed, virtual worlds. In B. G. Wilson (Ed.), *Constructivist learning environments: Case studies in instructional design* (pp. 165–175). Englewood Cliffs, NJ: Educational Technology Publications.

Dempsey, J. V., Lucassen, B. A., Haynes, L. L., & Casey, M. S. (1998). Instructional applications of computer games. In J. J. Hirschbuhl & D. Bishop (Eds.), Computer studies: Computers in education (8th ed., pp. 85-91). Guilford, CT: Dushkin/McGraw Hill.

Dempsey, J. V., Rasmussen, K., & Lucassen, B. (1996). *The instructional gaming literature: Implications and 99 sources* (Tech. Rep. No. 96-1). Mobile, AL: University of South Alabama, College of Education. Retrieved May 30, 2002, from http://www.coe.usouthal.edu/TechReports/TR96_1.PDF

Dick, W. (1987). A history of instructional design and its impact on educational psychology. In J. Glover, & R. Ronning (Eds.), *Historical foundations of educational psychology*. New York: Plenum. doi:10.1007/978-1-4899-3620-2_10

Funk, J. B., Hagan, J., & Schimming, J. (1999). Children and electronic games: A comparison of parents' and children's perceptions of children's habits and preferences in a United States Sample. *Psychological Reports, 85*(3), 883–888. PubMed doi:10.2466/PR0.85.7.883-888

Glaser, R. (1971). The design of instruction. In M. D. Merrill (Ed.), *Instructional design: Readings* (pp. 18–37). Englewood, NJ: Prentice-Hall.

Gordon, J., & Zemke, R. (2000). The attack on ISD. *Training Magazine, 37*(4), 42–53.

Gorriz, C. M., & Medina, C. (2000). Engaging girls with computers through software games. *Communications of the ACM, 43*(1), 42–49. doi:10.1145/323830.323843

Gredler, M. E. (1994). *Designing and evaluating games and simulations: A process approach.* Houston, TX: Gulf Publication Company.

Gredler, M. E. (1996). Educational games and simulations: A technology in search of a (research) paradigm. In D. H. Jonassen (Ed.), *Handbook of research for educational communications and technology* (pp. 521–539). New York: Macmillan.

Greenfield, P. M., deWinstanley, P., Kilpatrick, H., & Kaye, D. (1994). Action video games and informal education: Effects on strategies for dividing visual attention [Abstract]. *Journal of Applied Developmental Psychology, 15*(1), 105–123. doi:10.1016/0193-3973(94)90008-6

Gros, B., Elen, J., Kerres, M., Merriënboer, J., & Spector, M. (1997). Instructional design and the authoring of multimedia and hypermedia systems: Does a marriage make sense? *Educational Technology, 37*(1), 48–56.

Gustafson, K. L., & Branch, R. M. (1997). *Survey of instructional development models* (3rd ed.). Syracuse, NY: ERIC Clearinghouse on Information Resources.

Gustafson, K. L., & Branch, R. M. (1998). Revisioning models of instructional development. *Educational Technology Research and Development, 45*(3), 73–89. doi:10.1007/BF02299731

Heinich, R., Molenda, M., Russell, J. D., & Smaldino, S. E. (2002). *Instructional media and technologies for learning* (7th ed.). Upper Saddle River, NJ: Merrill Prentice Hall.

Hoffman, S. (1997). Elaboration theory and hypermedia: Is there a link? *Educational Technology, 37*(1), 57–64.

Jacobs, J. W., & Dempsey, J. V. (1993). Simulation and gaming: Fidelity, feedback and motivation. In J. V. Dempsey, & G. C. Sales (Eds.), *Interactive instruction and feedback* (pp. 197–228). Engelwood Cliffs, NJ: Educational Technology Publications.

Jonassen, D. H. (1990). Thinking technology: Chaos in instructional design. *Educational Technology, 30*(2), 32–34.

Jonassen, D. H., Hennon, R. J., Ondrusek, A., Samouilova, M., Spaulding, K. L., & Yueh, H. P. et al. (1997). Certainty, determinism, and predictability in theories of instructional design: Lessons from science. *Educational Technology, 37*(1), 27–33.

Kaptelinin, V., & Nardi, B. A. (1997). *Activity theory: Basic concepts and applications.* Retrieved October 11, 2003, from http://www.acm.org/sigchi/chi97/proceedings/tutorial/bn.htm/

Kember, D., & Murphy, D. (1995). The impact of student learning research and the nature of design on ID fundamentals. In B. Seels (Ed.), *Instructional design fundamentals: A reconsideration* (pp. 99–112). Englewood Cliffs, NJ: Educational Technology Publications.

Kirriemuir, J. (2002). Video gaming, education and digital learning technologies [Online]. *D-lib Magazine, 8*(2). Retrieved May 12, 2003, from http://www.dlib.dlib.org/february02/kirriemuir/02kirriemuir.html

Kozma, R. B. (1991). Learning with media. *Review of Educational Research, 61*(2), 179–211. doi:10.3102/00346543061002179

Kozma, R. B. (1994). Will media influence learning? Reframing the debate. *Educational Technology Research and Development, 42*(2), 7–19. doi:10.1007/BF02299087

Malone, T. W. (1980). *What makes things fun to learn? Heuristics for designing instructional computer games.* Paper presented at the Joint Symposium: Association for Computing Machinery Special Interest Group on Small Computers and Special Interest Group on Personal Computers, Palo Alto, CA.

Malone, T. W., & Lepper, M. R. (1987). Making learning fun: A taxonomy of intrinsic motivations for learning. In R. E. Snow, & M. J. Farr (Eds.), *Aptitude, learning, and instruction, III: Cognitive and affective process analysis* (pp. 223–253). Hillsdale, NJ: Lawrence Erlbaum Associates.

Massachusetts Institute of Technology (MIT). (2003). Design principles of next-generation digital gaming for education. *Educational Technology, 43*(5), 17–22.

McCombs, B. L. (1986). The instructional systems development (ISD) model: A review of those factors critical to its successful implementation. *Education Communication and Technology Journal, 34*(2), 67–81.

Media Analysis Laboratory. Simon Fraser University, B.C. (1998). Video game culture: Leisure and play of B.C. teens [Online]. Retrieved June 11, 2003, from http://www.mediaawareness.ca/eng/ISSUES/VIOLENCE/RESOURCE/reports/vgames.html

Merrill, M. D., Li, Z., & Jones, M. K. (1990). Limitations of first generations instructional design. *Educational Technology, 30*(1), 7–11.

Molenda, M., & Sullivan, M. (2003). Issues and trends in instructional technology: Treading water. In M. A. Fitzgerald, M. Orey, & R. M. Branch (Eds.), *Educational Media and Technology Yearbook 2003* (pp. 3–20). Englewood, CO: Libraries Unlimited.

Morrison, J. L., & Aldrich, C. (2003). Simulations and the learning revolution: An interview with Clark Aldrich. *The Technology Source.* Retrieved August 11, 2003, from http://64.124.14.173/default.asp?show=article&id=2032

Prensky, M. (2001). *Digital game-based learning.* New York: McGraw-Hill.

Price, R. V. (1990). *Computer-aided instruction: A guide for authors.* Pacific Grove, CA: Brooks/Cole Publishing Company.

Provenzo, E. F. (1992). The video generation. *The American School Board Journal, 179*(3), 29–32.

Reigeluth, C., & Schwartz, E. (1989). An instructional theory for the design of computer-based simulations. *Journal of Computer-Based Instruction, 16*(1), 1–10.

Reigeluth, C. M. (1983). Instructional design: What is it and why is it? In C. M. Reigeluth (Ed.), *Instructional-design theories and models: An overview of their current status* (pp. 3–36). Hillsdale, NJ: Lawrence Erlbaum Associates.

Reigeluth, C. M. (1996). A new paradigm of ISD? *Educational Technology, 36*(3), 13–20.

Reigeluth, C. M. (1997). Instructional theory, practitioner needs, and new directions: Some reflections. *Educational Technology, 37*(1), 42–47.

Reigeluth, C. M. (1999). What is instructional-design theory and how is it changing? In C. M. Reigeluth (Ed.), Instructional-design theories and models (Vol. II): A new paradigm of instructional theory (pp. 5-29). Mahwah, NJ: Lawrence Erlbaum Associates.

Reigeluth, C. M., & Frick, T. W. (1999). Formative research: Methodology for creating and improving design theories. In C. M. Reigeluth (Ed.), Instructional-design theories and models (Vol. II): A new paradigm of instructional theory (pp. 633-652). Mahwah, NJ: Lawrence Erlbaum Associates.

Reiser, R. A. (2001). A history of instructional design and technology: Part II: A history of instructional design. *Educational Technology Research and Development*, *49*(2), 57–67. doi:10.1007/BF02504928

Richey, C. (1995). Trends in instructional design: Emerging theory-based models. *Performance Improvement Quarterly*, *8*(3), 96–110. doi:10.1111/j.1937-8327.1995.tb00689.x

Richey, R. C. (1997). Agenda-building and its implications for theory construction in instructional technology. *Educational Technology*, *37*(1), 5–11.

Rieber, L. P. (1996). Seriously considering play: Designing interactive learning environments based on the blending of microworlds, simulations, and games. *Educational Technology Research and Development*, *44*(2), 43–58. doi:10.1007/BF02300540

Rosas, R., Nussbaum, M., & Cumsille, P. (2003). Beyond Nintendo: Design and assessment of educational video games for first and second grade students. *Computers & Education*, *40*(1), 71–94. doi:10.1016/S0360-1315(02)00099-4

Rowland, G. (1992). What do instructional designers actually do? An initial investigation of expert practice. *Performance Improvement Quarterly*, *5*(2), 65–86. doi:10.1111/j.1937-8327.1992.tb00546.x

Rowland, G., Parra, M. L., & Basnet, K. (1994). Educating instructional designers: Different methods for different outcomes. *Educational Technology*, *34*(6), 5–11.

Russell, T. (2003). *The "No Significant Difference Phenomenon."* Retreived December 10, 2005, from http://teleeducation.nb.ca/nosignificantdifference/

Schiffman, S. S. (1995). Instructional systems design: Five views of the field. In G. Anglin (Ed.), *Instructional Technology: Past, present, and future* (2nd ed., pp. 131–144). Engelwood, CO: Libraries Unlimited.

Schindelka, B. (2003). A framework of constructivist instructional design: Shiny, happy design. *Inoad e-Zine, 3*(1). Retrieved August 18, 2003, from http://www.inroad.net/shindelka0403.html

Seels, B., & Richie, R. (1994). *Instructional technology: The definitions and domains of the field*. Washington, DC: AECT.

Shrock, S. A. (1995). A brief history of instructional development. In G. Anglin (Ed.), *Instructional technology: Past, present, and future* (2nd ed., pp. 11–19). Engelwood, CO: Libraries Unlimited.

Subrahmanyam, K., Greenfield, P., Kraut, R., & Gross, E. (2001). The impact of computer use on children's and adolescents' development. *Applied Developmental Psychology*, *22*(1), 7–30. doi:10.1016/S0193-3973(00)00063-0

Tessmer, M., Jonassen, D. H., & Caverly, D. (1989). *Non-programmer's guide to designing instruction for microcomputers*. Littleton, CO: Libraries Unlimited.

Tessmer, M., & Richey, R. C. (1997). The role of context in learning and instructional design. *Educational Technology Research and Development*, *45*(2), 85–115. doi:10.1007/BF02299526

Thurman, R. A. (1993). Instructional simulation from a cognitive psychology viewpoint. *Educational Technology Research and Development, 41*(4), 75–79. doi:10.1007/BF02297513

Tripp, S. D., & Bichelmeyer, B. (1990). Rapid prototyping: An alternative instructional design strategy. *Educational Technology Research and Development, 38*(1), 31–44. doi:10.1007/BF02298246

Turkle, S. (1984). Video games and computer holding power. In *The second self: Computers and the human spirit* (pp. 64–92). New York: Simon and Schuster.

Wilson, B. G. (1997). The postmodern paradigm. In C. R. Dills, & A. A. Romiszowski (Eds.), *Instructional development paradigms* (pp. 63–80). Englewood Cliffs, NJ: Educational Technology Publications.

Wilson, B. G., Teslow, J., & Osman-Jouchoux, R. (1995). The impact of constructivism (and post-modernism) on instructional design fundamentals. In B. B. Seels (Ed.), *Instructional design fundamentals: A reconsideration* (pp. 137–157). Englewood Cliffs, NJ: Educational Technology Publications.

Winn, W. (1996). Cognitive perspectives in psychology. In D. H. Jonassen (Ed.), *Handbook of research for educational communications and technology: A project of the Association for Educational Communications and Technology* (pp. 79–112). New York: Macmillan Library Reference.

Winn, W. (1997). Advantages of a theory-based curriculum in instructional technology. *Educational Technology, 37*(1), 34–41.

Yelland, N., & Lloyd, M. (2001). Virtual kids of the 21st century: Understanding the children in schools today. *Information Technology in Childhood Education Annual, 13*(1), 175–192.

You, Y. (1993). What can we learn from Chaos theory? An alternative approach to instructional systems design. *Educational Technology Research and Development, 41*(3), 17–32. doi:10.1007/BF02297355

Zemke, R. E., & Rossett, A. (2002). A hard look at ISD. *Training (New York, N.Y.), 39*(2), 26–34.

KEY TERMS AND DEFINITIONS

Educational Game: Educational games are academic games, which aim to teach and provide practice, while motivating the learners, and life simulation games, which are context simulation games including strict rules in real-life contexts, or open-ended life simulation games including flexible rules and goals in social science contexts.

Game: A game is a competitive activity that is creative and enjoyable in its essence, which is bounded by certain rules and requires certain skills.

Game-Like Learning Environments: Game-like learning environments are authentic or simulated places, which have been gamified to provide learning that is fostered and supported, especially by the integration of gamification motivating factors, such as challenge, curiosity, and fantasy.

Instructional Design: an organized procedure that includes steps of analyzing, designing, developing, implementing, and evaluating instruction

Simulation: A simulation is an interactive abstraction or simplification of some real life, which may require gamification methods to achieve.

This work was previously published in Games and Simulations in Online Learning: Research and Development Frameworks, edited by David Gibson, Clark Aldrich, and Marc Prensky, pp. 1-20, copyright 2007 by Information Science Publishing (an imprint of IGI Global).

Compilation of References

A+ Education Ltd. (2012). *The early childhood environment rating scale – Revised (ECERS-R)*. Retrieved 14 March, 2013, from http://www.ecersuk.org/4.html

Aarhus, R., Grönvall, E., Larsen, S. B., & Wollsen, S. (2011). Turning training into play: Embodied gaming, seniors, physical training and motivation. *Gerontechnology (Valkenswaard)*, *10*(2), 110–120. doi:10.4017/gt.2011.10.2.005.00

Abt, C. (1970). *Serious games*. New York: Viking Press.

ACARA. (2012). *The Australian curriculum*. Retrieved 20 February, 2013, from www.australiancurriculum.edu.au

Action for Health Kids. (2003). Retrieved from http://actionforhealthykids.org

Adams, E. (2005). *Educational Games Don't Have to Stink!* Retrieved 31 January, 2009, from http://www.gamasutra.com/features/20050126/adams_01.shtml

Aggarwal, C. C. (2003). Towards systematic design of distance functions for data mining applications. In *Proceedings of the Ninth ACM SIGKDD International Conference on Knowledge Discovery and Data Mining*, (pp. 9-18). ACM.

Agle, B. R., Mitchell, R. K., & Sonnenfeld, J. A. (1999). Who matters to CEOs? An investigation of stakeholder attributes and salience, corporate performance, and CEO values. *Academy of Management Journal*, *42*, 507–525. doi:10.2307/256973

Akilli, G. K., & Cagiltay, K. (2006). An instructional design/development model for game-like learning environments: The FIDGE model. In M. Pivec (Ed.), *Affective and emotional aspects of human-computer interaction game-based and innovative learning approaches* (Vol. 1, pp. 93–112). Amsterdam, The Netherlands: IOS Press.

Alessi, S. M., & Trollip, S. R. (2001). *Multimedia for learning: Methods and development* (3rd ed.). Boston: Allyn and Bacon Publication.

Alexandrova, E. (2011). Metamorphoses of civil society and politics: From ganko's café to facebook. In G. Lozanov, & O. Spassov (Eds.), *Media and politics*. Konrad Adenauer Stiftung.

Amory, A. (2007). Game object model version II: A theoretical framework for educational game development. *Educational Technology Research and Development*, *55*, 55–77. doi:10.1007/s11423-006-9001-x

Amory, A., Naicker, K., Vincent, J., & Adams, C. (1999). The use of computer games as an educational tool: Identification of appropriate game types and game elements. *British Journal of Educational Technology*, *30*(4), 311–321. doi:10.1111/1467-8535.00121

Anderson, C. A., & Bushman, B. J. (2001). Effects of violent video games on aggressive behavior, aggressive cognition, aggressive affect, physiological arousal, and prosocial behavior: A meta-analytic review of the scientific literature. *Psychological Science*, *12*(5), 353–359. PubMed doi:10.1111/1467-9280.00366

Anderson, L. W., Krathwohl, D. R., Airasian, P. W., Cruikshank, K. A., Mayer, R. E., & Pintrich, P. R. et al. (2000). *A Taxonomy for Learning, Teaching, and Assessing: A Revision of Bloom's Taxonomy of Educational Objectives* (2nd ed.). Boston: Allyn & Bacon.

Ang, C. S., Zaphiris, P., & Mahmood, S. (2007). A model of cognitive loads in massively multiplayer online role playing games. *Interacting with Computers*, *19*, 167–179. doi:10.1016/j.intcom.2006.08.006

Asakawa, K. (2004). Flow experience and autotelic personality in Japanese college students: How do they experience challenges in daily life? *Journal of Happiness Studies, 5*(2), 123–154. doi:10.1023/B:JOHS.0000035915.97836.89

Asby, F., Isen, A., & Turken, A. (1999). A neuropsychological theory of positive affects and its influence on cognition. *Psychological Review, 106*, 529–550. doi:10.1037/0033-295X.106.3.529 PMID:10467897

Ashcroft, J. D., & Ashcroft, J. E. (2007). *Law for business*. South-Western Pub.

Asquith, A. (1998). Non-elite employees: Perceptions of organizational change in english local government. *International Journal of Public Sector Management, 11*(4), 262–280. doi:10.1108/09513559810225825

Austen, I. (2007, February 19). Canadian company offers nude photos via cellphone. *The New York Times*.

Avgerou, C., & Walsham. (2000). Introduction: IT in developing countries. In C. Avgerou & G. Walsham (Eds.), *Information technology in context: Studies from the perspective of developing countries*. Aldershot, UK: Ashgate.

Ažderska, T., & Jerman Blažič, B. (2011). A novel systemic taxonomy of trust in the online environment. *Towards a Service-Based Internet*, 122-133.

Bahl, R. W., & Linn. (1992). *Urban public finance in developing countries*. New York: Oxford University Press.

Bainbridge. (2004). *Introduction to computer law*. New York: Longman.

Bainbridge, D. I. (2007). *Introduction to information technology law*. Upper Saddle River, NJ: Prentice Hall.

Ball, D., & Forzani, F. (2007). What makes education research 'educational'? *Educational Researcher, 36*(9), 529–540. doi:10.3102/0013189X07312896

Bandura, A. (1977). Self-efficacy: Toward a unifying theory of behavioural change. *Psychological Review, 84*, 191–215. PubMed doi:10.1037/0033-295X.84.2.191

Bannock, G. Baxter, & Davis. (1987). The Penguin dictionary of economics (4th ed.). London: Penguin Books.

Barab, S. A., Hay, K., Barnett, M., & Keating, T. (2000). Virtual solar system: Building understanding through model building. *Journal of Research in Science Teaching, 37*(7), 719–756. doi:10.1002/1098-2736(200009)37:7<719::AID-TEA6>3.0.CO;2-V

Barab, S. A., Ingram-Goble, A., & Warren, S. (2008). Conceptual Playspaces. In R. Ferdig (Ed.), *Handbook on Research on Effective Electronic Gaming in Education*. Hershey, PA: IGI Global. doi:10.4018/978-1-59904-808-6.ch057

Barab, S. A., Sadler, T. D., Heiselt, C., Hickey, D., & Zuiker, S. (2007). Relating narrative, inquiry and inscriptions: Supporting consequential play. *Journal of Science Education and Technology, 16*(1), 59–82. doi:10.1007/s10956-006-9033-3

Barlett, C. P., Anderson, C. A., & Swing, E. L. (2009). Video games effects–Confirmed, suspected, and speculative. A review of the evidence. *Simulation & Gaming, 40*(3), 377–403. doi:10.1177/1046878108327539

Basak, C., Boot, W. R., Voss, M. W., & Kramer, A. F. (2008). Can training in a real-time strategy video game attenuate cognitive decline in older adults? *Psychology and Aging, 23*(4), 765–777. PubMed doi:10.1037/a0013494

Basu, S. (2004). E-government and developing countries: An overview. *International Review of Law Computers & Technology, 18*(1), 109–132. doi:10.1080/13600860410001674779

Bates, A. W. (2000). *Managing technological change. Strategies for college and university leaders*. San Francisco: Jossey-Bass.

Battersby, S. J. (2007, December). Serious games and the Wii™ – a technical report. In *Proceedings of the 27th BCS SGAI Workshop on Serious Games*, Cambridge, UK.

Baudrillard, J. (1983). *Simulations*. New York: Semiotext. [e]

Bauman, R., & Briggs, C. (1990). Poetics and performance as critical perspectives on language and social life. *Annual Review of Anthropology, 19*, 59–88. doi:10.1146/annurev.an.19.100190.000423

Becerra, M. (2009). *Theory of the firm for strategic management: Economic value analysis*. Cambridge, UK: Cambridge University Press. doi:10.1017/CBO9780511626524

BECTa. (2006). *Computer Games in Education: Findings Report*. Retrieved 19 February, 2008, from http://partners.becta.org.uk/index.php?section=rh&rid=13595

Beeland, W. (2002). Student engagement, visual learning and technology: Can interactive whiteboards help?. *Action Research Exchange, 1*.

Beenen, G., Ling, K., Wang, X., Chang, K., Frankowski, D., Resnick, P., & Kraut, R. E. (2004). Using social psychology to motivate contributions to online communities. *Proceedings of the 2004 ACM Conference on Computer Supported Cooperative Work*, (p. 221).

Beike, D., & Wirth-Beaumont, E. (2005). Psychological closure as a memory phenomenon. *Memory (Hove, England), 13*(6), 574–593. PubMed doi:10.1080/09658210444000241

Beinhocker, E. D. (2006). *The origin of wealth: Evolution, complexity and the radical remaking of economics*. Boston: Harvard Business School Press.

Bellamy, C. (2000). The politics of public information systems. In G. D. Garson (Ed.), *Handbook of public information systems*. New York: Marcel Dekker Inc.

Bellamy, C., & Taylor. (1994). Introduction: Exploiting IT in public administration – Towards the information polity? *Public Administration, 72*, 1–12. doi:10.1111/j.1467-9299.1994.tb00996.x

Bennett, W. L., Wells, C., & Rank, A. (2008). *Young citizens and civic learning: Two paradigms of citizenship in the digital age*. Seattle, WA: University of Washington.

Benware, C., & Deci, E. L. (1984). Quality of learning with an active versus passive motivational set. *American Educational Research Journal, 21*, 755–765. doi:10.3102/00028312021004755

Berliner, M. S. (1974). Montessori and social development. *The Educational Forum, 38*(3), 295–304. doi:10.1080/00131727409338116

Bernard, D. (2010). What is Wikileaks? *News.com*. Retrieved from http://www.voanews.com/english/news/what-is-wikileaks--99239414

Bernstein, D. E. (2003). You can't say that: Canadian thought police on the march. *National Review Online*. Retrieved from http://www.nationalreview.com/script/comment/ bernstein 200312020910

Bers, M. (2012). *Designing digital experiences for positive youth development: From playpen to playground*. New York: Oxford University Press. doi:10.1093/acprof:oso/9780199757022.001.0001

Bers, M. U. (2008). Civic identities, online technologies: From designing civic curriculum to supporting civic engagement. In W. L. Bennett (Ed.), *Civic Life Online*. Cambridge, MA: MIT Press.

Bhatnagar, S. (2005). *E-government: Opportunities and challenges*. Retrieved from http://siteresources.worldbank.org/INTEDEVELOPMENT/ Resources/559323-1114798035525/ 1055531-1114798256329/ 10555556- 1114798371392/ Bhatnagar1.ppt

Bhatnagar, S. (2004). *E-government: From vision to implementation*. Thousand Oaks, CA: Sage Publications.

Bingham, C. B., & Eisenhardt, K. M. (2011). Rational heuristics: The 'simple rules' that strategists learn from process experience. *Strategic Management Journal, 32*, 1437–1464. doi:10.1002/smj.965

Birch, H. (2013). *Motivational effects of gamificaiton of piano instruction and practice*. Paper presented at the Dean's Graduate Student Research Conference 2013. New York, NY.

Bishop, J. (2005). The role of mediating artifacts in the design of persuasive e-learning systems. *Proceedings of the First International Conferences on Internet Technologies and Applications*, University of Wales, NEWI, Wrexham, (pp. 548-558).

Bishop, J. (2009c). Increasing membership in online communities: The five principles of managing virtual club economies. *Proceedings of the 3rd International Conference on Internet Technologies and Applications - ITA09*, Glyndwr University, Wrexham.

Bishop, J. (2010). *Multiculturalism in intergenerational contexts: Implications for the design of virtual worlds.* Paper Presented to the Reconstructing Multiculturalism Conference, Cardiff, UK.

Bishop, J. (2011a). *The equatrics of intergenerational knowledge transformation in techno-cultures: Towards a model for enhancing information management in virtual worlds.* (Unpublished MScEcon Thesis). Aberystwyth University, Aberystwyth, UK.

Bishop, J. (2011b). *The role of the prefrontal cortex in social orientation construction: A pilot study.* Paper Presented to the BPS Welsh Conference on Wellbeing. Wrexham.

Bishop, J. (2011b). Transforming lurkers into posters: The role of the participation continuum. *Proceedings of the Fourth International Conference on Internet Technologies and Applications (ITA11),* Glyndwr University.

Bishop, J. (2012). Taming the chatroom bob: The role of brain-computer interfaces that manipulate prefrontal cortex optimization for increasing participation of victims of traumatic sex and other abuse online. In *Proceedings of the 13th International Conference on Bioinformatics and Computational Biology (BIOCOMP'12).* BIOCOMP.

Bishop, J. (2007). Increasing participation in online communities: A framework for human–computer interaction. *Computers in Human Behavior, 23*(4), 1881–1893. doi:10.1016/j.chb.2005.11.004

Bishop, J. (2007a). Ecological cognition: A new dynamic for human-computer interaction. In B. Wallace, A. Ross, J. Davies, & T. Anderson (Eds.), *The mind, the body and the world: Psychology after cognitivism* (pp. 327–345). Exeter, UK: Imprint Academic.

Bishop, J. (2008). Increasing capital revenue in social networking communities: Building social and economic relationships through avatars and characters. In C. Romm-Livermore, & K. Setzekorn (Eds.), *Social networking communities and eDating services: Concepts and implications.* Hershey, PA: IGI Global. doi:10.4018/978-1-60566-104-9.ch004

Bishop, J. (2009a). Enhancing the understanding of genres of web-based communities: The role of the ecological cognition framework. *International Journal of Web Based Communities, 5*(1), 4–17. doi:10.1504/IJWBC.2009.021558

Bishop, J. (2009d). Increasing the economic sustainability of online communities: An empirical investigation. In M. F. Hindsworth, & T. B. Lang (Eds.), *Community participation: Empowerment, diversity and sustainability.* New York, NY: Nova Science Publishers.

Bishop, J. (2011a). All's well that ends well: A comparative analysis of the constitutional and administrative frameworks of cyberspace and the united kingdom. In A. Dudley-Sponaugle, & J. Braman (Eds.), *Investigating cyber law and cyber ethics: Issues, impacts and practices.* Hershey, PA: IGI Global. doi:10.4018/978-1-61350-132-0.ch012

Bishop, J. (2011a). *The equatrics of intergenerational knowledge transformation in techno-cultures: Towards a model for enhancing information management in virtual worlds. Unpublished MScEcon.* Aberystwyth, UK: Aberystwyth University.

Bishop, J. (2013). The effect of deindividuation of the internet troller on criminal procedure implementation: An interview with a hater. *International Journal of Cyber Criminology, 7*(1), 28–48.

Bishop, J. (2013). The psychology of trolling and lurking: The role of defriending and gamification for increasing participation in online communities using seductive narratives. In J. Bishop (Ed.), *Examining the concepts, issues and implications of internet trolling* (pp. 106–123). Hershey, PA: IGI Global. doi:10.4018/978-1-4666-2803-8.ch009

Biswas, A., Licata, J. W., McKee, D., Pullig, C., & Daughtridge, C. (2000). The Recycling Cycle. *Journal of Public Policy & Marketing, 19*(1), 93–105. doi:10.1509/jppm.19.1.93.16950

Bixler, B. (2006). *Games and Motivation: Implications for Instructional Design.* Paper presented at the 2006 NMC Summer Conference.

Björke, K. (2003). Seizing power: Shaders and storytellers. In *Proceedings of ICVS 2003* (Vol. 2897, pp. 3–11). Berlin, Germany: Springer.

Blackshield, A. (2006). Constitutional issues affecting public private partnerships. *UNSWLJ, 29,* 302.

Blaiklock, K. (2011). Curriculum guidelines for early literacy: A comparison of New Zealand and England. *Australian Journal of Early Childhood, 36*(1), 62–68.

Bloom, B. S. (1956). *Taxonomy of Educational Objectives, Handbook I: The Cognitive Domain.* New York: David McKay Co. Inc.

Bloom, B. S., Engelhart, M. D., Furst, E. J., Hill, W. H., & Krathwohl, D. R. (1956). *Taxonomy of educational objectives: The cognitive domain.* New York: Longman.

Bobick, A., & Davis, J. W. (1996). Real-time recognition of activity using temporal templates. In *Proceedings of the 3rd IEEE Workshop on Computer Vision*, Sarasota, FL (pp. 39-42).

Bogost, I. (2005). *The rhetoric of exergaming.* The Georgia Institute of Technology. Retrieved August 11, 2011, from http://www.exergamefitness.com/pdf/The%20Rhetoric%20of%20Exergaming.pdf

Boulay, M., Benveniste, S., Boesplug, S., Jouvelot, P., & Rigaud, A. S. (2009). A pilot usability study of MINWii, a music therapy game for demented patients. [PubMed]. *Technology and Health Care, 19*(4), 233–246. PMID:21849735

Bowker, N., & Tuffin, K. (2002). Disability discourses for online identities. *Disability & Society, 17*(3), 327–344. doi:10.1080/09687590220139883

Boyle-Baise, L., & Goodman, J. (2009). The influence of Harold O. Rugg: Conceptual and pedagogical considerations. *Social Studies, 1*, 31–40. doi:10.3200/TSSS.100.1.31-40

Boyle, E., Connolly, T. M., & Hainey, T. (2011). The role of psychology in understanding the impact of computer games. *Entertainment Computing, 2*(2), 69–74. doi:10.1016/j.entcom.2010.12.002

Brammer, S., & Millington, A. (2004). The development of corporate charitable contributions in the UK: a stakeholder analysis. *Journal of Management Studies, 41*(8), 1411–1434. doi:10.1111/j.1467-6486.2004.00480.x

Bransford, J. D. L. B. A., & Crocking, R. R. (1999). How People Learn: Brain, Mind, Experience, and School. Washington, DC: National Academic Press.

Bransford, J., Brown, A., & Cocking, R. (2000). *How people learn: Brain, mind, and experience & school.* Washington, DC: National Academy Press.

Bredekamp, S. (1993). Reflections on Reggio Emilia. *Young Children, 49*(1), 13–17.

Bristow, D. N., & Mowen, J. C. (1998). The consumer resource exchange model: An empirical investigation of construct and predictive validity. *Marketing Intelligence & Planning, 16*(6), 375–386. doi:10.1108/02634509810237587

Broetz, D., Braun, C., Weber, C., Soekadar, S. R., Caria, A., & Birbaumer, N. (2010). Combination of brain-computer interface training and goal-directed physical therapy in chronic stroke: A case report. *Neurorehabilitation and Neural Repair, 24*(7), 674–679. PubMed doi:10.1177/1545968310368683

Broussard, J. (2012). Making the MMOst of your online class. In *Proceedings of 28th Annual Conference on Distance Education & Learning* (pp. 1–5). IEEE.

Brown, D., Standen, P., Barker, M., Battersby, S., Lewis, J., & Walker, M. (2009, October). Designing serious games using Nintendo™'s Wiimote controller for upper limb stroke rehabilitation. In *Proceedings of the 3rd European Conference on Games Based Learning*, Graz, Austria (pp. 68-77).

Brown, E., & Cairns, P. (2004). *A grounded investigation of game immersion.* Paper presented at the Conference on Human Factors in Computing Systems.

Brown, D. J., McHugh, D., Standen, P., Evett, L., Shopland, N., & Battersby, S. (2010b). Designing location based learning experiences for people with intellectual disabilities and additional sensory impairments. *Computers & Education, 56*(1), 11–20. doi:10.1016/j.compedu.2010.04.014

Brown, D. J., Standen, P., Evett, L., & Battersby, S. (2010a). Designing serious games for people with dual diagnosis: learning disabilities and sensory impairments. In P. Zemliansky, & D. M. Wilcox (Eds.), *Design and implementation of educational games: Theoretical and practical perspectives.* Hershey, PA: IGI Global. doi:10.4018/978-1-61520-781-7.ch027

Brown, J. S., Collins, A., & Duguid, P. (1989). Situated cognition and the culture of learning. *Educational Researcher, 18*(1), 32–42. doi:10.3102/0013189X018001032

Bruhat Bangalore Mahanagara Palike. (2000). Property tax self-assessment scheme: Golden jubilee year 2000. Mahanagara Palike Council Resolution No. 194/99-2000, Bangalore.

Bruhat Bangalore Mahanagara Palike. (2007). *Assessment and calculation of property tax under the capital value system (new SAS), 2007- 2008*. Unpublished.

Bruner, J. (1990). *Acts of meaning*. Cambridge, MA: Harvard University Press.

Bruner, J. S. (1960). *The Process of Education*. Cambridge, MA: Harvard University Press.

Bruner, J. S. (1961). The act of discovery. *Harvard Educational Review*, *31*(1), 21–32.

Brünken, R., Plass, J., & Leutner, D. (2003). Direct measurement of cognitive load in multimedia learning. *Educational Psychologist*, *38*, 53–61. doi:10.1207/S15326985EP3801_7

Buckley, K. E., & Anderson, C. A. (2006). A theoretical model of the effects and consequences of playing video games. In P. Vorderer, & J. Bryant (Eds.), *Playing video games: Motives, responses, and consequences*. Mahwah, NJ: Lawrence Erlbaum Associates.

Burgoon, J. K., Stoner, G., Bonito, J. A., & Dunbar, N. E. (2003). Trust and deception in mediated communication. In *Proceedings of the 36th Annual Hawaii International Conference*. IEEE.

Burke, J. W., McNeill, M., Charles, D., Morrow, P., Crosbie, J., & McDonough, S. (2009, March 23-24). Serious games for upper limb rehabilitation following stroke. In *Proceedings of the Conference in Games and Virtual Worlds for Serious Applications*, Coventry, UK (pp. 103-110).

Burton, S. J. (1980). Breach of contract and the common law duty to perform in good faith. *Harvard Law Review*, *94*, 369. doi:10.2307/1340584

Cagiltay, K. (2001). A design/development model for building electronic performance support systems. In M. Simonson & C. Lamboy (Eds.), *Annual Proceedings of Selected Research and Development [and] Practice Papers presented at the 24th National Convention of the Association for Educational Communications and Technology*, Atlanta, GA (pp. 433-440). Bloomington, IN: Association for Educational Communications and Technology. (ERIC Document Reproduction Service No: ED 470175).

Calvert, S. L., & Jordan, A. B. (2001). Children in the digital age. *Applied Developmental Psychology*, *22*(1), 3–5. doi:10.1016/S0193-3973(00)00062-9

Cambourne, B. (1988). *The whole story natural learning and the acquisition of literacy in the classroom*. London: Ashton Scholastic.

Carroll, J. M. (Ed.). (1998). *Minimalism Beyond the Nurnberg Funnel (Technical Communication, Multimedia and Information Systems)*. Cambridge, MA: MIT Press.

Carson, S., Kanchanaraksa, S., Gooding, I., Mulder, F., & Schuwer, R. (2012). Impact of OpenCourseWare publication on higher education participation and student recruitment. *International Review of Research in Open and Distance Learning*, *13*(4), 19–32.

Carstens, C. B., Huskins, E., & Hounshell, G. W. (1995). Listening to Mozart may not enhance performance on the revised Minnesota paper form board test. *Psychological Reports*, *77*, 111–114. doi:10.2466/pr0.1995.77.1.111 PMID:7501747

Carter, S. M. (2006). The interaction of top management group, stakeholder, and situational factors on certain corporate reputation management activities. *Journal of Management Studies*, *43*(5), 1145–1176. doi:10.1111/j.1467-6486.2006.00632.x

Casamassina, M. (2006). *Wii Sports review*. Retrieved from http://Wii™.ign.com/articles/745/745708p1.html

Casely, J. (2004, March 13). Public sector reform and corruption: CARD facade in andhra pradesh. *Economic and Political Weekly*, 1151–1156.

Cassity, H. D., Henley, T. B., & Markley, R. P. (2007). The Mozart effect: Musical phenomenon or musical preference? A more ecologically valid reconsideration. *Journal of Instructional Psychology*, *34*(1), 13–17.

Castaneda, V., & Navab, N. (2011). Time-of-flight and kinect imaging: Labcourse. Retrieved August 10, 2011, from http://campar.in.tum.de/twiki/pub/Chair/TeachingSs11Kinect/2011-DSensors_LabCourse_Kinect.pdf

Castel, A. D., Pratt, J., & Drummond, E. (2005). The effects of action video game experience on the time course of inhibition of return and the efficiency of visual search. *Acta Psychologica*, *119*, 217–230. PubMed doi:10.1016/j.actpsy.2005.02.004

Cerny, M., & John, M. (2002, June). Game development. Myth vs. method. *Game Developer Magazine*, 32-36.

Chadwick, A., & Howard. (Eds.). (2009). *The handbook of internet politics*. London: Routledge.

Chandler, P., & Sweller, J. (1991). Cognitive load theory and the format of instruction. *Cognition and Instruction*, *8*, 293–332. doi:10.1207/s1532690xci0804_2

Charter, D., & Richards, J. (2009). The European Union wants to block Internet searches for bomb recipes. In *Freedom of expression*. Farmington Hills, MI: Greenhaven Press.

Chen-Wishart, M. (2012). *Contract law*. Oxford, UK: Oxford University Press. doi:10.1093/he/9780199644841.001.0001

Chen, Y., Harper, F. M., Konstan, J., & Li, S. X. (2009). Group identity and social preferences. *The American Economic Review*, *99*(1). doi:10.1257/aer.99.1.431

Cherniack, E. P. (2011). Not just fun and games: Applications of virtual reality in the identification and rehabilitation of cognitive disorders of the elderly. *Disability and Rehabilitation. Assistive Technology, 6*(4), 283–289. PubMed doi:10.3109/17483107.2010.542570

Chin A Paw, M., Jacobs, W., Vaessen, E., Titze, S., & van Mechelen, W. (2008). The motivation of children to play an active video game. *Journal of Science and Medicine in Sport, 11*, 163–166. PubMed doi:10.1016/j.jsams.2007.06.001

Chou, T. J., & Ting, C. C. (2003). The role of flow experience in cyber-game addiction. *Cyberpsychology & Behavior*, *6*(6), 663–675. doi:10.1089/109493103322725469 PMID:14756934

Ciborra, C. (2005). Interpreting e-government and development: Efficiency, transparency or governance at a distance? *Information Technology & People*, *18*(3), 260–279. doi:10.1108/09593840510615879

Ciborra, C., & Navarra. (2005). Good governance, development theory and aid policy: Risks and challenges of e-government in Jordan. *Information Technology for Development*, *11*(2), 141–159. doi:10.1002/itdj.20008

Clarka, R., Bryanta, A., Puab, Y., Bennella, P. M. K., & Hunta, M. (2010). Validity and reliability of the Nintendo™ Wii mote balance board for assessment of standing balance. *Gait & Posture*, *31*(3), 307–310. doi: doi:10.1016/j.gaitpost.2009.11.012

Clark, R. C. (2002). The new ISD: Applying cognitive strategies to instructional design. *Performance Improvement Journal*, *41*(7), 8–14.

Clark, R. E. (1983). Reconsidering research on learning from media. *Review of Educational Research*, *53*(4), 445–459. doi:10.3102/00346543053004445

Clark, R. E. (1994a). Media will never influence learning. *Educational Technology Research and Development*, *42*(2), 21–29. doi:10.1007/BF02299088

Clark, R. E. (1994b). Media and method. *Educational Technology Research and Development*, *42*(3), 7–10. doi:10.1007/BF02298090

Clayton, D. M. (2010). *The presidential campaign of Barack Obama: A critical analysis of a racially transcendent strategy*. San Francisco: Taylor & Francis.

Cobb, T. (1997). Cognitive efficiency: Toward a revised theory of media. *Educational Technology Research and Development*, *45*, 1042–1062. doi:10.1007/BF02299681

Cohen, G. M. (2000). Implied terms and interpretation in contract law. Encyclopedia of Law and Economics, 3, 78-99.

Cohen, D. K., & Ball, D. L. (1990). Relations between policy and practice: A commentary. *Educational Evaluation and Policy Analysis*, *12*(3), 331–338.

Collins, H. (2003). *The law of contract*. London: Butterworths.

Colombo, R., Pisano, F., Mazzone, A., Delconte, C., Micera, S., Chiara Carrozza, M., et al. (2007). Design strategies to improve patient motivation during robot-aided rehabilitation. *Journal of Neuroengineering and Rehabilitation, 4*(3). PubMed

Colt, H. G., Crawford, S. W., & III, O. G. (2001). Virtual Reality Bronchoscopy Simulation*: A Revolution in Procedural Training. *Chest, 120*(4), 1333–1339. PubMed doi:10.1378/chest.120.4.1333

Comaniciu, D., & Meer, P. (1999). Mean shift analysis and applications. In. *Proceedings of the International Conference on Computer Vision, 2*, 1197–1203. doi:10.1109/ICCV.1999.790416

Cornillie, F., Lagatie, R., & Desmet, P. (2012). Language learners' use of automatically generated corrective feedback in a gamified and task-based tutorial CALL program: A pilot study. Paper presented at the EuroCALL 2012: CALL: Using, Learning, Knowing. London, UK.

Cornwell, B., Curry, & Schwirian. (2003). Revisiting norton long's ecology of games: A network approach. *City & Community, 2*(2), 121–142. doi:10.1111/1540-6040.00044

Courchesne, E. (1995, February). *An MRI study of autism: The cerebellum revisited. Journal of Autism and Developmental Disorders, 25*(1), 19–22.PubMed doi:10.1007/BF02178164

Crawford, C. (1982). *The art of computer game design*. Retrieved from http://www.vancouver.wsu.edu/fac/peabody/game-book/Coverpage.html

Creemers, B. P. M. (1994). *The Effective Classroom*. London: Cassell.

Crockett, M. J., Braams, B. R., Clark, L., Tobler, P. N., Robbins, T. W., & Kalenscher, T. (2013). Restricting temptations: Neural mechanisms of precommitment. *Neuron, 79*(2), 391–401. doi:10.1016/j.neuron.2013.05.028 PMID:23889938

Crozier, M., & Friedberg. (1980). *Actors and systems*. Chicago: University of Chicago Press.

Csikszentmihalyi, I., & Csikszentmihalyi, M. (1988). *Optimal experience: Psychological studies of flow in consciousness*. New York, NY: Cambridge University Press. doi:10.1017/CBO9780511621956

Csikszentmihalyi, M. (1975). *Beyond boredom and anxiety*. San Francisco, CA: Jossey-Bass.

Csikszentmihalyi, M. (1990). *Flow: The psychology of optimal experience*. New York, NY: Harper & Row.

Csikszentmihalyi, M., Abuhamdeh, S., & Nakamura, J. (2005). Flow. In C. S. Dweck, & A. J. Elliot (Eds.), *Handbook of competence and motivation* (pp. 598–608). New York, NY: Guilford Publications.

Cui, G. (1997). Marketing strategies in a multi-ethnic environment. *Journal of Marketing Theory and Practice*, 122-134.

D'Orlando, F. (2009). The demand for pornography. *Journal of Happiness Studies*.

Da Silva, C. M., Bermudez, I., Badia, S., Duarte, E., & Verschure, P. F. (2011). Virtual reality based rehabilitation speeds up functional recovery of the upper extremities after stroke: A randomized controlled pilot study in the acute phase of stroke using the rehabilitation gaming system.[Epub ahead of print]. *Restorative Neurology and Neuroscience*.

Dada, D. (2006). The failure of e-government in developing countries: A literature review. *The Electronic Journal on Information Systems in Developing Countries, 26*(7), 1–10.

Danilicheva, P., Klimenko, S., Baturin, Y., & Serebrov, A. (2009). Education in virtual worlds: Virtual storytelling. In *Proceedings of International Conference on CyberWorlds* (pp. 333–338). IEEE.

Darnall, N., Henriques, I., & Sadorsky, P. (2010). Adopting proactive environmental strategy: The influence of stakeholders and firm size. *Journal of Management Studies, 47*(6), 1072–1094. doi:10.1111/j.1467-6486.2009.00873.x

Das, D. A., Grimmer, K. A., Sparnon, A. L., McRae, S. E., & Thomas, B. H. (2005). The efficacy of playing a virtual reality game in modulating pain for children with acute burn injuries: A randomized controlled trial [ISRCTN87413556]. *BMC Pediatrics, 5*(1). PubMed doi:10.1186/1471-2431-5-1

Davis, S. (2008). With a little help from my online friends: The health benefits of internet community participation. *The Journal of Education, Community and Values, 8*(3).

Davis, M. H. (1994). *Empathy: A social psychological approach*. Dubuque, IA: Brown and Benchmark Publishers.

Dawson, C. R., Cragg, A., Taylor, C., & Toombs, B. (2007). Video Games Research to improve understanding of what players enjoy about video games, and to explain their preferences for particular games. London: British Board of Film Classification (BBFC).

de Crook, M. B. M., van Merriënboer, J. J. G., & Paas, F. G. W. C. (1998). High versus low contextual interference in simulation-based training of troubleshooting skills: Effects on transfer performance and invested mental effort. *Computers in Human Behavior*, *14*, 249–267. doi:10.1016/S0747-5632(98)00005-3

De Schutter, B. (2011). *De betekenis van digitale spellen voor een ouder publiek.* (Unpublished doctoral dissertation). Catholic University of Leuven, Leuven, Belgium.

De, R. (2007). Antecedents of corruption and the role of e-government systems in developing countries. In *Proceedings of Ongoing Research*. Retrieved from http://www.iimb.ernet.in/~rahulde/CorruptionPaperEgov07_RDe.pdf

Deci, E. L., & Ryan, R. M. (2004). *Intrinsic motivation inventory.* Retrieved from http://www.psych.rochester.edu/SDT/ measures/IMI_description.php

Deci, E. L., & Ryan, R. M. (1985). *Intrinsic motivation and self-determination in human behavior.* New York, NY: Plenum. doi:10.1007/978-1-4899-2271-7

Deci, E. L., Schwartz, A. J., Sheinman, L., & Ryan, R. M. (1981). An instrument to assess adults' orientations toward control versus autonomy with children: Reflections on intrinsic motivation and perceived competence. *Journal of Educational Psychology*, *73*, 642–650. doi:10.1037/0022-0663.73.5.642

Dede, C. (1996). The evolution of constructivist learning environments: Immersion in distributed, virtual worlds. In B. G. Wilson (Ed.), *Constructivist learning environments: Case studies in instructional design* (pp. 165–175). Englewood Cliffs, NJ: Educational Technology Publications.

DEEWR. (2009). *Belonging, being & becoming: The early years learning framework for Australia.* Retrieved 10 February, 2013, from http://deewr.gov.au/early-years-learning-framework

Dellarocas, C. (2006). Strategic manipulation of internet opinion forums: Implications for consumers and firms. *Management Science*, *52*(10), 1577–1593. doi:10.1287/mnsc.1060.0567

Dempsey, J. V., Lucassen, B. A., Haynes, L. L., & Casey, M. S. (1998). Instructional applications of computer games. In J. J. Hirschbuhl & D. Bishop (Eds.), Computer studies: Computers in education (8th ed., pp. 85-91). Guilford, CT: Dushkin/McGraw Hill.

Dempsey, J. V., Rasmussen, K., & Lucassen, B. (1996). *The instructional gaming literature: Implications and 99 sources* (Tech. Rep. No. 96-1). Mobile, AL: University of South Alabama, College of Education. Retrieved May 30, 2002, from http://www.coe.usouthal.edu/TechReports/TR96_1.PDF

Denis, G., & Jouvelot, P. (2005). *Motivation-Driven Educational Game Design: Applying Best Practice to Music Education.* Paper presented at the Advances in Computer Entertainment (ACE) 2005.

Denzin, N. K., & Lincoln, Y. S. (2000). The discipline and practice of qualitative research. In N. K. Denzin, & Y. S. Lincoln (Eds.), *Handbook of Qualitative Research* (2nd ed.). Thousand Oaks, CA: Sage.

Department for Education. (2012). *Statutory framework for the early years foundation stage.* Cheshire, UK: Department for Education.

Desiato, C. (2009). *The conditions of permeability: How shared cyberworlds turn into laboratories of possible worlds.* Paper presented at the International Conference on CyberWorlds. New York, NY.

Deterding, S., Dixon, D., Khaled, R., & Nacke, L. (2011). From game design elements to gamefulness: Defining gamification. In *Proceedings of the 15th International Academic MindTrek Conference: Envisioning Future Media Environments* (pp. 9–15). Tampere, Finland: ACM. doi:10.1145/2181037.2181040

Deterding, S., Sicart, M., Nacke, L., O'Hara, K., & Dixon, D. (2011). Gamification: Using game-design elements in non-gaming contexts. In *Proceedings of the 2011 Annual Conference on Human Factors in Computing Systems.* New York, NY: IEEE.

Deterding, S. (2012). Gamification: Designing for motivation. *Interaction*, *19*(4), 14–17. doi:10.1145/2212877.2212883

Deutsch, J. E., Borbely, M., Filler, J., Huhn, K., & Guarrera-Bowlby. (2008). Use of a low-cost, commercially available gaming console (Wii) for rehabilitation of an adolescent with cerebral palsy. *Physical Therapy*, *88*(10), 1196–1207. PubMed doi:10.2522/ptj.20080062

Dewey, J. (1909). *Moral Principles in Education.* New York: Houghton Mifflin.

Dewey, J. (1916). *Democracy and Education*. New York: Macmillan.

Dewey, J. (1956). *The child and the curriculum, and the school and society*. Chicago: University of Chicago Press.

Diamond, M., & Hopson, J. (1998). *Magic trees of the mind*. New York: Penguin.

Dick, W. (1987). A history of instructional design and its impact on educational psychology. In J. Glover, & R. Ronning (Eds.), *Historical foundations of educational psychology*. New York: Plenum. doi:10.1007/978-1-4899-3620-2_10

Dietz W.H., Bandini L.G., Morelli J.A., Peers K.F., Ching P.L. Effect of sedentary activities on resting metabolic rate. *American Journal of Clinical Nutition, 59*, 556–559.

Difede, J., & Hoffman, H. G. (2002). Virtual reality exposure therapy for World Trade Center post-traumatic stress disorder: A case report. *Cyberpsychology & Behavior, 5*(6), 529–535. PubMed doi:10.1089/109493102321018169

Dillinger, W. (1988). *Urban property taxation in developing countries*. Retrieved from http://ideas.repec.org/p/wbk/wbrwps/41.html

Doh, J. P., & Guay, T. R. (2006). Corporate social responsibility, public policy, and NGO activism in Europe and the United States: An institutional-stakeholder perspective. *Journal of Management Studies, 43*(1), 47–73. doi:10.1111/j.1467-6486.2006.00582.x

Domínguez, A., Saenz-de-Navarrete, J., De-Marcos, L., Fernández-Sanz, L., Pagés, C., & Martínez-Herráiz, J. (2013). Gamifying learning experiences: Practical implications and outcomes. *Computers & Education, 63*, 380–392. doi:10.1016/j.compedu.2012.12.020

Donaldson, T. (1985). Multinational decision-making: Reconciling international norms. *Journal of Business Ethics, 4*(4), 357–366. doi:10.1007/BF00381779

Dorling, A., & McCaffery, F. (2012). The gamification of SPICE. In *Software Process Improvement and Capability Determination* (pp. 295–301). Academic Press. doi:10.1007/978-3-642-30439-2_35

Downes-Le Guin, T., Baker, R., Mechling, J., & Ruylea, E. (2012). Myths and realities of respondent engagement in online surveys. *International Journal of Market Research, 54*(5), 1–21. doi:10.2501/IJMR-54-5-613-633

Downs, A. (1964). *Inside bureaucracy*. Boston: Little Brown.

Doyle, T. (2004). Should web sites for bomb-making be legal? *Journal of Information Ethics, 13*(1), 34–37. doi:10.3172/JIE.13.1.34

Dubbink, W. (2004). The fragile structure of free-market society: The radical implications for corporate social responsibility. *Business Ethics Quarterly, 14*(1), 23–46. doi:10.5840/beq20041412

Dunleavy, P. Margetts, Bastow, & Tinkler. (2006). Digital era governance: IT corporations, the state and e-government. Oxford, UK: Oxford University Press.

Durkin, K., & Barber, B. (2002). Not so doomed: Computer game play and positive adolescent development. *Applied Developmental Psychology, 23*(4), 373–392. doi:10.1016/S0193-3973(02)00124-7

Dutton, W. H. Schneider, & Vedel. (2011). Large technical systems as ecologies of games: Cases from telecommunications to the internet. In J. Bauer, A. Lang, & V. Schneider (Eds.), Innovation policy and governance in high-tech industries: The complexity of coordination. Berlin: Springer.

Dutton, W. H. (1992). The ecology of games shaping telecommunications policy. *Communication Theory, 2*(4), 303–324. doi:10.1111/j.1468-2885.1992.tb00046.x

Dutton, W. H. (1999). *Society on the line: Information politics in the digital age*. Oxford, UK: Oxford University Press.

Dweck, C. S. (1986). Motivational processes affecting learning. *The American Psychologist, 41*, 1040–1048. doi:10.1037/0003-066X.41.10.1040

Economist. (2007, October 13). International: The tongue twisters, civil liberties: Freedom of speech. *The Economist*.

Economist. (2010, December 2). Wikileaks unpluggable: How WikiLeaks embarrassed and enraged America, gripped the public and rewrote the rules of diplomacy. *The Economist*.

Edwards, C., Gandini, L., & Forman, G. (1995). Introduction. In C. Edwards, L. Gandini, & G. Forman (Eds.), *The hundred languages of children: The Reggio Emilia approach to early childhood education* (pp. 3–18). Norwood, NJ: Ablex Publishing Corporation.

Edwards, S., & Cutter-Mackenzie, A. (2011). Environmentalising early childhood education curriculum through pedagogies of play. *Australian Journal of Early Childhood, 36*(1), 51–59.

Efimova, L. (2009). Weblog as a personal thinking space. *Proceedings of the 20th ACM Conference on Hypertext and Hypermedia,* (pp. 289-298).

Eliot, J., & Smith, I. M. (1983). *An international directory of spatial tests*. Windsor, UK: NFER-Nelson.

Emanuel, S. (2010). *Emanuel law outlines: Contracts*. Aspen Law & Business.

Entertainment Software Association. (2006). *Essential Facts about the computer and video game industry.*

Erenli, K. (2012). *The impact of gamification: A recommendation of scenarios for education.* Paper presented at the 15th International Conference on Interactive Collaborative Learning. Villach, Austria.

Espeland, P. (2003). *Life lists for teens: Tips, steps, hints, and how-tos for growing up, getting along, learning, and having fun.* Minneapolis, MN: Free Spirit Publishing.

Esposito, J. J. (2010). Creating a consolidated online catalogue for the university press community. *Journal of Scholarly Publishing, 41*(4), 385–427. doi:10.3138/jsp.41.4.385

Evett, L., Battersby, S., Ridley, A., & Brown, D. (2009). An interface to virtual environments for people who are blind using Wii technology – mental models and navigation. *Journal of Assistive Technologies, 3*, 26–34. doi:10.1108/17549450200900013

Fardinpour, A., Reiners, T., & Dreher, H. (2013). Action-based learning assessment: Evaluating goal-oriented actions in virtual training environments. In *Proceedings of ASCILITE 2013*. ASCILITE.

Faria, A. J. (1998). Business Simulation Games: Current Usage Levels - An Update. *Simulation & Gaming, 29*(3), 295–308. doi:10.1177/1046878198293002

Farné, R. (2005). Pedagogy of play. *Topoi, 24*(2), 169–181. doi:10.1007/s11245-005-5053-5

Fathi, N. (2010, February 11). Iran disrupts Internet communications. *The New York Times.*

Federation of American Scientists. (2006). Harnessing the power of video games for learning [report]. *Summit on Educational Games*. Federation of American Scientists, Washington, DC. Retrieved from http://fas.org/gamesummit/Resources/Summit%20on%20Educational%20Games.pdf

Fehr, E., & Gachter. (1998). Reciprocity and economics: The economic implications of homo reciprocans. *European Economic Review, 42*(3), 845–859. doi:10.1016/S0014-2921(97)00131-1

Fehr, E., & Gachter. (2002). Altruistic punishment in humans. *Nature, 415*, 137–145. doi:10.1038/415137a PMID:11805825

Feng, J., Spence, I., & Pratt, J. (2007). Playing an action video game reduces gender differences in spatial cognition. *Psychological Science, 18*(10), 850–855. PubMed doi:10.1111/j.1467-9280.2007.01990.x

Ferguson, C. J. (2007). The good, the bad and the ugly: A meta-analytic review of positive and negative effects of violent video games. *The Psychiatric Quarterly, 78*, 309–316. PubMed doi:10.1007/s11126-007-9056-9

Ferguson, C. J. (2010). Blazing Angels or Resident Evil? Can violent video games be a force for good? *Review of General Psychology, 14*(2), 68–81. doi:10.1037/a0018941

Ferguson, C. J., Cruz, A., & Rueda, S. (2008). Gender, video game playing habits and visual memory tasks. *Sex Roles, 58*, 279–286. doi:10.1007/s11199-007-9332-z

Ferguson, C. J., & Rueda, S. M. (2010). Violent video game exposure effects on aggressive behavior, hostile feelings, and depression. *European Psychologist, 15*(2), 99–108. doi:10.1027/1016-9040/a000010

Fernández-Calvo, B., Rodríguez-Pérez, R., Contador, I., Rubio-Santorum, A., & Ramos, F. (2011). Eficacia del entrenamiento cognitivo basado en nuevas tecnologías en pacientes con demencia tipo Alzheimer.[PubMed]. *Psicothema, 23*(1), 44–50. PMID:21266141

Festinger, L. (1957). *A theory of cognitive dissonance.* Evanston, IL: Row, Peterson.

Feys, H. M., De Weerdt, W. J., Selz, B. E., Cox Steck, G. A., Spichiger, R., & Vereeck, L. E. et al. (1998). Effect of a therapeutic intervention for the hemiplegic upper limb in the acute phase after stroke: A single-blind, randomized, controlled multicenter trial. *Stroke, 29*, 785–792. doi: doi:10.1161/01.STR.29.4.785

Fineman, S., & Clarke, K. (1996). Green stakeholders: Industry interpretations and response. *Journal of Management Studies, 33*(6), 716–730. doi:10.1111/j.1467-6486.1996.tb00169.x

Finestone, H. M., & Greene-Finestone, L. S. (2002). The role of nutrition and diet in stroke rehabilitation. *Topics in Stroke Rehabilitation, 6*, 46–66.

Fink, C., & Kenny. (2003). W(h)ither the digital divide? *Info: The Journal of Policy. Regulation and Strategy for Telecommunications, 5*(6), 15–24. doi:10.1108/14636690310507180

Firestone, W. A. (1989). Educational policy as an ecology of games. *Educational Researcher, 18*(7), 18–24.

Fisch, S. M. (2005). *Making Educational Computer Games "Educational"*. Paper presented at the 4th International Conference on Interaction Design and Children (IDC2005).

Fit, V. I. (2010). *VI Fit, exergames for users who are visually impaired or blind.* Retrieved from http://vifit.org

Fleer, M. (2008). A cultural-historical perspective on play: Play as a leading activity across cultural communities. In I. Pramling-Samuelsson (Ed.), *Play and learning in early childhood settings*. Dordrecht, The Netherlands: Springer. doi:10.1007/978-1-4020-8498-0_1

Forrester Research, Inc. (2004). *Accessible technology in computing: Examining awareness, use, and future potential* (pp. 22–41). Cambridge, MA: Forrester Research, Inc.

Foulds, L. R. (1983). The heuristic problem-solving approach. *The Journal of the Operational Research Society, 34*(10), 927–934.

Freud, S. (1933). *New introductory lectures on psychoanalysis*. New York, NY: W.W. Norton & Company, Inc.

Friedman, M. (2002). Capitalism and freedom (40th Anniversary Ed.). Chicago: The University of Chicago Press.

Friedman, A. L., & Miles, S. (2002). Developing stakeholder theory. *Journal of Management Studies, 39*(1), 1–21. doi:10.1111/1467-6486.00280

Friedman, T. (2005). *The world is flat*. New York: Farrar, Straus and Giroux.

Fujimoto, R. (2012). *Games and failure*. Retrieved from http://shoyulearning.wordpress.com/2012/05/20/games-and-failure/

Funk, J. B., Hagan, J., & Schimming, J. (1999). Children and electronic games: A comparison of parents' and children's perceptions of children's habits and preferences in a United States Sample. *Psychological Reports, 85*(3), 883–888. PubMed doi:10.2466/PR0.85.7.883-888

Gagne, R. M. (1970). *The Conditions of Learning and Theory of Instruction* (2nd ed.). New York: Holt, Rinehart & Winston.

Gagné, R. M., & Driscoll, D. (1988). *Essentials of learning for instruction*. Englewood Cliffs, NJ: Prentice-Hall.

Gajadhar, B. J., Nap, H. H., de Kort, Y. A. W., & IJsselsteijn, W. A. (2010). Out of sight, out of mind: Co-player effects on seniors' player experience. In *Proceedings of Fun and Games '10 the 3rd International Conference on Fun and Games*, (pp. 74-83).

Gajadhar, B. J., de Kort, Y. A. W., & IJsselsteijn, W. A. (2008). Shared fun is doubled fun: Player enjoyment as a function of social setting. In P. Markopoulos, B. de Ruyter, W. IJsselsteijn, & D. Rowland (Eds.), *Fun and games* (pp. 106–117). New York, NY: Springer. doi:10.1007/978-3-540-88322-7_11

Gajadhar, B. J., de Kort, Y. A. W., & IJsselsteijn, W. A. (2009). Rules of engagement: Influence of social setting on player involvement in digital games. *International Journal of Gaming and Computer-Mediated Simulations, 1*(3), 14–27. doi:10.4018/jgcms.2009070102

Galt, V. (2007, February 21). Telus hangs up on mobile porn service. *The Globe and Mail*, p. A1.

Games, G. M. A. (2008). *Lone wolf for Windows – A submarine simulation, GMA games*. Retrieved from http://www.gmagames.com/lonewolf.shtml

Gandini, L. (1993). Fundamentals of the Reggio Emilia approach to early childhood education. *Young Children, 49*(1), 4–8.

Gartner. (2011). *Gartner predicts over 70 percent of global 2000 organisations will have at least one gamified application by 2014.* Retrieved 12 February, 2012, from http://www.gartner.com/newsroom/id/1844115

Gee, J. P. (2005). *Game-like learning: An example of situated learning and implications for the opportunity to learn* [Electronic Version]. Retrieved September 6, 2006 from http://www.academiccolab.org/resources/documents/Game-Like%20Learning.rev.pdf

Gee, J. P. (2003). *What video games have to teach us about learning and literacy.* New York: Palgrave Macmillan. doi:10.1145/950566.950595

Generation Fit. (2007). [Video File]. Video posted to http://www.generation-fit.com

Gergen, M. P. (2011). *Negligent misrepresentation as contract.* Berkeley, CA: University of California.

Gerjets, P., Scheiter, K., & Catrambone, R. (2004). Designing instructional examples to reduce intrinsic cognitive load: Molar versus modular presentation of solution procedures. *Instructional Science, 32*, 33–58. doi:10.1023/B:TRUC.0000021809.10236.71

Gibson, C., & Levine, P. (2003). *The Civic Mission of Schools.* New York: Carnegie Corporation.

Gibson, J. J. (1986). *The ecological approach to visual perception.* Lawrence Erlbaum Associates.

Giere, R. (1991). *Understanding Scientific Reasoning* (3rd ed.). Fort Worth, TX: Holt, Rinehardt and Winston.

Gintis, H. (2006). *The economy as a complex adaptive system - A review of Eric D. Beinhocker, the origins of wealth: Evolution, complexity, and the radical remaking of economics.* MacArthur Research Foundation. Retrieved from http://www.umass.edu/preferen/Class%20Material/Readings%20in %20Market%20Dynamics/ Complexity%20Economics.pdf

Gintis, H. (2000). *Game theory evolving.* Princeton, NJ: Princeton University Press.

Girot, C. (2001). *User protection in IT contracts: A comparative study of the protection of the user against defective performance in information technology.* Boston: Kluwer Law International.

Glaser, R. (1971). The design of instruction. In M. D. Merrill (Ed.), *Instructional design: Readings* (pp. 18–37). Englewood, NJ: Prentice-Hall.

Glover, D., & Miller, D. (2001). Running with technology: The pedagogic impact of the large-scale introduction of interactive whiteboards in one secondary school. *Journal of Information Technology for Teacher Education, 10*(3), 257–278. doi:10.1080/14759390100200115

Goffin, S., & Wilson, C. (2001). *Curriculum models and early childhood education: Appraising the relationship.* Upper Saddle River, NJ: Prentice-Hall Inc.

Goldberg, E. (2005). *The wisdom paradox: How your mind can grow stronger as your brain grows older.* New York, NY: Penguin Group.

Goode, M. M. H., & Moutinho, L. A. C., C. (1996). Structural equation modelling of overall satisfaction and full use of services for ATMs. *International Journal of Bank Marketing, 14*(7), 4–11. doi:10.1108/02652329610151331

Gopal Jaya, N. (2006). Introduction. In N. Gopal Jaya, A. Prakash, & P. K. Sharma (Eds.), *Local governance in India: Decentralisation and beyond.* New Delhi: Oxford University Press.

Gordon, J., & Zemke, R. (2000). The attack on ISD. *Training Magazine, 37*(4), 42–53.

Gorgan, C., & Brett, J. (2006). *Google and the government of China: A case study of cross-cultural negotiations.* Kellogg School of Management, Northwestern University.

Gorriz, C. M., & Medina, C. (2000). Engaging girls with computers through software games. *Communications of the ACM, 43*(1), 42–49. doi:10.1145/323830.323843

Grammenos, D., & Savidis, A. (2006). *Unified design of universally accessible games (say what?).* Retrieved from http://www.gamasutra.com/features/20061207/grammenos_01.shtml

Grammenos, D., Savidis, A., & Stephanidis, C. (2005, July). UA-Chess: A universally accessible board game. In *Proceedings of the 3rd International Conference on Universal Access in Human-Computer Interaction*, Las Vegas, NV.

Graves, L. E., Ridgers, N. D., Williams, K., Stratton, G., Atkinson, G., & Cable, N. T. (2010). The physiological cost and enjoyment of Wii Fit in adolescents, young adults, and older adults.[PubMed]. *Journal of Physical Activity & Health*, *7*(3), 393–401. PMID:20551497

Gray, J. R., Braver, T. S., & Raichle, M. E. (2002, March 19). Integration of emotion and cognition in the lateral prefrontal cortex. *Proceedings of the National Academy of Sciences of the United States of America, 99*(6), 4115–4020. PubMed doi:10.1073/pnas.062381899

Gredler, M. (1994). *Designing and evaluating games and simulations: A process approach.* Houston, TX: Gulf Publishing Company.

Gredler, M. E. (1996). Educational games and simulations: A technology in search of a (research) paradigm. In D. H. Jonassen (Ed.), *Handbook of research for educational communications and technology* (pp. 521–539). New York: Macmillan.

Green, C. S., & Bavelier, D. (2003). Action video game modifies visual selective attention. *Nature, 423*(6939), 534–538. PubMed doi:10.1038/nature01647

Green, C. S., & Bavelier, D. (2006). The cognitive neuroscience of video games. In P. Messaris, & L. Humphreys (Eds.), *Digital media: Transformations in human communication* (pp. 211–223). New York, NY: Peter Lang.

Greenfield, P. M., deWinstanley, P., Kilpatrick, H., & Kaye, D. (1994). Action video games and informal education: Effects on strategies for dividing visual attention[Abstract]. *Journal of Applied Developmental Psychology, 15*(1), 105–123. doi:10.1016/0193-3973(94)90008-6

Greenley, G. E., & Foxall, G. R. (1997). Multiple stakeholder orientation in UK companies and the implications for company performance. *Journal of Management Studies, 34*(2), 259–284. doi:10.1111/1467-6486.00051

Gregory, S., Lee, M. J. W., Dalgarno, B., & Tynan, B. (2013). *Virtual worlds in online and distance education.* Athabasca, Canada: Athabasca University Press.

Gregory, S., Reiners, T., & Tynan, B. (2010). Alternative realities: Immersive learning for and with students. In H. Song (Ed.), *Distance learning technology, current instruction, and the future of education: Applications of today, practices of tomorrow* (pp. 245–272). Hershey, PA: IGI Global.

Greitemeyer, T., & Osswald, S. (2009). Prosocial video games reduce aggressive cognitions. *Journal of Experimental Social Psychology*, *45*(4), 896–900. doi:10.1016/j.jesp.2009.04.005

Grieshaber, S. (2008). Interrupting stereotypes: Teaching and the education of young children. *Early Childhood Education and Development*, *19*(3), 505–518. doi:10.1080/10409280802068670

Griffiths, M. (2002). The educational benefits of videogames. *Education for Health, 20*(3), 47–51.

Grodal, T. (2000). Video games and the pleasures of control. In D. Zillmann, & P. Vorderer (Eds.), *Media entertainment: The psychology of its appeal.* Mahwah, NJ: Lawrence Erlbaum Associates.

Grolnick, W. S., & Ryan, R. M. (1987). Autonomy in children's learning: An experimental and individual difference investigation. *Journal of Personality and Social Psychology, 52*, 890–898. doi:10.1037/0022-3514.52.5.890 PMID:3585701

Gros, B., Elen, J., Kerres, M., Merriënboer, J., & Spector, M. (1997). Instructional design and the authoring of multimedia and hypermedia systems: Does a marriage make sense? *Educational Technology*, *37*(1), 48–56.

Guardian. (2011). *PopCap games survey claims sharp rise for social gaming in US and UK.* Retrieved 10 February, 2013, from http://www.guardian.co.uk/technology/appsblog/2011/nov/14/popcap-social-games-survey

Guillaumin, J. (1987). Entre blessure et cicatrice: Le destin du négatif dans la psychanalyse. Ed.s Champ Vallon.

Gully, A. (2012). It's only a flaming game: A case study of arabic computer-mediated communication. *British Journal of Middle Eastern Studies, 39*(1), 1–18. doi:10.1080/13530194.2012.659440

Gunter, G. A., Kenny, R. F., & Vick, E. H. (2008). Taking educational games seriously: Using the RETAIN model to design endogenous fantasy into standalone educational games. *Educational Technology Research and Development, 56*, 511–537. doi:10.1007/s11423-007-9073-2

Gustafson, K. L., & Branch, R. M. (1997). *Survey of instructional development models* (3rd ed.). Syracuse, NY: ERIC Clearinghouse on Information Resources.

Gustafson, K. L., & Branch, R. M. (1998). Re-visioning models of instructional development. *Educational Technology Research and Development, 45*(3), 73–89. doi:10.1007/BF02299731

Habermas, J. (1984). *The theory of communicative action.* Boston: Beacon Press.

Hacker, K. L., & van Dijk. (2000). Introduction: What is digital democracy?. In K. L. Hacker & J. van Dijk (Eds.), *Digital democracy: Issues of theory and practice.* London: Sage Publications.

Hall, P. A. & Taylor. (1996). Political science and the three new institutionalisms. *MPIFG Discussion Paper 96/9.*

Hamari, J., & Koivisto, J. (2013). Social motivations to use gamification: An empirical study of gamifying exercise. In *Proceedings of the 21st European Conference on Information Systems.* IEEE.

Hamilton, J. B., Knousse, S. B., & Hill, V. (2009). Google in China: A manager friendly heuristic-model for resolving cross cultural ethical conflicts. *Journal of Business Ethics, 86*, 143–157. doi:10.1007/s10551-008-9840-y

Hargreaves, I. (2005). The ethics of journalism: A summing-up for lord hutton. *Ethics, Law, and Society, 1*, 153.

Harp, S. F., & Mayer, R. E. (1998). How seductive details do their damage: A theory of cognitive interest in science learning. *Journal of Educational Psychology, 90*(3), 414–434. doi:10.1037/0022-0663.90.3.414

Harris, L. C., & Goode, M. M. (2004). The four levels of loyalty and the pivotal role of trust: A study of online service dynamics. *Journal of Retailing, 80*(2), 139–158. doi:10.1016/j.jretai.2004.04.002

Harris, L. C., & Goode, M. M. H. (2010). Online servicescapes, trust, and purchase intentions. *Journal of Services Marketing, 24*(3), 230–243. doi:10.1108/08876041011040631

Harshaw, T. (2010). The hunt for Julian Assange. *Opinionator, NYTBlogs.* Retrieved from http://opinionator.blogs.nytimes.com/2010/12/03/the-hunt-for-julian-assange/

Hart, S. G., & Staveland, L. E. (1988). Development of NASA-TLX (Task Load Index): Results of experimental and theoretical research. In P. A. Hancock, & N. Meshkati (Eds.), *Human mental workload* (pp. 139–183). Amsterdam, The Netherlands: North Holland. doi:10.1016/S0166-4115(08)62386-9

Hattie, J. A. (2003). Teachers make a difference: What is the research evidence? 2003 Australian Council for Educational Research Conference. Melbourne.

Hattie, J. A. (2005). *What is the nature of evidence that makes a difference to learning?* Paper presented at the Research Conference 2005 VIC: Australian Council for Educational Research.

Hebb, D. O. (1955). Drives and the C.N.S. (conceptual nervous system). *Psychological Review, 62*, 243–254. doi:10.1037/h0041823 PMID:14395368

Hechter, M., & Kanazawa. (1997). Sociological rational choice theory. *Annual Review of Sociology, 23*(1), 191–214. doi:10.1146/annurev.soc.23.1.191

Heeks, R. (2003). Most eGovernment-for-development projects fail: How can the risks be reduced? *i-Government Working Paper Series,* Paper No. 14, IDPM.

Heeks, R. (2005). eGovernment as a carrier of context. *Journal of Public Policy, 25*(1), 51–74. doi:10.1017/S0143814X05000206

Heeks, R. (2006). *Implementing and managing eGovernment – An international text.* New Delhi: Vistar Publications.

Heinich, R., Molenda, M., Russell, J. D., & Smaldino, S. E. (2002). *Instructional media and technologies for learning* (7th ed.). Upper Saddle River, NJ: Merrill Prentice Hall.

Heish, N. (2004). The obligations of transnational corporations: Rawlsian justice and duty of assistance. *Business Ethics Quarterly, 14*, 643–661.

Heron, S. (2009). Online privacy and browser security. *Network Security,* (6): 4–7. doi:10.1016/S1353-4858(09)70061-3

Hewett, V. (2001). Examining the Reggio Emilia approach to early childhood education. *Early Childhood Education Journal, 29*(2), 95–100. doi:10.1023/A:1012520828095

Higginbotham, A. (2010). *Dragging accessible games into the 21st century, BBC – Ouch! (disability).* Retrieved from http://www.bbc.co.uk/ouch/features/dragging_accessible_computer_games_into_.shtml

Hill, C. W. L., & Jones, T. M. (1992). Stakeholder-agency theory. *Journal of Management Studies, 29*(2), 131–154. doi:10.1111/j.1467-6486.1992.tb00657.x

Hoffman, M. (2000). *Empathy & moral development: Implications for caring and justice.* New York: Cambridge University Press. doi:10.1017/CBO9780511805851

Hoffman, S. (1997). Elaboration theory and hypermedia: Is there a link? *Educational Technology, 37*(1), 57–64.

Hogan, R. (1969). Development of an empathy scale. *Journal of Consulting and Clinical Psychology, 33,* 307–316. PubMed doi:10.1037/h0027580

Hokkanen, L., Kataja, E., Kaukomies, O., Linturi, A., Pylkkö, I., & Salohalla, L. (2011). The future of gaming: technology, behaviour & gamification. In A. G. Paterson (Ed.), *They are everywhere* (pp. 46–82). Retrieved from http://booki.cc/they-are-everywhere/

Holmes, E. A., James, E. L., Coode-Bate, T., & Deeprose, C. (2011). Can playing the computer game "Tetris" reduce the build-up of flashbacks for trauma? A proposal from cognitive science.[PubMed]. *PLoS ONE, 4*(1), e1453–e1458. PMID:19127289

Homma, K., & Takenaka, E. (1985). An image processing method for feature extraction of space-occupying lesions. *Journal of Nuclear Medicine, 26*(12), 1472–1477.

Honey, P., & Mumford, A. (1982). *Manual of Learning Styles.* London: P. Honey.

Horton, S. (2006). *Access by design: A guide to universal usability for web designers.* Berkeley, CA: New Riders.

Howard, T. W. (2010). *Design to thrive: Creating social networks and online communities that last.* Morgan Kaufmann.

Hsu, S. H., Chang, J., & Lee, C. (2013). *Designing attractive gamification features for collaborative storytelling websites.* Cyberpsychology, Behavior, and Social Networking. doi:10.1089/cyber.2012.0492

Huang, W. D., & Johnson, J. (2009). Let's get serious about E-games: A design research approach towards emergence perspective. In B. Cope, & M. Kalantzis (Eds.), *Ubiquitous learning.* Champaign, IL: University of Illinois Press.

Huang, W. D., & Johnson, T. (2008). Instructional game design using Cognitive Load Theory. In R. Ferdig (Ed.), *Handbook of research on effective electronic gaming in education.* Hershey, PA: Information Science Reference. doi:10.4018/978-1-59904-808-6.ch066

Huber, M., Rabin, B., Docan, C., Burdea, G. C., Abdelbaky, M., & Golomb, M. R. (2010). *IEEE Transactions on Information Technology in Biomedicine, 14*(2), 526–534. PubMed doi:10.1109/TITB.2009.2038995

Huk, T., Bieger, S., Ohrmann, S., & Weigel, B. (2004). Computer animations in science education: Is background music beneficial or detrimental? In *Proceedings of the World Conference on Educational Multimedia, Hypermedia and Telecommunications* (pp. 4227-4234).

Hull, C. L. (1951). *Essentials of behavior.* New Haven, CT: Yale University Press.

Hu, M. K. (1962). Visual pattern recognition by moment invariants. *I.R.E. Transactions on Information Theory, 8,* 179–187. doi: doi:10.1109/TIT.1962.1057692

Hunt, D., Atkin, D., & Krishnan, A. (2012). The influence of computer-mediated communication apprehension on motives for facebook use. *Journal of Broadcasting & Electronic Media, 56*(2), 187–202. doi:10.1080/08838151.2012.678717

Hurkmans, H. L., Ribbers, G. M., Streur-Kranenburg, M. F., Stam, H. J., & Van den Berg-Emons, R. J. (2011). Energy expenditure in chronic stroke patients playing Wii Sports: A pilot study. *Journal of Neuroengineering and Rehabilitation, 8*(1), 38. PubMed doi:10.1186/1743-0003-8-38

Hurley, P. J. (2000). *A concise introduction to logic.* Belmont, CA: Wadsworth.

IJsselsteijn, W. A., Nap, H. H., de Kort, Y. A. W., & Poels, K. (2007). Digital game design for elderly users. In B. Kapralos, & M. Katchabaw (Eds.), *Proceedings of Future Play '07 the 2007 conference on Future Play* (pp. 17–22). doi:10.1145/1328202.1328206

Imagawa, K., Lu, S., & Igi, S. (1998, April 14-16). Color-based hands tracking system for sign language recognition. In *Proceedings of the 3rd International Conference on Face & Gesture Recognition*, Nara, Japan (p. 462).

Immordino-Yang, M. H., & Damasio, A. (2008). *We feel, therefore we learn: The relevance of affective and social neuroscience to education. In the brain and learning. The Jossey-Bass reader.* San Francisco: Jossey-Bass.

Institute for Creative Technologies. (2009). *Full spectrum warrior*. Retrieved from http://ict.usc.edu/projects/full_spectrum_warrior

Internet Watch Foundation (IWF). (2008). *IWF statement regarding Wikipedia webpage*. Retrieved from http://www.iwf.org.uk/about-iwf/news/post/251-iwf-statement-regarding-wikipedia-webpage

Isaac-Henry, K. (1997). Development and change in the public sector. In K. Isaac-Henry, C. Painter, & C. Barnes (Eds.), *Management in the public sector: Challenge and change*. London: International Thomson Business Press.

ISTE. (2008). *National educational technology standards for teachers*. Retrieved 12 February, 2013, from http://www.iste.org/standards/nets-for-teachers/nets-for-teachers-2008

Jacobs, J. W., & Dempsey, J. V. (1993). Simulation and gaming: Fidelity, feedback and motivation. In J. V. Dempsey, & G. C. Sales (Eds.), *Interactive instruction and feedback* (pp. 197–228). Engelwood Cliffs, NJ: Educational Technology Publications.

Jamal, A., & Goode, M. M. H. (2001). Consumers and brands: A study of the impact of self-image congruence on brand preference and satisfaction. *Marketing Intelligence & Planning*, *19*(7), 482–492. doi:10.1108/02634500110408286

Jansz, J., & Martens, L. (2005). Gaming at a LAN event: The social context of playing video games. *New Media & Society*, *7*(3), 333–355. doi:10.1177/1461444805052280

Jenkins, H. (2006). *Confronting the challenges of participatory culture: Media education for the 21st Century*. Chicago, IL: John D. and Catherine A. MacArthur Foundation.

Jennings, S., & Gersie, A. (1987). *Drama therapy with disturbed adolescents* (pp. 162–182).

Jha, S.N., & Mathur. (1999). *Decentralization and local politics*. New Delhi: Sage Publications.

Jha, G. (1983). Area basis of valuation of property tax: An evaluation. In A. Datta (Ed.), *Property taxation in India*. New Delhi: Centre for Urban Indian Studies – The Indian Institute of Public Administration.

Jick, T. D. (1979). Mixing qualitative and quantitative methods: Triangulation in action. *Administrative Science Quarterly*, *24*(4), 602–611. doi:10.2307/2392366

Johnson, R. N., & Libecap. (1994). *The federal civil service system and the problem of bureaucracy*. Chicago: University of Chicago Press.

Johnson, J., Christie, H., & Wardle, F. (2005). *Play, development, and early education*. Boston: Pearson Education, Inc.

Johnson-Laird, P. N., & Oatley, K. (2000). Cognitive & social construction in emotions. In M. Lewis, & J. M. Haviland-Jones (Eds.), *Handbook of emotions* (2nd ed.). New York: The Guildford Press.

Jonassen, D. H. (1990). Thinking technology: Chaos in instructional design. *Educational Technology*, *30*(2), 32–34.

Jonassen, D. H., Hennon, R. J., Ondrusek, A., Samouilova, M., Spaulding, K. L., & Yueh, H. P. et al. (1997). Certainty, determinism, and predictability in theories of instructional design: Lessons from science. *Educational Technology*, *37*(1), 27–33.

Jones, M. A. (1995). Liability for psychiatric Illness—More principle, less subtlety? *Web Journal of Current Legal Issues Yearbook*, *258*, 259.

Jordan & Henderson. (1995). Interaction analysis: Foundations and practice. *Journal of the Learning Sciences*, *4*(1), 39–104. doi:10.1207/s15327809jls0401_2

Julian, S. D., Ofori-Dankwa, J. C., & Justis, R. T. (2008). Understanding strategic responses to interest group pressures. *Strategic Management Journal*, *29*, 963–984. doi:10.1002/smj.698

Kahn, F. R. (2007). Representational approaches matter. *Journal of Business Ethics*, *73*, 77–89. doi:10.1007/s10551-006-9199-x

Kalra, D., Gertz, R., Singleton, P., & Inskip, H. M. (2006). Confidentiality of personal health information used for research. *British Medical Journal*, *333*(7560), 196. doi:10.1136/bmj.333.7560.196 PMID:16858053

Kalyuga, S., Ayres, P., Chandler, P., & Sweller, J. (2003). The expertise reversal effect. *Educational Psychologist*, *38*, 23–31. doi:10.1207/S15326985EP3801_4

Kapp, K. (2012). *The gamification of learning and instruction: Game-based methods and strategies for training and education*. San Francisco, CA: John Wiley & Sons.

Kaptelinin, V., & Nardi, B. A. (1997). *Activity theory: Basic concepts and applications*. Retrieved October 11, 2003, from http://www.acm.org/sigchi/chi97/proceedings/tutorial/bn.htm/

Kato, P. M., Cole, S. W., Bradlyn, A. S., & Pollock, B. H. (2008). A video game improves behavioral outcomes in adolescents and young adults with cancer: A randomized trial. *Pediatrics*, *122*, e305–e317. PubMed doi:10.1542/peds.2007-3134

Kaufman, D. (2009). How does the use of interactive electronic whiteboards affect teaching and learning? *Distance Learning*, *6*(2), 23–33.

Keenan, D. J., & Smith, K. (2007). *Smith and keenan's english law*. Harlow, UK: Longman Publishing Group.

Ke, F. (2009). A Qualitative Meta-Analysis of Computer Games as Learning Tools. In R. E. Ferdig (Ed.), *Handbook of Research on Effective Electronic Gaming in Education* (Vol. 1-32). Hershey, PA: Information Science Reference.

Kelly, H., Howell, K., Glinert, E., Holding, L., Swain, C., & Burrowbridge, A. et al. (2007). How to build serious games. *Communications of the ACM*, *50*(7), 44–49. doi:10.1145/1272516.1272538

Kember, D., & Murphy, D. (1995). The impact of student learning research and the nature of design on ID fundamentals. In B. Seels (Ed.), *Instructional design fundamentals: A reconsideration* (pp. 99–112). Englewood Cliffs, NJ: Educational Technology Publications.

Kenman, H. (1996). Konkordanzdemokratie und korporatismus aus der perspektive eines rationalen institutionalismus. *Politische Vierteljahresschrift*, *37*, 494–515.

Khalil, M., Paas, F., Johnson, T. E., & Payer, A. (2005). Design of interactive and dynamic anatomical visualizations: The implication of cognitive load theory. *Anatomical Record. Part B, New Anatomist*, *286B*, 15–20. doi:10.1002/ar.b.20078 PMID:16177992

Kil, S. H. (2010). Telling stories: The use of personal narratives in the social sciences and history. *Journal of Ethnic and Migration Studies*, *36*(3), 539–540. doi:10.1080/13691831003651754

Kim, A. J. (2000). *Community building on the web: Secret strategies for successful online communities*. Berkeley, CA: Peachpit Press.

Kirriemuir, J. (2002). Video gaming, education and digital learning technologies [Online]. *D-lib Magazine*, *8*(2). Retrieved May 12, 2003, from http://www.dlib.dlib.org/february02/kirriemuir/02kirriemuir.html

Kjeldsen, R., & Kender, J. (1996, October 14-16). Finding skin in color images. In *Proceedings of the 2nd International Conference on Automatic Face and Gesture Recognition*, Killington, VT (pp. 312).

Klopfer, E., Osterweil, S., & Salen, K. (2009). *Moving learning games forward: The education arcade*. Cambridge, MA: Massachusetts Institute of Technology.

Knewton. (2012). *The gamification of education infographic*. Retrieved 12 February, 2012, from http://www.knewton.com/gamification-education/

Knowles, M. (1996). Androgogy: An emerging techology for adult learning. In R. Edwards, A. Hanson, & P. Raggatt (Eds.), *Boundaries of Adult Learning* (pp. 82–96). London: Routledge.

Kolb, D. A. (1984). *Experiential Learning*. Englewood Cliffs, NJ: Prentice-Hall.

Kozma, R. B. (1991). Learning with media. *Review of Educational Research*, *61*(2), 179–211. doi:10.3102/00346543061002179

Kozma, R. B. (1994). Will media influence learning? Reframing the debate. *Educational Technology Research and Development*, *42*(2), 7–19. doi:10.1007/BF02299087

Krathwohl, D. R., Bloom, B. S., & Masia, B. B. (1964). *Taxonomy of Educational Objectives: Classification of Educational Goals, Handbook II: Affective Domain.* New York: David McKay Co., Inc.

Krepki, R., Curio, G., Blankertz, B., & Müller, K. R. (2007). Berlin brain-computer interface-the HCI communication channel for discovery. *International Journal of Human-Computer Studies, 65*(5), 460–477. doi:10.1016/j.ijhcs.2006.11.010

Kuhn, T. E. (1962). *The structure of scientific revolutions.* Chicago: University of Chicago Press.

Lady Chatterley's Lover. (n.d.). Retrieved from http://en.wikipedia.org/wiki/LadyChatterley'Lover's Plot

Lambert, K. A. (2008). Online identities unmasked. *Litig. News, 34,* 10.

Lampe, C., & Resnick, P. (2004). Slash (dot) and burn: Distributed moderation in a large online conversation space. *Proceedings of the SIGCHI Conference on Human Factors in Computing Systems,* (pp. 543-550).

Landers, R. N., & Callan, R. C. (2011). Casual social games as serious games: The psychology of gamification in undergraduate education and employee training. In *Serious Games and Edutainment Applications* (pp. 399–423). London: Springer. Retrieved from. doi:10.1007/978-1-4471-2161-9_20

Lange, B., Flynn, S., Proffitt, R., Chang, C. Y., & Rizzo, A. S. (2010). Development of an interactive game-based rehabilitation tool for dynamic balance training. *Topics in Stroke Rehabilitation, 17*(5), 345–352. PubMed doi:10.1310/tsr1705-345

Lange, B., Flynn, S. M., & Rizzo, A. (2009). Game-based telerehabilitation.[PubMed]. *European Journal of Physical and Rehabiliation Medicine, 45*(1), 143–151. PMID:19282807

Lanningham-Foster, L., Jensen, T., Foster, R. C., & Redmond, A. B. (2006, December). Energy expenditure of sedentary screen time compared with active screen time for children. *Pediatrics, 118*(6).PubMed doi:10.1542/peds.2006-1087

Lave, J., & Wenger, E. (1991). *Situated learning: Legitimate peripheral participation.* Cambridge, UK: Cambridge University Press. doi:10.1017/CBO9780511815355

Laver, M., & Schofield. (1990). *Multiparty government: The politics of coalition in Europe.* Oxford, UK: Oxford University Press.

Le Menestrel, M., Hunter, M., & de Bettignies, H. (2002). Internet e-ethics in confrontation with an activists' agenda. *Journal of Business Ethics, 39,* 135–144. doi:10.1023/A:1016348421254

Leiner, A. C., Leiner, H., & Noback, C. R. (1997). *Cerebellar Communications with the Prefrontal Cortex: Their Effect on Human Cognitive Skills.* Palo Alto, CA: Channing House.

Lenat, D. B. (1982). The nature of heuristics. *Artificial Intelligence, 19,* 189–249. doi:10.1016/0004-3702(82)90036-4

Lenhart, A., Kahne, J., Middaugh, E., Macgill, A. R., Evans, C., & Vitak, J. (2008). *Teens, Video Games and Civics: Teens' gaming experiences are diverse and include significant social interaction and civic engagement.* Washington, DC: Pew Internet and American Life Project.

Lessing, D. (2006, July 15). Testament of love. *The Guardian.*

Leung, C. H. (2010). Critical factors of implementing knowledge management in school environment: A qualitative study in Hong Kong. *Research Journal of Information Technology, 2*(2), 66–80. doi:10.3923/rjit.2010.66.80

Lewis, A. (1982). *The psychology of taxation.* Oxford, UK: Martin Robertson & Company.

Lieberman, D. A., Chamberlin, B., Medina, E., Franklin, B. A., Sanner, B. M., & Vafiadis, D. K. (2011). The power to play: Innovations in getting active summit 2011. A science panel proceedings report from the American Heart Association. *Circulation, 123*(21), 2507–2516. PubMed doi:10.1161/CIR.0b013e318219661d

Lieberman, D. A. (2006). Can we learn from playing interactive games? In P. Vorderer, & J. Bryant (Eds.), *Playing video games: Motives, responses, & consequences.* Mahwah, NJ: Lawrence Erlbaum Associates.

Lieten, G. K., & Srivatsava. (1999). *Unequal partners: Power relations, devolution and development in uttar Pradesh.* New Delhi: Sage Publications.

Lincoln, Y. S., & Guba, E. G. (1985). *Naturalistic Inquiry.* Newbury Park, CA: Sage.

Lindblom, C. (1959). The science of muddling through. *Public Administration Review, 19*(2), 79–88. doi:10.2307/973677

Long, N. E. (1958). The local community as an ecology of games. *American Journal of Sociology, 64*(3), 251–261. doi:10.1086/222468

Loots, P. C., & Charrett, D. (2009). *Practical guide to engineering and construction contracts*. CCH Australia Limited.

Low, R., Jin, P., & Sweller, J. (2010). Learner's cognitive load when using education technology. In R. Van Eck (Ed.), *Interdisciplinary models and tools for serious games: Emerging concepts and future directions*. Hershey, PA: IGI.

Ludwig, J. L. (2011). Protections for virtual property: A modern restitutionary approach: Why would anyone pay 50,000 for a virtual property. *Loy. LA Ent.L.Rev., 32*, 1.

MacInnis, L. (2009). U.N. body adopts resolution on religious defamation. *Reuters*. Retrieved from http://www.reuters.com/article/2009/03/26/us-religion-defamation-idUSTRE52P60220090326

Mackay, J., & Mensah, G. A. (2004). *The atlas of heart disease and stroke*. Geneva, Switzerland: World Health Organization.

Madon, S. (1993). Introducing administrative reform through the application of computer-based information systems: A case study in India. *Public Administration and Development, 13*, 37–48. doi:10.1002/pad.4230130104

Madon, S. (1997). Information-based global economy and socio-economic development: The case of Bangalore. *The Information Society, 13*, 227–243. doi:10.1080/019722497129115

Madon, S. (2004). Evaluating the developmental impact of e-governance initiatives: An exploratory framework. *Electronic Journal of Information Systems in Developing Countries, 20*(5), 1–13.

Madon, S., & Bhatnagar. (2000). Institutional decentralised information systems for local level planning: Comparing approaches across two states in India. *Journal of Global Information Technology Management, 3*(4), 45–59.

Madon, S., & Krishna, S., & Michael. (2010). Health information systems, decentralisation and democratic accountability. *Public Administration and Development, 30*(4), 247–260. doi:10.1002/pad.571

Madon, S., & Sahay, S., & Sahay. (2004). Implementing property tax reforms in Bangalore: An actor-network perspective. *Information and Organization, 14*, 269–295. doi:10.1016/j.infoandorg.2004.07.002

Maitra, I., & McGowan, M. K. (2007). The limits of free speech: Pornography and the question of coverage. *Legal Theory, 13*, 41–68. doi:10.1017/S1352325207070024

Malaguzzi, L. (1993). For an education based on relationships. *Young Children, 49*(1), 9–12.

Malik, K. (2009). A marketplace of outrage. *New Statesman (London, England), 138*(4940), 40–42.

Malone, T. W. (1980). *What makes things fun to learn? heuristics for designing instructional computer games*. Paper presented at the Proceedings of the 3rd ACM SIGSMALL symposium and the first SIGPC symposium on Small systems.

Malone, T. W., & Lepper, M. R. (1987). Making learning fun: A taxonomy of intrinsic motivations for learning. In R. E. Snow, & M. J. Farr (Eds.), *Aptitude, learning, and instruction, III: Cognitive and affective process analysis* (pp. 223–253). Hillsdale, NJ: Lawrence Erlbaum Associates.

Maloney, A. (2007). *Generation-Fit, a Pilot Study of Youth in Maine Middle Schools Using an "Exerlearning" Dance Video Game to Promote Physical Activity During School*. Retrieved from http://clinicaltrials.gov/ct2/show/NCT00424918

Mantovani, G. (1996a). *New communication environments: From everyday to virtual*. London: Taylor & Francis.

Mantovani, G. (1996b). Social context in HCI: A new framework for mental models, cooperation, and communication. *Cognitive Science, 20*(2), 237–269. doi:10.1207/s15516709cog2002_3

March, J. G., & Olsen. (1989). *Rediscovering institutions: The organisational basis of politics*. New York: The Free Press.

March, J.G., & Olsen. (1984). The new institutionalism: Organisational factors in political life. *The American Political Science Review, 78*(3), 734–749. doi:10.2307/1961840

Margetts, H. (1998). *Information technology in government: Britain and America*. London: Routledge.

Margetts, H. (2006). Transparency and digital government. In C. Hood, & D. Heald (Eds.), *Transparency: The key to better governance?* London: The British Academy. doi:10.5871/bacad/9780197263839.003.0012

Martens, R. L., Gulikers, J., & Bastiaens, T. (2004). The impact of intrinsic motivation on e-learning in authentic computer tasks. *Journal of Computer Assisted Learning, 20*, 368–376. doi:10.1111/j.1365-2729.2004.00096.x

Martínez Alemán, A. M. (2001). The ethics of democracy: Individuality and educational policy. *Educational Policy, 15*(3), 379–403. doi:10.1177/0895904801015003003

Martin, K. E. (2008). Internet technologies in China: Insights on the morally important influence of managers. *Journal of Business Ethics, 27*(4), 315–324.

Marzano, R. J. (2009). *MRL Meta-Analysis Database Summary*. Retrieved 5th July 2009: http://files.solution-tree.com/MRL/documents/strategy_summary_6_10_09.pdf

Marzano, R. J., & Kendall, J. S. (2006). *The New Taxonomy of Educational Objectives* (2nd ed.). Thousand Oaks, CA: Corwin Press.

Maslow, A. H. (1943). A theory of motivation. *Psychological Review, 50*(4), 370–396. doi:10.1037/h0054346

Maslow, A. H. (1946). A Theory of Human Motivation. In P. L. Harriman (Ed.), *Twentieth Century Psychology: Recent Developments in Psychology* (pp. 22–48). New York: The Philosophical Library.

Massachusetts Institute of Technology (MIT). (2003). Design principles of next-generation digital gaming for education. *Educational Technology, 43*(5), 17–22.

Maxwell, J. A., & Miller, B. A. (2008). Categorizing and connecting strategies in qualitative data analysis. In P. Leavy, & S. Hesse-Biber (Eds.), *Handbook of emergent methods* (pp. 461–477).

Mayer, R. E. (2001). *Multimedia learning*. New York: Cambridge University Press. doi:10.1017/CBO9781139164603

Mayer, R. E., & Moreno, R. (2003). Nine ways to reduce cognitive load in multimedia learning. *Educational Psychologist, 38*, 43–52. doi:10.1207/S15326985EP3801_6

Mayo, M. J. (2009). Video games: A route to large-scale STEM education? *Science, 323*, 79–82. PubMed doi:10.1126/science.1166900

McCombs, B. L. (1986). The instructional systems development (ISD) model: A review of those factors critical to its successful implementation. *Education Communication and Technology Journal, 34*(2), 67–81.

McCubbins, M. D., & Sullivan. (1987). *Congress: Structure and policy*. Cambridge, UK: Cambridge University Press.

McDarby, G., Condron, J., & Sharry, J. (2003). Affective Feedback–Learning skills in the virtual world for use in the real world. Retrieved August 15, 2011, from http://medialabeurope.org/mindgames/publications/publicationsAAATEAffectiveFeedback2003.pdf

McGonigal, J. (2011). *Reality is broken, why games make us better and how they can change the world*. London: Random House.

McKelvie, P., & Low, J. (2002). Listening to Mozart does not improve children's spatial ability: Final curtains for the Mozart effect. *The British Journal of Developmental Psychology, 20*(2), 241–258. doi:10.1348/026151002166433

McKenzie, A. (2001). Liability for information provision. In A. Scammell (Ed.), *Handbook of information management* (8th ed.). ASLIB Information Managment.

McMahon, M. J. (2012). Relationship between premium finance agency and insurance company is not sufficient to sustain a cause of action for negligent misrepresentation. *St. John's Law Review, 56*(2), 9.

McNamara, D., Kintsch, E., Songer, N. B., & Kintsch, W. (1996). Are good texts always better? Interactions of text coherence, background knowledge, and levels of understanding in learning from text. *Cognition and Instruction, 14*, 1–43. doi:10.1207/s1532690xci1401_1

Mead, L., Sagar, D., & Bampton, K. (2009). *Cima official learning system fundamentals of ethics, corporate governance and business law*. Cima Pub.

Media Analysis Laboratory. Simon Fraser University, B.C. (1998). Video game culture: Leisure and play of B.C. teens [Online]. Retrieved June 11, 2003, from http://www.mediaawareness.ca/eng/ISSUES/VIOLENCE/RESOURCE/reports/vgames.html

Mehta, J. (2010). Ideas and politics: Towards a second generation. *Perspectives on Politics*. Retrieved from http://www.allacademic.com//meta/p_mla_apa_research_citation/ 0/2/2/1/1/pages22111/p22111-1.php

Melià-Seguí, J., Zhang, R., Bart, E., Price, B., & Brdiczka, O. (2012). Activity duration analysis for context-aware services using foursquare check-ins. In *Proceedings of the 2012 International Workshop on Self-Aware Internet of Things,* (pp. 13-18). IEEE.

Merrill, M. D., Li, Z., & Jones, M. K. (1990). Limitations of first generations instructional design. *Educational Technology, 30*(1), 7–11.

Michigan State University. (2004). *Children spend more time playing video games than watching TV*. Retrieved from http://www.newsroom.msu.edu/site/indexer/1943/content.htm

Microsoft Corporation. (2009). *Engineering software for accessibility*. Redmond, WA: Microsoft Press.

Microsoft. (2010). *XNA game studio*. Retrieved from http://msdn.microsoft.com/ en-us/library/bb200104.aspx

Miller, G. A. (2003). The Magical Number Seven, Plus or Minus Two: Some Limits on Our Capacity for Processing Information. In B. J. Baars, W. P. Banks, & J. B. Newman (Eds.), *Essential Sources in the Scientific Study of Consciousness* (pp. 357–372). Cambridge, MA: MIT Press.

Ministry of Education. (1996). *Te whāriki: He whāriki mātauranga mō ngā mokopuna o aotearoa early childhood curriculum*. Wellington, New Zealand: Learning Media Limited.

Minke, A. G. (2013). *Conducting transatlantic business: Basic legal distinctions in the US and Europe*. London: Bookboon.

Mintzberg, H. (1985). The organisation as political arena. *Journal of Management Studies, 22*(2), 133–154. doi:10.1111/j.1467-6486.1985.tb00069.x

Misra, S. (2005). eGovernance: Responsive and transparent service delivery mechanism. In A. Singh (Ed.), Administrative reforms: Towards sustainable practices. New Delhi: Sage Publications.

Molenda, M., & Sullivan, M. (2003). Issues and trends in instructional technology: Treading water. In M. A. Fitzgerald, M. Orey, & R. M. Branch (Eds.), *Educational Media and Technology Yearbook 2003* (pp. 3–20). Englewood, CO: Libraries Unlimited.

Moneta, G. B., & Csikszentmihalyi, M. (1996). The effect of perceived challenges and skills on the quality of subjective experience. *Journal of Personality, 64*, 275–310. doi:10.1111/j.1467-6494.1996.tb00512.x PMID:8656320

Montessori, M., & Gutek, G. L. (2004). The Montessori method: The origins of an educational innovation: Including an abridged and annotated Ed. of Maria Montessori's The Montessori Method. Oxford, UK: Rowman & Littlefield Publishers, Inc.

Mook, D. G. (1987). *Motivation: The organization of action*. London, UK: W.W. Norton & Company Ltd.

Moon, M. J. (2002). The evolution of e-government among municipalities: Rhetoric or reality? *Public Administration Review, 62*(4), 424–433. doi:10.1111/0033-3352.00196

Moore Jackson, M., & Mappus, R. (2010). Applications for brain-computer interfaces. In D. S. Tan, & A. Nijholt (Eds.), *Brain-computer interfaces*. London, UK: Springer-Verlag. doi:10.1007/978-1-84996-272-8_6

Moore, M. G., & Anderson, W. G. (2012). *Handbook of distance education*. London: Routledge.

Moran, J. (2007). Generating more heat than light? Debates on civil liberties in the UK. *Policing, 1*(1), 80. doi:10.1093/police/pam009

Morelli, T., Foley, J., Columna, L., Lieberman, L., & Folmer, E. (2010, June). VI-Tennis: A vibrotactile/audio exergame for players who are visually impaired. In *Proceedings of the Foundations of Digital Interactive Games Conference*, Monterey, CA (pp. 147-154).

Moreno-Ger, P., Blesius, C., Currier, P., Sierra, J. L., & Fernández-Manjón, B. (2008). Online Learning and Clinical Procedures: Rapid Development and Effective Deployment of Game-Like Interactive Simulations. In Z. Pan, A. D. Cheok, W. Müller, & A. E. Rhabili (Eds.), *Transactions on Edutainment I* (Vol. 5080, pp. 288–304). Berlin, Heidelberg: Springer Verlag. doi:10.1007/978-3-540-69744-2_22

Moreno, R., & Mayer, R. E. (2000). A coherency effect in multimedia learning: The case for minimizing irrelevant sounds in the design of multimedia instructional messages. *Journal of Educational Psychology*, *92*, 117–125. doi:10.1037/0022-0663.92.1.117

Morozov, E. (2011). *The net delusion: How not to liberate the world.* New York: Penguin.

Morphy, E. (2008, December 8). British ISPs block Wikipedia page, reigniting 30-year-old child porn controversy. *ECT Newsnet.*

Morrell, R. W., Mayhorn, C. B., & Bennett, J. (2002). Older adults online in the internet century. *Older Adults, Health Information, and the World Wide Web,* 43-57.

Morrison, J. L., & Aldrich, C. (2003). Simulations and the learning revolution: An interview with Clark Aldrich. *The Technology Source.* Retrieved August 11, 2003, from http://64.124.14.173/default.asp?show=article&id=2032

Morrison, G. R., Ross, S. M., & Kemp, J. E. (2006). *Designing Effective Instruction* (5th ed.). London: Wiley.

Morsillo, R. (2011). One down, two to go: Public policy in service of an available, affordable and accessible national broadband network for people with disability. *Telecommunications Journal of Australia, 61*(2).

Mouawad, M. R., Doust, C. G., Max, M. D., & McNulty, P. A. (2011). Wii-based movement therapy to promote improved upper extremity function post-stroke. A pilot study. *Journal of Rehabilitation Medicine, 43*(6), 527–533. PubMed doi:10.2340/16501977-0816

Muntean, C. (2011). *Raising engagement in e-learning through gamification.* Paper presented at the 6th International Conference on Virtual Learning ICVL 2011. Bucharest, Romania.

Murray, J. H. (1997). *Hamlet on the Holodeck: The future of narrative in cyberspace.* Cambridge, MA: MIT Press.

NAEYC. (2009). *Developmentally appropriate practice in early childhood programs serving children from birth through age 8.* Retrieved 14 March, 2013, from www.naeyc.org/files/naeyc/file/positions/position%20statement%20Web.pdf

NAEYC. (2012). *Technology and interactive media as tools in early childhood programs serving children from birth through age 8.* Retrieved 2 March, 2013, from http://www.naeyc.org/files/naeyc/file/positions/PS_technology_WEB2.pdf

Nap, H. H., de Kort, Y. A. W., & IJsselsteijn, W. A. (2009). Senior gamers: Preferences, motivations and needs. *Gerontechnology (Valkenswaard)*, *8*(4), 247–262. doi:10.4017/gt.2009.08.04.003.00

National Institute of Urban Affairs. (2004). *Reforming the property tax system.* New Delhi: NIUA Press.

Neal, L. (2001). Storytelling at a distance. *elearn Magazine.* Retrieved from http://elearnmag.acm.org/featured.cfm?aid=566979

Nelson, B., Ketelhut, D., Clarke, J., Bowman, C., & Dede, C. (2005, November). Design based research strategies for developing a science inquiry curriculum in a multi-user virtual environment. *Educational Technology.*

Nemeth, K., & Simon, F. (2012). *Preschool curriculum and technology crosswalk.* Paper presented at the EETC Early Education and Technology for Children. Budapest, Hungary.

Neulight, N., Kafai, Y. B., Kao, L., Foley, B., & Galas, G. (2007, February). Children's participation in a virtual epidemic in the science classroom: Making connections to natural infectious diseases. *Journal of Science Education and Technology, 16*(1), 47–58. doi:10.1007/s10956-006-9029-z

Nicholson, S. (2012). *A user-centered theoretical framework for meaningful gamification.* Paper presented at the Games+Learning+Society 8.0. New York, NY.

Nieborg, D. B. (2004). *America's Army: More Than a Game.* Paper presented at the Transforming Knowledge into Action through Gaming and Simulation, Munchen: SAGSAGA.

Nielsen, J. (1993). *Usability engineering.* San Francisco: Morgan Kaufman.

Nijholt, A., Plass-Oude Bos, D., & Reuderink, B. (2009). Turning shortcomings into challenges: Brain-computer interfaces for games. *Entertainment Computing, 1*(2), 85–94. doi:10.1016/j.entcom.2009.09.007

Nintendo. (2007). *Supplementary information about earnings release.* Retrieved from http://www.nintendo.co.jp/ir/pdf/2010/100507e.pdf#page=6

Nintendo. (2009). *Wii Sports Resort.* Retrieved http://Wiisportsresort.com/en/#/home

Nintendo. (2010). *Wii Sports.* Retrieved from http://www.nintendo.co.uk/NOE/en_GB/games/Wii/Wii_sports_2781.html

Nitsche, M., & Thomas, M. (2003). Stories in space: The concept of the story map. In *Proceedings of ICVS 2003* (Vol. 2897, pp. 85–93). Berlin, Germany: Springer.

Nitz, J. C., Kuys, S., Isles, R., & Fu, S. (2010). Is the Wii Fit a new-generation tool for improving balance, health and well-being? A pilot study. *Climacteric, 13*(5), 487–491. PubMed doi:10.3109/13697130903395193

Noddings, N. (2007). *When school reform goes wrong.* New York: Teachers College Press.

Norman, D. A. (1991). Cognitive artifacts. In J. M. Carroll (Ed.), *Designing interaction: Psychology at the human-computer interface* (pp. 17–38). New York, NY: Cambridge University Press.

North, A. C., & Hargreaves, D. J. (1999). Music and driving game performance. *Scandinavian Journal of Psychology, 40*, 285–292. doi:10.1111/1467-9450.404128

Nunziato, D. C. (2009). *Virtual freedom: Net neutrality and free speech in the internet age.* Stanford, CA: Stanford Law Books, Stanford University Press.

OECD. (2006). *Executive summary: Starting strong II: Early childhood education and care.* Retrieved from http://www.oecd.org/edu/school/startingstrongiiearly-childhoodeducationandcare.htm

Oliver, C., & Shinal, J. (2006). Google will censor new China service. *MarketWatch.* Retrieved from http://www.marketwatch.com/story/google-builds-censorship-into-china-search-service

Ostrom, E. Gardner, & Walker. (1994). Rules, games and common-pool resources. Ann Arbor, MI: University of Michigan Press.

Oughton, D. W., & Davis, M. (1996). *Sourcebook on contract law.* Cavendish.

Owen, A. M., Hampshire, A., Grahn, J. A., Stenton, R., Dajani, S., & Burns, A. S. et al. (2010, April 20). Putting brain training to the test. *Nature.* doi:10.1038/nature09042 PMID:20407435

Paas, F., Tuovinen, J. E., Tabbers, H., & van Gerven, P. W. M. (2003). Cognitive load measurement as a means to advance cognitive load theory. *Educational Psychologist, 38*, 63–71. doi:10.1207/S15326985EP3801_8

Page, S. E. (1999). Computational models from A to Z. *Complexity, 5*(1), 35–41. doi:10.1002/(SICI)1099-0526(199909/10)5:1<35::AID-CPLX5>3.0.CO;2-B

Pajunen, K. (2006). Stakeholder influences in organizational survival. *Journal of Management Studies, 43*(6), 1261–1288. doi:10.1111/j.1467-6486.2006.00624.x

Pan, P. P. (2006, February 19). The click that broke a government's grip. *Washington Post.*

Papert, S. (1980). *Mindstorms: Children, Computers, and Powerful Ideas.* New York: Basic Books.

Paras, B., & Bizzocchi, J. (2005). *Game, Motivation, and Effective Learning: An Integrated Model for Educational Game Design.* Paper presented at the DiGRA 2005 – the Digital Games Research Association's 2nd International Conference, Simon Fraser University, Burnaby, BC, Canada.

Parker, S. G., Oliver, P., Pennington, M., Bond, J., Jagger, C., & Enderby, P. et al. (2009). Rehabilitation of older patients: Day hospital compared with rehabilitation at home. A randomised controlled trial.[PubMed]. *Health Technology Assessment, 13*(39), 1–168. PMID:19712593

Pasquale, F. (2006). Rankings, reductionism, and responsibility. *Cleveland State Law Review, 54*, 115–140.

Patel, A., Schieble, T., Davidson, M., Tran, M. C. J., Schoenberg, C., Delphin, E., & Bennett, H. (2006). Distraction with a hand-held video game reduces pediatric preoperative anxiety. *Paediatric Anaesthesia, 16*, 1019–1027. PubMed doi:10.1111/j.1460-9592.2006.01914.x

Pearce, C. (2006). Productive Play: Game Culture From the Bottom Up. *Games and Culture, 1*(1), 17–24. doi:10.1177/1555412005281418

Pearce, C. (2008). The truth about baby boomer gamers: A study of over-forty. *Games and Culture*, *3*(2), 142–174. doi:10.1177/1555412008314132

Pederson, E. M. (1995). Storytelling and the art of teaching. *Forum*, *33*(1).

Peloza, J., & Falkenberg, L. (2009). The role of collaboration in achieving corporate social responsibility objectives. *California Management Review*, *51*(3), 95–113. doi:10.2307/41166495

Perlin, K. (2003). Building virtual actors who can really act. In *Proceedings of ICVS 2003* (Vol. 2897, pp. 127–134). Berlin, Germany: Springer.

Peters, B. G. (2000). *Institutional theory in political science: The new institutionalism*. London: Continuum.

Peters, L. H., O'Connor, E. J., & Eulberg, J. R. (1985). Situational constraints: Sources, consequences, and future considerations. *Research in Personnel and Human Resources Management*, *3*, 79–114.

Pfeffer, J., & Salancik, G. R. (1978). *The external control of organizations: A resource dependence perspective*. New York: Harper & Row Publishers.

Pfeil, U., & Zaphiris, P. (2009). *Theories and methods for studying online communities for people with disabilities and older people*. Academic Press. doi:10.1201/9781420064995-c42

Pieter van Foreest. (2011). *Digitaal revalideren bij Fysiotherapie De Naaldhorst*. Retrieved August 10, 2011, from http://www.pietervanforeest.nl/PieterVanForeest/Over+Pieter+Van+Foreest/Actueel/Persberichten/Digitaal+revalideren+bij+Fysiotherapie+De+Naaldhorst.htm

Pino-Silva, J., & Mayora, C. A. (2010). English teachers' moderating and participating in OCPs. *System*, *38*(2). doi:10.1016/j.system.2010.01.002

Pivec, M., Dziabenko, O., & Schinnerl, I. (2003). *Aspects of Game-Based Learning*. Paper presented at the Third International Conference on Knowledge Management (IKNOW 03), Graz, Austria.

Pollock, E., Chandler, P., & Sweller, J. (2002). Assimilating complex information. *Learning and Instruction*, *12*, 61–86. doi:10.1016/S0959-4752(01)00016-0

Poor, N. (2005). Mechanisms of an online public sphere: The website slashdot. *Journal of Computer-Mediated Communication*, *10*(2).

PopCap Games. (2008). Survey: *'Disabled gamers' comprise 20% of casual video games audience*. Retrieved from http://popcap.mediaroom.com/index.php?s=43&item=30

Pope, A. T., & Palsson, O. S. (2001). *Helping video games 'rewire our minds'*. Paper presented at Playing by the Rules Conference, Chicago, IL.

Popescu, V. G., Burdea, G. C., Bouzit, M., & Hentz, V. R. (2000). A virtual-reality-based telerehabilitation system with force feedback. *IEEE Transactions on Information Technology in Biomedicine*, *4*(1), 45–51. PubMed doi:10.1109/4233.826858

Popper, K. (1959). *The logic of scientific discovery*. London: Hutchinson.

Powazek, D. M. (2002). *Design for community: The art of connecting real people in virtual places*. New Riders.

Preece, J. (2001). *Online communities: Designing usability, supporting sociability*. Chichester, UK: John Wiley & Sons.

Preece, J., Nonnecke, B., & Andrews, D. (2004). The top 5 reasons for lurking: Improving community experiences for everyone. *Computers in Human Behavior*, *2*(1), 42.

Prensky, M. (2001). *Digital game-based learning*. New York: McGraw-Hill Publishing Company.

Prensky, M. (2002). The Motivation of Gameplay or the REAL 21st century learning revolution. *Horizon*, *10*, 1–14. doi:10.1108/10748120210431349

Prensky, M. (2005). Computer games and learning: Digital game-based learning. In J. Raessens, & J. Goldstein (Eds.), *Handbook of computer game studies* (pp. 97–122). Cambridge, MA: MIT Press.

Price, R. V. (1990). *Computer-aided instruction: A guide for authors*. Pacific Grove, CA: Brooks/Cole Publishing Company.

Pritchett, J. E. (2009). *Identification of situational constraints in middle school business information technology programs. (Doctor of Education)*. Partial, Georgia: The University of Georgia.

Provenzo, E. F. (1992). The video generation. *The American School Board Journal, 179*(3), 29–32.

Qiu, Q., Ramirez, D. A., Saleh, S., Fluet, G. G., Parikh, H. D., Kelly, D., & Adamovich, S. V. (2009). The New Jersey Institute of Technology robot-assisted virtual rehabilitation (NJIT-RAVR) system for children with cerebral palsy: A feasibility study. *Journal of Neuroengineering and Rehabilitation, 6*(40). PubMed

Rademaker, A., van der Linden, S., & Wiersinga, J. (2010). SilverFit, a virtual rehabilitation system. *Gerontechnology (Valkenswaard), 8*(2), 119.

Raja, Y., McKenna, S. J., & Gong, S. (1998, April 14-16). Tracking and segmenting people in varying lighting conditions using colour. In *Proceedings of the Third IEEE International Conference on Automatic Face and Gesture Recognition*, Nara, Japan (pp. 228-233).

Ramachander, S. (2008). Internet filtering in Europe. In *Access denied: The practice and policy of global internet filtering*. Cambridge, MA: The MIT Press.

Rao, A. S., & Georgeff, M. P. (1998). Decision procedures for BDI logics. *Journal of Logic and Computation, 8*(3), 293. doi:10.1093/logcom/8.3.293

Rashid, A. M., Ling, K., Tassone, R. D., Resnick, P., Kraut, R., & Riedl, J. (2006). Motivating participation by displaying the value of contribution. *Proceedings of the SIGCHI Conference on Human Factors in Computing Systems,* (p. 958).

Ratey, J. (2008). *Spark: The revolutionary new science of exercise and the brain.* New York: Little, Brown and Company.

Rauscher, F. H., Shaw, G. L., & Ky, K. N. (1993). Music and spatial task performance. *Nature, 365*, 611. doi:10.1038/365611a0 PMID:8413624

Raybourn, E. M. (2007). Applying simulation experience design methods to creating serious game-based adaptive training systems. *Interacting with Computers, 19*, 206–214. doi:10.1016/j.intcom.2006.08.001

Ready, D. J., Gerardi, R. J., Backscheider, A. G., Mascaro, N., & Rothbaum, B. O. (2010). Comparing virtual reality exposure therapy to present-centered therapy with 11 U.S. Vietnam veterans with PTSD. *Cyberpsychology, Behavior, and Social Networking, 13*(1), 49–54. PubMed doi:10.1089/cyber.2009.0239

Reigeluth, C. M. (1999). What is instructional-design theory and how is it changing? In C. M. Reigeluth (Ed.), Instructional-design theories and models (Vol. II): A new paradigm of instructional theory (pp. 5-29). Mahwah, NJ: Lawrence Erlbaum Associates.

Reigeluth, C. M., & Frick, T. W. (1999). Formative research: Methodology for creating and improving design theories. In C. M. Reigeluth (Ed.), Instructional-design theories and models (Vol. II): A new paradigm of instructional theory (pp. 633-652). Mahwah, NJ: Lawrence Erlbaum Associates.

Reigeluth, C. M. (1983). Instructional design: What is it and why is it? In C. M. Reigeluth (Ed.), *Instructional-design theories and models: An overview of their current status* (pp. 3–36). Hillsdale, NJ: Lawrence Erlbaum Associates.

Reigeluth, C. M. (1996). A new paradigm of ISD? *Educational Technology, 36*(3), 13–20.

Reigeluth, C. M. (1997). Instructional theory, practitioner needs, and new directions: Some reflections. *Educational Technology, 37*(1), 42–47.

Reigeluth, C. M., Merrill, M. D., Wilson, B. G., & Spille, R. T. (1980). The elaboration theory of instruction: A model for sequencing and synthesizing instruction. *Instructional Science, 9*(3), 195–219. doi:10.1007/BF00177327

Reigeluth, C., & Schwartz, E. (1989). An instructional theory for the design of computer-based simulations. *Journal of Computer-Based Instruction, 16*(1), 1–10.

Reiners, T., Dreher, C., & Dreher, H. (2012). Transforming ideas to innovations: A methodology for 3D systems development. In A. Hebbel-Segger, T. Reiners, & D. Schäfer (Eds.), *Alternate realities: Emerging Technologies in education and economics*. Berlin: Springer.

Reiners, T., Gregory, S., & Knox, V. (2012). Virtual bots, their influence on learning environments and how they increase immersion. In S. Gregory, M. J. W. Lee, B. Dalgarno, & B. Tynan (Eds.), *Virtual worlds in open and distance education*. Athabasca, Canada: Athabasca University Press.

Reiners, T., & Wood, L. C. (2013). Immersive virtual environments to facilitate authentic education in logistics and supply chain management. In Y. Kats (Ed.), *Learning management systems and instructional design: Metrics, standards, and applications* (pp. 323–343). Hershey, PA: IGI Global. doi:10.4018/978-1-4666-3930-0.ch017

Reiser, R. A. (2001). A history of instructional design and technology: Part II: A history of instructional design. *Educational Technology Research and Development*, *49*(2), 57–67. doi:10.1007/BF02504928

Reiss, S. (2004). Multifaceted nature of intrinsic motivation: The theory of 16 basic desires. *Review of General Psychology*, *8*(3), 179–193. doi:10.1037/1089-2680.8.3.179

Rheault, M., & Moghaed, D. (2008, Summer). Gallup presents...cartoon and controversy: Free expression or Muslim exceptionalism in Europe? *Harvard International Review*, 68–71.

Rheinberg, F. (2004, July 5-8). *Motivational competence andflow-experience.* Paper presented at the 2nd European Conference on Positive Psychology, Verbania, Italy.

Rheinberg, F., Vollmeyer, R., & Engeser, S. (2003). Die Erfassung des Flow-Erlebens. In J. Stiensmeier-Pelser, & F. Rheinberg (Eds.), *Diagnostic von Selbstkonzept, Lernmotivation und Selbstregulation* (pp. 261–279). Göttingen, Germany: Hogrefe.

Rheingold, H. (2000). *The virtual community: Homesteading on the electronic frontier* (2nd ed.). London, UK: MIT Press.

Richards, D., Fassbender, E., Bilgin, A., & Thompson, W. F. (2008). An investigation of the role of background music in IVWs for learning. *ALT-J Research in Learning Technology*, *16*(3), 231–244. doi:10.1080/09687760802526715

Richey, C. (1995). Trends in instructional design: Emerging theory-based models. *Performance Improvement Quarterly*, *8*(3), 96–110. doi:10.1111/j.1937-8327.1995.tb00689.x

Richey, R. C. (1997). Agenda-building and its implications for theory construction in instructional technology. *Educational Technology*, *37*(1), 5–11.

Ricks, V. (2011). *Assent is not an element of contract formation (No. SSRN 1898824)*. Social Science Research Network.

Rideout, V. J., Vandewater, E. A., & Wartella, E. A. (2003). *Zero to Six: Electronic Media in the Lives of Infants, Toddlers and Preschoolers*. Menlo Park, CA: Kaiser Family Foundation.

Rieber, L. P. (1996). Seriously considering play: Designing interactive learning environments based on the blending of microworlds, simulations, and games. *Educational Technology Research and Development*, *44*(2), 43–58. doi:10.1007/BF02300540

Riedl, M. O., & Young, M. (2003). Character-focused narrative generation for execution in virtual worlds. In *Virtual storytelling: Using virtual reality technologies for storytelling* (Vol. 2897, pp. 47–56). Berlin: Springer. doi:10.1007/978-3-540-40014-1_6

Rinaldi, C. (1995). The emergent curriculum and social constructivism: An interview with Lella Gandini. In C. Edwards, L. Gandini, & G. Forman (Eds.), *The hundred languages of children: The Reggio Emilia approach to early childhood education* (pp. 101–111). Norwood, NJ: Ablex Publishing Corporation.

Rollings, A., & Adams, E. (2003). *Andrew Rollings and Ernest Adams on Game Design*. St. Carmel, IN: New Riders Publishing.

Romm, C. T., & Setzekom, K. (2008). *Social network communities and E-dating services: Concepts and implications*. London, UK: Information Science Reference. doi:10.4018/978-1-60566-104-9

Ronaghan, S. A. (2002). Benchmarking e-government: A global perspective. New York: The United Nations Division for Public Economics and Public Administration (DPEPA) Report.

Roosendaal, A. (2007). Elimination of anonymity in regard to liability for unlawful acts on the internet. *International Journal of Technoloy Transfer and Commercialisation*, *6*(2), 184–195. doi:10.1504/IJTTC.2007.017805

Rosas, R., Nussbaum, M., & Cumsille, P. (2003). Beyond Nintendo: Design and assessment of educational video games for first and second grade students. *Computers & Education, 40*(1), 71–94. doi:10.1016/S0360-1315(02)00099-4

Rose, F. (2006, February 19). Why I published those cartoons. *The Washington Post*. Retrieved from http://www.washingtonpost.com/wp-dyn/content/article/2006/02/17/AR2006021702499.html

Rose-Ackerman, S. (1978). *Corruption: A study in political economy*. New York: Academic Press.

Rosenberg, D., Depp, C. A., Vahia, I. V., Reichstadt, J., Palmer, B. W., Kerr, J., et al. (2010). Exergames for subsyndromal depression in older adults: A pilot study of a novel intervention. *The American Journal of Geriatric Psychiatry, 18*(3), 221–226. PubMed doi:10.1097/JGP.0b013e3181c534b5

Rosengard, J. K. (1998). *Property tax reform in developing countries*. Boston: Kluwer Academic Publications. doi:10.1007/978-1-4615-5667-1

Rose, R. (2005). A global diffusion model of e-governance. *Journal of Public Policy, 25*(1), 5–28. doi:10.1017/S0143814X05000279

Roth, W. M., & McGinn, M. K. (1998). Inscriptions: Toward a theory of representing as social practice. *Review of Educational Research, 68*(1), 35–59. doi:10.3102/00346543068001035

Roulstone, A., Thomas, P., & Balderston, S. (2011). Between hate and vulnerability: Unpacking the British criminal justice system's construction of disablist hate crime. *Disability & Society, 26*(3), 351–364. doi:10.1080/09687599.2011.560418

Rowland, G. (1992). What do instructional designers actually do? An initial investigation of expert practice. *Performance Improvement Quarterly, 5*(2), 65–86. doi:10.1111/j.1937-8327.1992.tb00546.x

Rowland, G., Parra, M. L., & Basnet, K. (1994). Educating instructional designers: Different methods for different outcomes. *Educational Technology, 34*(6), 5–11.

Royal College of Physicians. (2004). *National clinical guidelines for stroke* (2nd ed.). London, UK: Royal College of Physicians.

Rufer-Bach, K. (2009). *The second life grid: The official guide to communication, collaboration, and community engagement*. Sybex.

Ruiz, S., York, A., Truong, H., Tarr, A., Keating, N., Stein, M., et al. (2006). *Darfur is Dying*. Thesis Project. Retrieved from http://interactive.usc.edu/projects/games/20070125-darfur_is_.php

Russell, T. (2003). *The "No Significant Difference Phenomenon."* Retreived December 10, 2005, from http://teleeducation.nb.ca/nosignificantdifference/

Rutter, C. E., Dahlquist, L. M., & Weiss, K. E. (2009). Sustained efficacy of virtual reality distraction. *The Journal of Pain, 10*(4), 391–397. PubMed doi:10.1016/j.jpain.2008.09.016

Ryan, R. M., & Deci, E. L. (2000). Self-determination theory and the facilitation of intrinsic motivation, social development and well-being. *The American Psychologist, 55*(1), 68–78. PubMed doi:10.1037/0003-066X.55.1.68

Ryan, R. M. (1982). Control and information in the intrapersonal sphere: An extension of cognitive evaluation theory. *Journal of Personality and Social Psychology, 43*, 450–461. doi:10.1037/0022-3514.43.3.450

Ryan, R. M., & Deci, E. L. (2000a). Intrinsic and extrinsic motivations: Classic definitions and new directions. *Contemporary Educational Psychology, 25*, 54–67. doi:10.1006/ceps.1999.1020 PMID:10620381

Sadler, T. D., Barab, S. A., & Scott, B. (2007). What do students gain by engaging in socioscientific inquiry? *Research in Science Education, 37*, 371–391. doi:10.1007/s11165-006-9030-9

Salen, K., & Zimmerman, E. (2003). *Rules of Play: Game Design Fundamentals*. Cambridge, MA: The MIT Press.

Sallis, J. F., Hovell, M. F., Hofstetter, C. R., & Barrington, E. (1992). Explanation of vigorous physical activity during two years using social learning variables. *Social Science & Medicine, 34*, 25–32. doi:10.1016/0277-9536(92)90063-V PMID:1738853

Salthouse, T. A. (2006). Mental exercise and mental aging: Evaluating the validity of the "use it or lose it" hypothesis. *Perspectives on Psychological Science, 1*, 68–87. doi:10.1111/j.1745-6916.2006.00005.x

Salzedo, S., & Brunner, P. (1995). *Briefcase on contract law*. Routledge-Cavendish.

Sandham, J. (2005). *Medical thermography*. Retrieved from http://www.ebme.co.uk/arts/ thermog/

Saposnik, G., & Levin, M. (2011). Virtual reality in stroke rehabilitation: A meta-analysis and implications for clinicians. *Stroke, 42*(5), 1380–1386. PubMed doi:10.1161/STROKEAHA.110.605451

Sara, G., Burgess, P., Harris, M., Malhi, G., Whiteford, H., & Hall, W. (2012). Stimulant use disorders: Characteristics and comorbidity in an australian population sample. *The Australian and New Zealand Journal of Psychiatry*. doi:10.1177/0004867412461057 PMID:22990432

Sawyer, B., & Smith, P. (2008). *Serious Games Taxonomy*. Retrieved 26 March, 2008, from http://www.dmill.com/presentations/serious-games-taxonomy-2008.pdf

Scharpf, F. W. (1997). *Games real actors play: Actor-centered institutionalism in policy research*. Oxford, UK: Westview Press.

Schech, S. (2005). Wired for change: The links between ICTs and development discourses. *Journal of International Development, 14*(1), 13–23. doi:10.1002/jid.870

Scheffler, S. (2001). *Boundaries and allegiances: Problems of justice and responsibility in liberal thought*. Oxford, UK: Oxford University Press.

Schellenberg, E. G., & Hallam, S. (2005). Music and cognitive abilities in 10- and 11-year-olds: The blur effect. *Annals of the New York Academy of Sciences, 1060*, 202–209. doi:10.1196/annals.1360.013 PMID:16597767

Schell, J. (2005). Understanding entertainment: story and gameplay are one.[CIE]. *Computers in Entertainment, 3*(1), 6. doi:10.1145/1057270.1057284

Schiffman, S. S. (1995). Instructional systems design: Five views of the field. In G. Anglin (Ed.), *Instructional Technology: Past, present, and future* (2nd ed., pp. 131–144). Engelwood, CO: Libraries Unlimited.

Schiller, H. I. (1981). *Who knows: Information in the age of the fortune 500*. Norwood, NJ: Ablex Publishing Corp.

Schindelka, B. (2003). A framework of constructivist instructional design: Shiny, happy design. *Inoad e-Zine, 3*(1). Retrieved August 18, 2003, from http://www.inroad.net/shindelka0403.html

Scholing, A., & Emmelkamp, P. M. G. (1993). Exposure with and without cognitive therapy for generalized social phobia: Effects of individual and group treatment. *Behaviour Research and Therapy, 31*(7), 667–681. PubMed doi:10.1016/0005-7967(93)90120-J

Schon, D. A. (1985). *The design studio: An exploration of its traditions and potentials*. London: RIBA Publications.

Schon, D. A. (1987). *Educating the reflective practitioner: Toward a new design for teaching and learning in the professions*. San Francisco: Jossey-Bass.

Schwabach, A. (2006). *Internet and the law: Technology, society, and compromises*. Abc-clio.

Schweinle, A., Meyer, D. K., & Turner, J. C. (2006). Striking the right balance: Students' motivation and affect in elementary mathematics. *The Journal of Educational Research, 99*(5), 271–293. doi:10.3200/JOER.99.5.271-294

Schwimmer, J. B., Burwinkle, T. M., & Varni, J. W. (2003, April 9). Health-related Quality of Life of Severely Obese Children and Adolescents. *Journal of the American Medical Association*, 289. PMID:12684360

Seels, B., & Richie, R. (1994). *Instructional technology: The definitions and domains of the field*. Washington, DC: AECT.

Segovia, K. Y., Bailenson, J. N., & Monin, B. (2009). Morality in tele-immersive environments. In *Proceedings of the 2nd International Conference on Immersive Telecommunications*. IEEE.

Serious Games Initiative. (2010). Retrieved on January 14, 2010, from http://www.seriousgames.org/about2.html

Shaffer, D. W. (2004). Pedagogical praxis: The professional models for post-industrial education. *Teachers College Record, 106*(7). doi:10.1111/j.1467-9620.2004.00383.x

Shaffer, D. W. (2006). *How Computer Games Help Children Learn*. New York: Palgrave. doi:10.1057/9780230601994

Shasek, J. (2009). *Brainy Stuff* [PowerPoint slides]. Retrieved from http://www.slideshare.net/invenTEAM/brainy-stuff-1078097

Shepard, R. J. (1997). Curricular Physical Activity and Academic Performance. *Pediatric Exercise Science.*

Sherry, J. L. (2001). The effects of violent video games on aggression. A Meta-analysis. *Human Communication Research, 27*(3), 409–431.

Sherwin, A. (2006, April 3). A family of Welsh sheep - The new stars of Al-Jazeera. *Times (London, England),* (n.d), 7.

Shrock, S. A. (1995). A brief history of instructional development. In G. Anglin (Ed.), *Instructional technology: Past, present, and future* (2nd ed., pp. 11–19). Engelwood, CO: Libraries Unlimited.

Sicart, M. (2005, 16-20 June 2005). *The Ethics of Computer Game Design.* Paper presented at the DiGRA 2005 - the Digital Games Research Association's 2nd International Conference, Simon Fraser University, Burnaby, BC, Canada.

Silva, M. C. d., Conti, C. L., Klauss, J., & Alves, L. G., Cavalcante, Henrique Mineiro do Nascimento, Fregni, F., et al. (2013). Behavioral effects of transcranial direct current stimulation (tDCS) induced dorsolateral prefrontal cortex plasticity in alcohol dependence. *Journal of Physiology, Paris.* PMID:23891741

Silver, E. A. (2004). An overview of heuristic solution methods. *The Journal of the Operational Research Society, 55,* 936–956. doi:10.1057/palgrave.jors.2601758

Simmons, L. L., & Clayton, R. W. (2010). The impact of small business B2B virtual community commitment on brand loyalty. *International Journal of Business and Systems Research, 4*(4), 451–468. doi:10.1504/IJBSR.2010.033423

Simon, F., & Nemeth, K. (2012). *Digital decisions: Choosing the right technology tools for early childhood.* Lewisville, NC: Gryphon House.

Simon, H. A. (1955). A behavioural model of rational choice. *The Quarterly Journal of Economics, 69*(1), 99–118. doi:10.2307/1884852

Simon, J. G., Powers, C. W., & Gunnemann, J. P. (1983). The responsibilities of corporations and their owners. In *Ethical theory and business* (2nd ed.). Englewood Cliffs, NJ: Prentice-Hall, Inc.

Simpson, E. J. (1972). *The classification of educational objectives in the psychomotor domain: The psychomotor domain* (Vol. 3). Washington, DC: Gryphin House.

Sin, C. H., Hedges, A., Cook, C., Mguni, N., & Comber, N. (2011). Adult protection and effective action in tackling violence and hostility against disabled people: Some tensions and challenges. *Journal of Adult Protection, 13*(2), 63–75.

Sinclair, B. (2009). *Visually impaired gamer sues Sony Online.* Retrieved from http://uk.gamespot.com/news/6239339.html

Sinha, K. P. (1981). *Property taxation in a developing economy.* New Delhi: Puja Publications.

Skinner, B. F. (1935). Two Types Of Conditioned Reflex And A Pseudo Type. *The Journal of General Psychology, 12,* 66–77. doi:10.1080/00221309.1935.9920088

Sklar, M., Shaw, B., & Hogue, A. (2012). Recommending interesting events in real-time with foursquare check-ins. In *Proceedings of the Sixth ACM Conference on Recommender Systems,* (pp. 311-312). ACM.

Slegers, K., & Donoso, V. (2012). The impact of paper prototyping on card sorting: A case study. *Interacting with Computers, 24*(5), 351–357.

Smith, G. J. H. (1996). Building the lawyer-proof web site. Paper presented at the *Aslib Proceedings, 48,* (pp. 161-168).

Smith, H., Higgins, S., Wall, K., & Miller, J. (2005). Interactive whiteboards: Boon or bandwagon? A critical review of the literature. *Journal of Computer Assisted Learning, 21*(2), 91–101. doi:10.1111/j.1365-2729.2005.00117.x

Smith, R. E. (2011). *Elementary information security.* Jones & Bartlett Learning.

Soler, J., & Miller, L. (2003). The struggle for early childhood curricula: A comparison of the English foundation stage curriculum, Te Wha¨riki and Reggio Emilia. *International Journal of Early Years Education, 11*(1).

Sousa, D. (2005). *How the brain learns.* Thousand Oaks, CA: Corwin Press.

Spencer, R. (2009). Muhammad and Aisha, a love story. *Middle East Quarterly.* Retrieved from http://www.me-forum.org/2010/the-jewel-of-medina

Spiegelman, A. (2006, June). Drawing blood: Outrageous cartoons and the art of outrage. *Harper's,* 43–51.

Spierling, U. (2002). Digital storytelling. *Computers & Graphics, 26*(1), 1–66. doi:10.1016/S0097-8493(01)00172-8

Squire, K. (2005). Changing the game: What happens when video games enter the classroom? *Innovate: Journal of Online Education, 1*(6). Retrieved from http://innovateonline.info.proxy.bc.edu/index.php?view=article&id=82

Squire, K. (2006). From content to context: Videogames as designed experience. *Educational Researcher, 35*(8), 19–29. doi:10.3102/0013189X035008019

Staiano, A. E., & Calvert, S. L. (2011). Exergames for physical education courses: Physical, social, and cognitive benefits. *Child Development Perspectives, 5*(2), 93–98. PubMed doi:10.1111/j.1750-8606.2011.00162.x

Stanforth, C. (2006). *Analysing eGovernment in developing countries using actor-network theory.* iGovernment Working Paper Series – Paper no. 17.

Stark, T. (2002). *Negotiating and drafting contract boilerplate.* Incisive Media, LLC.

Steele, K. M. (2000). Arousal and mood factors in the "Mozart Effect". *Perceptual and Motor Skills, 91,* 188–190. doi:10.2466/pms.2000.91.1.188 PMID:11011888

Steele, K. M., Bass, K., & Crook, M. (1999). The mystery of the Mozart effect: Failure to replicate. *Psychological Science, 10,* 366–369. doi:10.1111/1467-9280.00169

Štogr, J. (2011). Surveillancebased mechanisms in MUVEs (MultiUser virtual environments) used for monitoring, data gathering and evaluation of knowledge transfer in VirtuReality. *Journal of Systemics. Cybernetics & Informatics, 9*(2), 24–27.

Stone, C. D. (1975). *Where the law ends: The social control of corporate behavior.* New York: Harper & Row Publishers.

Stough, C., Kerkin, B., Bates, T., & Mangan, G. (1994). Music and spatial IQ. *Personality and Individual Differences, 17,* 695. doi:10.1016/0191-8869(94)90145-7

Stovers, S. (2011). *Play's progress? Location play in the educationalisation of early childhood in Aotearoa New Zealand. (Doctor of Philosophy).* Auckland, New Zealand: Auckland University of Techology.

Subrahmanyam, K., Greenfield, P., Kraut, R., & Gross, E. (2001). The impact of computer use on children's and adolescents' development. *Applied Developmental Psychology, 22*(1), 7–30. doi:10.1016/S0193-3973(00)00063-0

Suits, B. (1978). *The grasshopper: Games, life, and utopia.* Ontario, CA: University of Toronto Press.

Sumsion, J., Barnes, S., Cheeseman, S., Harrison, L., Kennedy, A., & Stonehouse, A. (2009). Insider perspectives on developing: Belonging, being & becoming: The early years learning framework for Australia. *Australian Journal of Early Childhood, 34*(4), 4–13.

Sweller, J. (1994). Cognitive load theory, learning difficulty, and instructional design. *Learning and Instruction, 4*(4), 295–312. doi:10.1016/0959-4752(94)90003-5

Sweller, J., van Merriënboer, J. J. G., & Paas, F. G. W. C. (1998). Cognitive architecture and instructional design. *Educational Psychology Review, 10,* 251–296. doi:10.1023/A:1022193728205

Synodi, E. (2010). Play in the kindergarten: The case of Norway, Sweden, New Zealand and Japan. *International Journal of Early Years Education, 18*(3), 185–200. doi:10.1080/09669760.2010.521299

Szewczyk, J. (2003). Difficulties with the novices' comprehension of the computer-aided design (CAD) interface: Understanding visual representations of CAD tools. *Journal of Engineering Design, 14*(2), 169–185. doi:10.1080/0954482031000091491

Tang, S. Martin Hanneghan, & El-Rhalibi, A. (2009). Introduction to Games-Based Learning. In T. M. Connolly, M. H. Stansfield & L. Boyle (Eds.), Games-Based Learning Advancements for Multi-Sensory Human Computer Interfaces: Techniques and Effective Practices (pp. 1-17). Hershey, PA: Idea-Group Publishing.

Tang, S., Hanneghan, M., & El-Rhalibi, A. (2007). *Pedagogy Elements, Components and Structures for Serious Games Authoring Environment.* Paper presented at the 5th International Game Design and Technology Workshop (GDTW 2007), Liverpool, UK.

Tanzi, V. (1987). Quantitative characteristics of the tax systems of developing countries. In D. Newbery, & N. Stern (Eds.), *The theory of taxation for developing countries*. New York: Oxford University Press.

Tapscott, D., & Williams, A. D. (2010). *Macrowikinomics: Rebooting business and the world.* Canada: Penguin Group.

Telfer, R. (1993). *Aviation Instruction and Training.* Aldershot, UK: Ashgate.

Tessmer, M., Jonassen, D. H., & Caverly, D. (1989). *Non-programmer's guide to designing instruction for microcomputers*. Littleton, CO: Libraries Unlimited.

Tessmer, M., & Richey, R. C. (1997). The role of context in learning and instructional design. *Educational Technology Research and Development, 45*(2), 85–115. doi:10.1007/BF02299526

Tettegah, S. (2007). Pre-service teachers, victim empathy, and problem solving using animated narrative vignettes. *Technology, Instruction. Cognition and Learning, 5*, 41–68.

Tettegah, S., & Anderson, C. (2007). Pre-service teachers' empathy and cognitions: Statistical analysis of text data by graphical models. *Contemporary Educational Psychology, 32*, 48–82. doi:10.1016/j.cedpsych.2006.10.010

Tettegah, S., & Neville, H. (2007). Empathy among Black youth: Simulating race-related aggression in the classroom. *Scientia Paedagogica Experimentalis, XLIV, 1*, 33–48.

Thelwall, M. (2008). Social networks, gender, and friending: An analysis of MySpace member profiles. *Journal of the American Society for Information Science and Technology, 59*(8), 1321–1330. doi:10.1002/asi.20835

Thelwall, M. (2009). Social network sites: Users and uses. In M. Zelkowitz (Ed.), *Advances in computers: Social networking and the web* (p. 19). London, UK: Academic Press. doi:10.1016/S0065-2458(09)01002-X

Thomas, R., Cahill, J., & Santilli, L. (1997). Using an interactive computer game to increase skill and self-efficacy regarding safer sex negotiation: Field test results. *Health Education & Behavior, 24*(1), 71–86. PubMed doi:10.1177/109019819702400108

Thomas, J., Andrysiak, T., Fairbanks, L., Goodnight, J., Sarna, G., & Jamison, K. (2006). Cannabis and cancer chemotherapy: A comparison of oral delta-9-thc and prochlorperazine. *Cancer, 50*(4), 636–645. doi:10.1002/1097-0142(19820815)50:4<636::AID-CNCR2820500404>3.0.CO;2-4

Thorndike, E. L. (1933). A proof of the law of effect. *Science, 77*, 173–175. PubMed doi:10.1126/science.77.1989.173-a

Thurman, R. A. (1993). Instructional simulation from a cognitive psychology viewpoint. *Educational Technology Research and Development, 41*(4), 75–79. doi:10.1007/BF02297513

Time. (1930). *National affairs: Decency squabble.* Retrieved from http://www.time.com/time/magazine/article/0,9171,738937,00.html

Tompson, W. F., Schellenberg, E. G., & Husain, G. (2001). Arousal, mood, and the Mozart effect. *Psychological Science, 12*, 248–251. doi:10.1111/1467-9280.00345 PMID:11437309

Tootell, H., Plumb, M., Hadfield, C., & Dawson, L. (2013). *Gestural interface technology in early childhood education: A framework for fully-engaged communication.* Paper presented at the Hawaii International Conference on System Sciences (HICSS). Maui, HI.

Trevisan, F. (2010). *More barriers? Yes please: Strategies of control, co-optation and hijacking of online disability campaigns in scotland.* Glasgow, UK: Social Science Research Network. doi:10.2139/ssrn.1667427

Tripp, S. D., & Bichelmeyer, B. (1990). Rapid prototyping: An alternative instructional design strategy. *Educational Technology Research and Development, 38*(1), 31–44. doi:10.1007/BF02298246

Tsebelis, G. (1990). *Nested games: Rational choice in comparative politics.* Berkeley, CA: University of California Press.

Turja, L., Endepohls-ulpe, M., & Chatoney, M. (2009). A conceptual framework for developing the curriculum and delivery of technology education in early childhood. *International Journal of Technology and Design Education, 19*(4), 353–365. doi:10.1007/s10798-009-9093-9

Turkle, S. (1984). Video games and computer holding power. In *The second self: Computers and the human spirit* (pp. 64–92). New York: Simon and Schuster.

Turner Bell, D. (2009, March). *Exercise Gives The Brain A Workout, Too.* Retrieved from http://www.cbsnews.com/stories/2009/01/30/earlyshow/health/main4764523.shtml

UK Department for Education. (2012). *Statutory framework for the early years foundation stage.* Retrieved 10 February, 2013, from https://www.education.gov.uk/publications/eOrderingDownload/00267-2008BKT-EN.pdf

Um, E., Song, H., & Plass, J. L. (2007). *The effect of positive emotions on multimedia learning.* Paper presented at the World Conference on Educational Multimedia, Hypermedia & Telecommunications (ED-MEDIA 2007) in Vancouver, Canada, June 25–29, 2007.

UN Global Compact/Office of the United Nations High Commissioner for Human Rights (UNHCHR). (2004). *Embedding human rights in business practice.* New York: UN.

UN. (2008). *World population prospects, the 2008 revision.* United Nations Department of Economic and Social Affairs, Population Division. Retrieved November 17, 2011, from http://www.un.org/esa/population/publications/wpp2008/wpp2008_highlights.pdf

UNESCO. (2011). *ICT in education.* Retrieved 23 April, 2012, from http://www.unesco.org/new/en/unesco/themes/icts/

Upshaw, L. B., & Taylor, E. L. (2000). *The masterbrand mandate: The management strategy that unifies companies and multiplies value.* Hoboken, NJ: Wiley.

Valas, H., & Sovik, N. (1993). Variables affecting students' intrinsic motivation for school mathematics: Two empirical studies based on Deci and Ryan's theory of motivation. *Learning and Instruction, 3*, 281–298. doi:10.1016/0959-4752(93)90020-Z

Van den Hoogen, W. M., IJsselsteijn, W. A., & de Kort, Y. A. W. (2009). Yes Wii can! Using digital games as a rehabilitation platform after stroke-The role of social support. In *Proceedings of 2009 Virtual Rehabilitation International Conference* (pp. 195).

Van der Eerden, W. J., Otten, E., May, G., & Even-Zohar, O. (1999). CAREN-computer assisted rehabiliation environment.[PubMed]. *Studies in Health Technology and Informatics, 62*, 373–378. PMID:10538390

van Gerven, P. W. M., Paas, F., van Merriënboer, J. J. G., & Schmidt, H. G. (2006). Modality and variability as factors in training the elderly. *Applied Cognitive Psychology, 20*, 311–320. doi:10.1002/acp.1247

van Merriënboer, J. J. G., Clark, R. E., & de Croock, M. B. M. (2002). Blueprints for complex learning: The 4C/ID-model. *Educational Technology Research and Development, 50*, 39–64. doi:10.1007/BF02504993

Vanacken, L., Notelaers, S., Raymaekers, C., Coninx, K., Van den Hoogen, W., IJsselsteijn, W., & Feys, F. (2010). Game-based collaborative training for arm rehabilitation of MS patients: A proof-of-concept game. *Proceedings of GameDays, 2010*, 65–75.

Vedel, T. (1989). Télématique et configurations d'acteurs: Une perspective européenne. *Reseaux, 7*(37), 9–28.

Viola, P., & Jones, M. (2002). *Robust real-time object detection.* Paper presented at the Second International Workshop on Statistical and Computational Theories of Vision – Modeling, Learning, Computing, and Sampling, Vancouver, BC, Canada.

Virkar, S. (2011b). *The politics of implementing e-government for development: The ecology of games shaping property tax administration in Bangalore City.* (Unpublished Doctoral Thesis). University of Oxford, Oxford, UK.

Virkar, S. (2011a). Exploring property tax administration reform through the use of information and communication technologies: A study of e-government in Karnataka, India. In J. Steyn, & S. Fahey (Eds.), *ICTs and sustainable solutions for global development: Theory, practice and the digital divide.* Hershey, PA: IGI Global.

Vygotsky, L. S. (1930). *Mind in society*. Cambridge, MA: Waiton, S. (2009). Policing after the crisis: Crime, safety and the vulnerable public. *Punishment and Society*, *11*(3), 359.

Wade, R. H. (1985). The market for public office: Why the Indian state is not better at development. *World Development*, *13*(4), 467–497. doi:10.1016/0305-750X(85)90052-X

Waldkirch, R. W., Meyer, M., & Homann, K. (2009). Accounting for the benefits of social security and the role of business: Four ideal types and their different heuristics. *Journal of Business Ethics*, *89*, 247–267. doi:10.1007/s10551-010-0392-6

Wall Street Journal. (2008, April 24). Google Brazil turns in user data amid child-pornography inquiry. *The Wall Street Journal*, p. B9.

Wang, V., Tucker, J. V., & Rihll, T. E. (2011). On phatic technologies for creating and maintaining human relationships. *Technology in Society*. doi:10.1016/j.techsoc.2011.03.017

Weber, C., & Matthews, H. S. (2008). Quantifying the global and distributional aspects of American Household carbon footprints. *Ecological Economics*, *66*, 379–391. doi:10.1016/j.ecolecon.2007.09.021

Weiler, J. H. H. (2002). A constitution for Europe? Some hard choices. *Journal of Common Market Studies*, *40*(4), 563–580. doi:10.1111/1468-5965.00388

Weiner, B. (1979). A Theory of Motivation for Some Classroom Experiences. *Journal of Educational Psychology*, *71*(1), 3–25. PubMed doi:10.1037/0022-0663.71.1.3

Westheimer, J., & Kahne, J. (2004). What kind of citizen? The politics of educating for democracy. *American Educational Research Journal*, *41*(2), 237–269. doi:10.3102/00028312041002237

Westin, T. (2004). Game accessibility case study: Terraformers – a real-time 3D graphic game. In *Proceedings of the 5th International Conference on Disability, Virtual Reality & Associates Technology*, Oxford, UK (pp. 95-100).

Wettstein, F. (2010). For better or for worse: Corporate responsibility beyond do not harm. *Business Ethics Quarterly*, *20*(2), 275–283. doi:10.5840/beq201020220

Wikileaks. (n.d.). *What is Wikileaks*. Retrieved from http://wikileaks.org/About.html

Wikipedia. (2011). *Blocking of Wikipedia by the People's Republic of China*. Retrieved from http://en.wikipedia.org/wiki/Censorship_of_Wikipedia#China

Wikipedia. (n.d.). *Wikipedia: About*. Retrieved from http://en.wikipedia.org/wiki/Wikipedia:About

Wilkinson, N., Ang, R. P., & Goh, D. H. (2008). Online video game therapy for mental health concerns: A review. *The International Journal of Social Psychiatry*, *54*(4), 370–382. PubMed doi:10.1177/0020764008091659

Wilkinson, G. (2006). Commercial breaks: An overview of corporate opportunities for commercializing education in US and English schools. *London Review of Education*, *4*(3), 253–269. doi:10.1080/14748460601043932

Williams, D., & Skoric, M. (2005). Internet fantasy violence: A test of aggression in an online game. *Communication Monographs*, *72*(2), 217–233. doi:10.1080/03637750500111781

Willis, S. L., Tennstedt, S. L., Marsiske, M., Ball, K., Elias, J., Koepke, K. M.... Wright, E., for the ACTIVE Study Group. (2006). Long-term effects of cognitive training on everyday functional outcomes in older adults. *Journal of the American Medical Association*, *296*(23), 2805–2814. PubMed doi:10.1001/jama.296.23.2805

Wilson, A. D., Member, S., Society, I. C., Bobick, A. F., & Society, I. C. (1999). Parametric hidden Markov models for gesture recognition. *IEEE Transactions on Pattern Analysis and Machine Intelligence*, *21*, 884–900. doi: doi:10.1109/34.790429

Wilson, B. G. (1997). The postmodern paradigm. In C. R. Dills, & A. A. Romiszowski (Eds.), *Instructional development paradigms* (pp. 63–80). Englewood Cliffs, NJ: Educational Technology Publications.

Wilson, B. G., Teslow, J., & Osman-Jouchoux, R. (1995). The impact of constructivism (and post-modernism) on instructional design fundamentals. In B. B. Seels (Ed.), *Instructional design fundamentals: A reconsideration* (pp. 137–157). Englewood Cliffs, NJ: Educational Technology Publications.

Winn, W. (1996). Cognitive perspectives in psychology. In D. H. Jonassen (Ed.), *Handbook of research for educational communications and technology: A project of the Association for Educational Communications and Technology* (pp. 79–112). New York: Macmillan Library Reference.

Winn, W. (1997). Advantages of a theory-based curriculum in instructional technology. *Educational Technology, 37*(1), 34–41.

Wispé, L. (1987). History of the concept of empathy. In N. Eisenberg, & J. Strayer (Eds.), *Empathy and its development*. Cambridge, UK: Cambridge University Press.

Wolters, C. A., & Pintrich, P. R. (1998). Contextual differences in student motivation and self-regulated learning in mathematics, English, and social studies classrooms. *Instructional Science, 26*, 27–47. doi:10.1023/A:1003035929216

Wood, L. C., Teräs, H., Reiners, T., & Gregory, S. (2013). *The role of gamification and game-based learning in authentic assessment within virtual environments*. Paper presented at the HERDSA 2013. New York, NY.

Woodfill, W. (2009, October 1). The transporters: Discover the world of emotions. *School Library Journal Reviews*, 59.

World Bank - Global Information and Communication Technologies Department. (2002). *The networking revolution: Opportunities and challenges for developing countries*. Washington, DC: World Bank.

Wriedt, S., Reiners, T., & Ebeling, M. (2008). How to teach and demonstrate topics of supply chain management in virtual worlds. In *Proceedings of ED-MEDIA 2008: World Conference on Educational Multimedia, Hypermedia & Telecommunications* (pp. 5501–5508). Chesapeake, VA: AACE.

Wu, E. (2006). Liberating Wikipedia in China (almost). *Fortune International, 154*(9), 14–16.

Xie, X., Wang, Q., & Chen, A. (2012). Analysis of competition in chinese automobile industry based on an opinion and sentiment mining system. *Journal of Intelligence Studies in Business, 2*(1).

Yelland, N., & Lloyd, M. (2001). Virtual kids of the 21st century: Understanding the children in schools today. *Information Technology in Childhood Education Annual, 13*(1), 175–192.

Yerkes, R. M., & Dodson, J. D. (1908). The relation of strength of stimulus to rapidity of habit-formation. *The Journal of Comparative Neurology and Psychology, 18*, 459–482. doi:10.1002/cne.920180503

Yin, R. K. (2003). *Case study research: Design and methods*. London: Sage Publications.

Yin, S. (2011). *Anonymous v anonymous: This is why we can't have nice things*. PM Magazine.

Yong Joo, L., Soon Yin, T., Xu, D., Thia, E., Pei Fen, C., Kuah, C. W. K., & Kong, K. H. (2010). A feasibility study using interactive commercial off-the-shelf computer gaming in upper limb rehabilitation in patients after stroke. *Journal of Rehabilitation Medicine, 42*(5), 437–441. PubMed doi:10.2340/16501977-0528

You, Y. (1993). What can we learn from Chaos theory? An alternative approach to instructional systems design. *Educational Technology Research and Development, 41*(3), 17–32. doi:10.1007/BF02297355

Yukawa, J. (2005). Story-lines: A case study of online learning using narrative analysis. *Proceedings of the 2005 Conference on Computer Support for Collaborative Learning: Learning 2005: The Next 10 Years!* (p. 736).

Zakaria, F. (2008). *The post-American world*. New York: W.W. Norton & Company.

Zehnder, S., Igoe, L., & Lipscomb, S. D. (2003, June). *Immersion factor – sound: A study of influence of sound on the perceptual salience of interactive games*. Paper presented at the Conference of the Society for Music Perception & Cognition, Las Vegas, NV.

Zehnder, S., & Lipscomb, S. (2006). The role of music in video games. In P. Vorderer, & J. Bryant (Eds.), *Playing video games: Motives, responses, and consequences* (pp. 241–258). Mahwah, NJ: Lawrence Erlbaum.

Zemke, R. E., & Rossett, A. (2002). A hard look at ISD. *Training (New York, N.Y.), 39*(2), 26–34.

Zhang, P., Ma, X., Pan, Z., Li, X., & Xie, K. (2010). Multi-agent cooperative reinforcement learning in 3D virtual world. In *Advances in swarm intelligence* (pp. 731–739). London, UK: Springer. doi:10.1007/978-3-642-13495-1_90

Zikmund, W. G., & Scott, J. E. (1973). A multivariate analysis of perceived risk, self-confidence and information sources. *Advances in Consumer Research. Association for Consumer Research (U. S.)*, *1*(1), 406–416.

Zyda, M. (2005). From Visual Simulation to Virtual Reality to Games. *Computer*, *38*, 25–32. doi:10.1109/MC.2005.297

About the Contributors

Jonathan Bishop is an information technology executive, researcher, and writer. Having played video games on systems as early as the Atari VGS2600, before he finished high school he had programmed his first non-joystick operated video game on an Atari STE using a chair controller. This made his step into EEG operated gamification systems the natural research direction to take. Jonathan has 4 degrees, namely in multimedia studies, e-learning, law, and information systems. He has over 30 publications in journals, conference proceedings, and books. A fellow of the Royal Anthropological Institute and BCS – The Chartered Institute for IT, Jonathan has always been more interested in the human side of computing than the technology side. He has an increasing amount of research understanding the implications of Internet trolling, including understanding behavioural and social patterns. He is regularly interviewed by the media, including on the effect gamification can have in improving human understanding and interaction.

* * *

Göknur Kaplan Akilli completed her undergraduate degree on Mathematics Education at Hacettepe University, Turkey, in 2001, and ranked first in graduating Class of Faculty of Education the same year. In 2004, she earned her master's degree from Middle East Technical University (METU), with her thesis "A Proposal of Instructional Design/Development Model for Game-like Learning Environments: The FID^{2}GE Model," which is nominated to many national and international awards. Currently, she is pursuing a Ph.D. degree in Instructional Systems Program at Penn State University.

Dietrich Albert is professor of psychology at University of Graz, senior scientist at Graz University of Technology, Knowledge Management Institute and key researcher at the Know-Center Graz. Since 1993 Dietrich is the head of the Cognitive Science Section at the University of Graz, the Department of Psychology's largest working group. In the preceding years, he was with the Universities of Göttingen, Marburg, Heidelberg, and Hiroshima. His research topics cover several areas, including learning and memory, psychometrics, anxiety and performance, psychological decision theory, computer-based tutorial systems, values, and behaviour. Dietrich's actual focus is on knowledge and competence structures, their applications, and empirical research. By working with psychologists, computer scientists, and mathematicians, several academic disciplines are represented within his research team. Beside national activities, his expertise in European research and development projects is documented by several successful European projects.

Janice L. Anderson is an assistant professor of science education at the University of North Carolina at Chapel Hill. Prior to joining the faculty at UNC-Chapel Hill, she taught biology and anatomy in Ohio and worked in elementary classrooms in Massachusetts. Anderson received her Ph.D. in Curriculum and Instruction from Boston College with a focus on Science and Technology. Her dissertation research explored the use of a 3D virtual world (Quest Atlantis) to teach concepts related to water quality and ecosystems to urban fifth-grade students. Her research considers the impact of gender and learning outcomes on how students engaged with the game. The catalyst for her professional efforts has been the notion of improving students' engagement with science and technology particularly among populations that are underrepresented in science, based on both gender and race.

Steven Battersby is a doctoral student in the Interactive Systems Research Group in the Computing and Technology Team at NTU. He has worked on many national and European research and development projects using computer games and games technology for the education and rehabilitation of people with disabilities. Two of his latest projects include the Virtual Cane; allowing blind people to develop spatial maps using virtual environments and Wii technology, and the Stroke rehabilitation project developing serious games and data gloves for home based rehabilitation.

Thomasina Borkman received her Ph.D. from Columbia University, NYC in 1969 in sociology. Sociology Professor at George Mason University since 1974 who combines teaching and research. Her research and consulting is in her specialty area of health, illness and disability, especially with self-help/ mutual aid groups and nonprofit organizations nationally and cross-nationally. Between 1997-98 she co-taught the team aspects of an online course titled Taming the Electronic Frontier designed by Brad Cox; the course won the Paul Allen Distance Education national award in 1998 ($25,000 prize). Her major recent book is Understanding Self-Help/Mutual Aid: Experiential Learning in the Commons, Rutgers University Press, 1999.

David Brown is Professor of Interactive Systems for Social Inclusion at Nottingham Trent University, and Director of the Interactive System Research Group with the Computing and technology Team. He is EU Project Coordinator for several projects concerning games based learning and location based services for the education and rehabilitation of people with disabilities: RECALL Project http://recall-project.eu (Location based services for people with disabilities); GOET Project http://goet-project.eu (Serious games for people with learning disabilities); GOAL Project http://goal-net.eu (Serious games for people with learning disabilities); Game On Project http://gameon.europole.org (Serious games for prisoners with disabilities). David is a member of the International Steering Committees of the International Conference of Disability, Virtual Reality and Associated Technology (ICDVRAT), Interactive Technologies and Games (ITAG) and the European Conference of Games Based Learning (ECGBL).

Andy Burton is a research fellow at Nottingham Trent University in the Interactive Systems Research Group. He joined this team in April 2010 initially assisting Steven Battersby with the development of the IR Glove hardware and CLAHRC stroke rehabilitation games software. Previously to this he was a research fellow/associate at The University of Nottingham. Most recently Dr. Burton was 'research fellow for 3D graphics and virtual reality' in the Spatial Literacy in Teaching (SPLINT) Centre for Excellence in Teaching and Learning (CETL). In this role he developed stereoscopic applications for augmented 3D environments, to aid teaching and learning in both specialised lecture rooms and in fieldwork scenarios, and recorded their pedagogic impact. His previous research was in applications of virtual reality and 3D graphics in teaching, simulation, forensics, psychology and health and safety training in the Computer Sciences and Mining Engineering departments at The University of Nottingham.

Unai Diaz-Orueta (PhD) is Psychology graduate (2000) and PhD (2006) at Deusto University (Spain). Clinical Psychologist at Crownsville Hospital Center (USA, 2000-2001), Bermeo Hospital, (Spain, 2001-2002), La Loma Geriatric Residence (Spain, 2003-2005), and Zutitu (Spain, 2005-2006). His doctoral dissertation "Effects of psychological intervention in cognitive decline of residentialized elderly people" was published by UMI Dissertation Publishing, Ann Arbor, MI (USA). From 2007-2008 he developed workshops of cognitive training, wellbeing, and laugh-therapy for older people, within IPACE Ltd, Vitoria (Spain). Since 2008, he works as a research psychologist in Fundación INGEMA, in projects related to ageing and physical disability. He is author and co-author of papers and articles related to ageing and application of technological aids for elderly people. In 2010, he published two books related to his doctoral dissertation: "The ageing process. A comprehensive perspective" and "Memory Gym: A cognitive stimulation program for elderly people based on objectives," published by Deusto Publicaciones (Bilbao, Spain).

Jon Dron is a member of the Technology Enhanced Knowledge Research Institute and an Associate Professor in the School of Computing and Information Systems, Athabasca University (Canada's open university), where he teaches various graduate and undergraduate courses. He is also an Honorary Faculty Fellow in the Centre for Learning and Teaching, University of Brighton, UK. Jon has received both national and local awards for his teaching, is author of various award-winning research papers and is a regular keynote speaker at international conferences. Jon's research in learning technologies is highly cross-disciplinary, including social, pedagogical, technological, systemic and philosophical aspects of technology and learning design and management.

Lindsay Evett is a lecturer in the Computing and Technology Team. Her research is on accessibility and assistive technology, especially with respect to Serious Games, and Web-based content. She is a lecturer in Artificial Intelligence, and a member of Nottingham Trent University's working group on accessibility. She is a co-investigator on the Recall European project on route-learning systems and location-based services for people with cognitive and sensory disabilities. She is a member of the ETNA European thematic network on assistive information and communications technologies.

Loren Falkenberg has been a faculty in the Haskayne School of Business for over 20 years. Her Ph.D. is from the University of Illinois and her MBA is from Queen's University. She is currently the Associate Dean Research in Haskayne. She has been teaching and researching in the area of business ethics for over 20 years, and taught one of the first full semester business ethics courses in Canada. She is an editor for the Journal of Business Ethics. She has published in the *Journal of Business Ethics, Academy of Management Review, Journal of Management*, and *California Management Review*.

Joel Foreman is an associate professor in the English Department at George Mason University. He began teaching distance courses in 1996 and subsequently developed expertise in building and assessing Web based learning environments. As a member of GMU's Program on Social and Organizational

Learning from 1995 – 2001, he performed organizational learning studies sponsored by Hughes Information Technology Corporation, DynCorp, and Media General. He has been researching computerized instructional media since the 1980s and his applied experience includes documentaries he produced for NBC, public television, the Discovery Channel, and others. His current research is focused on game based learning and mobile learning.

Alison Freeman is an Honorary Fellow at the University of Wollongong. She has extensive experience in community informatics, policy development, project management and e-learning across both industry and academia.

Mark M. H. Goode is Professor of Marketing, Director of the Executive MBA and Chair of the '*Marketing and Retailing*' field group at Cardiff School of Management at Cardiff Metropolitan University. He also has two degrees in Economics and a PhD in Marketing. His past research has looked at the effect on the gaming environment on human behaviour and the effect online environments have on factors such as trust, perceived risk and behavioural intentions. Mark holds a distinguished teaching award from Swansea University and has also held posts at Cardiff University (Lecturer in Quantitative Methods and Lecturer in Operations Management), Swansea University (Senior Lecturer in Marketing and later Reader in Marketing). Mark has over 25 years teaching experience in Welsh Universities and has taught over 20,000 university students.

Martin Hanneghan (BSc Hons, PhD) is a Principal Lecturer in Computing and Head of Enterprise at Liverpool John Moores University in the UK where he teaches on undergraduate and postgraduate courses in Computer Games Technology. He has served as a member of the programme and technical committees for a number of games conferences around the world including Cybergames, GAME-ON, GDTW, and SBGames. His research interests include serious game applications and software engineering for games.

Wen-Hao David Huang is an Assistant Professor of E-Learning in the Department of Human Resource Education at University of Illinois at Urbana-Champaign. His academic background, consisting of material science and engineering, educational technology, and executive business administration, has enabled him to conduct interdisciplinary projects for instructional and research purposes for years. Dr. Huang currently teaches Learning Technologies and Instructional Design in the context of human resource development and E-Learning. His research interests include (1) design of game-based learning environments, (2) design and evaluation of E-Learning systems for adult learners, (3) Web 2.0 emerging technologies and their impact on teaching and learning, (4) measurement and manipulation of cognitive load in multimedia learning environment.

Liz Keating is a fully qualified personal assistant who used to run a team of nurses in the NHS. She then went on to work towards a degree in history, carrying out research into Anglo Saxon stone sculpture. She became interested in assistive technology due to being registered blind as a way of maintaining her independence, and to find new ways to keep interacting with people so that she does not become socially isolated.

Stephanie B. Linek graduated from the University of Wuerzburg (Germany), in 1997 with a diploma (MS) in Psychology. From 1998 to 1999, she worked as a postgraduate researcher at the University of Heidelberg (Germany). After postgraduate studies at the University of Koblenz-Landau (Germany), in 2002 she received the Certificate "Media- and Communication Psychologist." From 2003 to 2006, she was a postgraduate researcher at the Knowledge Media Research Center in Tuebingen (Germany) and worked on her PhD thesis on "Gender-specific design of narrated animations: Speaker/Gender Effect and the schema-incongruity of information." In 2007 she received her Dr. rer. nat (D.Sc.) by the University of Tuebingen. From 2006 to 2010, she worked in the Cognitive Science Section at the University of Graz (Austria) in several EC-research projects and as university assistant. Since 2011, she is expert for usability evaluation at the ZBW – Leibniz Information Centre for Economics in Kiel (Germany). Her research interests are in several areas of social media and game-based learning as well as in usability, evaluation and methodology.

Hao Liu is a research associate of Computer Science department of Nottingham Trent University, where he is working for a NHS sponsored serious games project for stroke rehabilitation. His work focuses on designing a makerless tracking system to control a series of 3D games by user's hand gestures and body languages. Hao obtained his PhD from University College London, where he examined the problem of flexible and transparent distributed computing. During that time, Hao was also working for various projects including a NASA leaded project, which was to calculate the lunar surface temperature using parallel processing technology. Hao has published more than 10 journal and conference papers.

Birgit Marte has a Diploma in Psychology from the University of Graz (Austria). From April 2004 to April 2008, she was working at the Cognitive Science Section (http://css.uni-graz.at) of the University of Graz on several European Commission-funded R&D projects that focused on e-learning. Her research addresses the representation and assessment of knowledge and competences as well as the evaluation of the effectiveness of e-learning.

Patrick Merritt is currently a student at Nottingham Trent University studying for a BSc in Computer Science (Games Technology). As part of his degree, he undertook a work placement at the university as a research assistant, working on several assistive technology and games projects. He is currently completing his final year, including a major project on virtual reality technology and its benefits to disabled user groups.

Henk Herman Nap (MSc PhD) received an MSc degree in Cognitive Ergonomics at Utrecht University (2002) and a PhD degree (2008) at Eindhoven University of Technology (EUT). The research during his PhD focused on stress in senior computer interaction. After his PhD, he worked as a postdoctoral research fellow on senior gamers and distributed gaming at the Game Experience Lab of the Human-Technology Interaction group at the EUT (2008-2010). He published journal and conference papers on gaming research and Gerontechnology and is active as a reviewer and project coordinator. Since 2011, he works as a project leader at Stichting Smart Homes, in European and national projects related to rehabilitation gaming, persuasive technologies, eHealth, and smart living.

Oleksiy (Olex) Osiyevskyy is a PhD candidate in Strategy and Global Management area at the Haskayne School of Business, University of Calgary. His research interests lie at the intersection of strategy, entrepreneurship and innovations studies. In 2012 Oleksiy was awarded the prestigious Izaak Walton Killam Memorial Scholarship (becoming a Killam Scholar), and the Eyes High International Doctoral Scholarship from the University of Calgary. His papers were recognized with best paper awards at the United States Association for Small Business and Entrepreneurship (USASBE) conference two years in a row (2012, 2013). The results of Oleksiy's research were presented at top academic conferences, and were accepted for publication in the *Journal of Small Business Management* and *Entrepreneurial Practice Review*.

Torsten Reiners is Senior Lecturer in Logistics at the Curtin University, Australia. His research and teaching experiences are in the areas of operations research, but include instructional design, development of adaptive learning environments, distant collaboration, and mobile learning, which is also manifested in his PhD Thesis about adaptive learning material in the field of operations research. Dr Reiners is the co-founder of the Second Life Island University of Hamburg and Students@work, an initiative to promote education in Web 3D. He participated in multiple projects to use 3D spaces for learning support; i.e. to improve the authenticity of learning in classes about production and simulation. In his latest project VirtualPREX, he explored professional experience for pre-service teacher in 3D spaces replicating authentic environments by using bots as actors in the role-play. After being an international research student at the University of Texas in Austin, he worked 9 month as a research fellow at the University of California, Davis, U.S. (Diploma Thesis). Since then, he had multiple research visits with universities in Australia, invited talks and over 20 international co-authors on his publications.

Allan Ridley was awarded an MRes with distinction in Computer Science by Nottingham Trent University in 2008. He is a PhD student carrying out research into accessible interactive systems. He is an assistive technology specialist has worked as a trainer in a blind and visually impaired person's resource centre. He is registered blind.

Judy Shasek, M.S. ExerLearning, is poised at the intersection of fitness, education, and technology. The author contributes vital expertise and resources in each of these key areas. Judy Shasek has 17 years of experience as a fitness/education consultant and 12 years as a public school teacher, curriculum designer, teacher trainer, and grant writer. By assimilating a massive amount of research and drawing on the invention and energy of many educators, researchers and fitness leaders around the country, ExerLearning was first delivered via Generation FIT. It is a program that developed organically over five years– in real schools with diverse students.

Nasser Sherkat received a B.Sc Honours degree in Mechanical Engineering from University of Nottingham in 1985. He received a Ph.D. in high speed geometric processing for continuous path generation, from the Nottingham Trent University in 1989. He is currently Associate Dean of Science and Technology at The Nottingham Trent University. His interests are use of intelligent pattern recognition in facilitating 'natural' human computer interaction and multimodal biometrics.

Nick Shopland has over 10 years experience in design, implementing and evaluation virtual training environments, serious games, location-based services, and assistive technology for people at risk of social exclusion, including people with a learning disability, people at risk of unemployment and young people at risk of social exclusion. He has a particular experience in the development of virtual training environments to travel train young people with learning disabilities in accessing work-based opportunities; to train people with low-level skills in entering the Care Industry, and is currently extending the travel training and route learning work to mobile devices.

Penny Standen is Professor in Health Psychology and Learning Disabilities at the University of Nottingham. Her main area of research is developing and evaluating virtual environments and interactive software for people with intellectual disabilities and from this grew an interest in the value of using serious games to help the acquisition of independence skills and improve cognition. She is currently collaborating with colleagues on the use of VR in stroke rehabilitation. She is a member of the programme committee for the International Conference Series on Disability, Virtual Reality and Associated Technologies an Associate member of the British Psychological Society and on the editorial board of the *Journal of Health Psychology*.

Stephen Tang (BSc, MSc) is a lecturer in Computer Games at Liverpool John Moores University (LJMU) in the UK. Prior to joining LJMU he was a lecturer at Tunku Abdul Rahman College (TARC) in Malaysia where he taught on undergraduate courses in multimedia and computer games design and technologies. Stephen has also served as a member of programme and technical committee members for game conferences such as Asian Game Developers Summit, GDTW and CyberGames. He is a technical reviewer of the *International Journal of Computer Games Technology*. Stephen is currently a PhD candidate at LJMU. His research interests include game-based learning, serious games design and development, and model driven engineering.

Sharon Tettegah is an Associate Professor in the Department of Curriculum and Instruction, at the University of Illinois, Urbana Champaign. Dr. Tettegah holds a doctorate degree in Educational Psychology, and also degrees in Curriculum and Supervision and Philosophy. In addition, she holds an appointment at the Beckman Institute where she is currently in the Division of Biotechnology, Cognitive Neuroscience Group. Her research focuses on the use of technologies to enhance teaching and learning with an emphasis on simulations and empathy.

Holly Tootell is a Senior Lecturer in the School of Information Systems and Technology at the University of Wollongong, Australia. She has extensive links across the Information Systems and Early Years Education disciplines, drawing together her academic knowledge with strong communication skills to develop practical and useful solutions for educators.

Shefali Virkar is research student at the University of Oxford, UK, currently reading for a D.Phil. in Politics. Her doctoral research seeks to explore the growing use of Information and Communication Technologies (ICTs) to promote better governance in the developing world, with special focus on the political and institutional impacts of ICTs on local public administration reform in India. Shefali holds an M.A. in Globalisation, Governance and Development from the University of Warwick, UK. Her Master's thesis analysed the concept of the Digital Divide in a globalising world, its impact developing countries and the ensuing policy implications. At Oxford, Shefali is a member of Keble College.

Marion Walker is Professor in Stroke Rehabilitation at the University of Nottingham. She is an occupational therapist and has played a leading role in developing a research culture within her own profession. She has served as President of the Society for Rehabilitation Research, the only multidisciplinary rehabilitation research group in the United Kingdom and was also Chairman of UK Stroke Forum from 2006-2008. She is Associate Director (Rehabilitation Lead) of the UK Stroke Research Network. Marion has both led and steered many stroke rehabilitation research projects including service evaluations of stroke patients not admitted to hospital, centre care for young stroke patients and therapy provision in nursing homes. Her research experience also encompasses evaluations of specific components of stroke therapy including: leisure, dressing, outdoor mobility, and behavioural approaches in patients with aphasia and depression. Marion is a strong advocate of patient partnership and co-chairs the Nottingham Stroke Research Consumer Group.

Lincoln Wood is a Senior Lecturer and researcher in operations, logistics, and supply chain management at Curtin University of Technology in Perth, Australia. He received the CSCMP's Young Researcher Award in 2009 in the USA and later earned his PhD at the University of Auckland. He draws on a range of industry experience in distribution companies and an international consulting company. While at Auckland, he developed a strong interest in effective supply chain education, and in 2010, he received the Outstanding Research Award at the International Higher Education Conference. He has published in leading international journals including *Transportation Research Part B: Methodological*, *International Journal of Operations & Production Management*, *The Service Industries Journal*, and *Habitat International*.

Index

A

B

C

D

E

F

G

H

I

K

L

M

N

O

P

Q

R

S

T

U

V

W

Y